Conductive Polymers and Their Composites

Yanmin Wang · Wei Feng

Conductive Polymers and Their Composites

Springer

Yanmin Wang
College of Material Science
and Engineering
Shandong University of Science
and Technology
Qingdao, China

Wei Feng
School of Materials Science
and Engineering
Tianjin University
Tianjin, China

ISBN 978-981-19-5362-0 ISBN 978-981-19-5363-7 (eBook)
https://doi.org/10.1007/978-981-19-5363-7

This Springer imprint is published by the registered company Springer Nature Singapore Pte Ltd.
The registered company address is: 152 Beach Road, #21-01/04 Gateway East, Singapore 189721, Singapore

Contents

Chapter 1
Introduction of Conductive Polymers

The people's traditional concept of insulating polymer materials was broken by the discovery of polyacetylene in the 1970s, and the new era of conductive polymers (CPs) was opened. In general, conductive polymers own the π-conjugated system with alternating single (σ) and double (π) bonds, resulting in their inherent electrical/electronic, electrochemical, and optical properties. These physical properties are influenced by the conjugation length, crystallinity, and intrachain and interchain interactions. Compared with their inorganic counterpart, CPs possess the advantages such as low density, chemical diversity, flexibility, tunable conductivity, easy-to-control shape, and morphology [14]. Therefore, they have potential applications in large-area optoelectronic devices [8], microwave absorbing materials, energy storage engineering, biological field, anticorrosive coating, various sensors, etc. Besides polyacetylene, the widely investigated CPs consist of polyaniline (PANI), polypyrrole (PPy), polythiophene (PTh), poly(3,4-ethylenedioxythiophene) (PEDOT), and poly(p-phenylene vinylene) (PPV). Their chemical structures are shown in Fig. 1.1a and their formula, abbreviation, electrical conductivity, and applications are illustrated in Table 1.1.

1.1 Electrical Properties

The inherent molecular structure of the CPs is composed of alternative single bonds and double bonds. Both single and double bonds include strong localized σ-bond, while the double bonds also include weaker localized π-bond. The π-bond between each two carbon atoms is transferred to the next two carbon atoms so that the electrons flow along the carbon skeleton (Fig. 1.1b). However, the pristine polymers are not highly conductive. Their conductivity of 10^{-6}–10^{-10} $S \cdot cm^{-1}$ lies in the region between insulator and semiconductor. Upon doping, their conductivity is increased by 10 or more orders of magnitude to the metallic regime (Fig. 1.1c). The reason can be explained as follows: When an electron was removed from a delocalized bonding

Y. Wang and W. Feng, *Conductive Polymers and Their Composites*,
https://doi.org/10.1007/978-981-19-5363-7_1

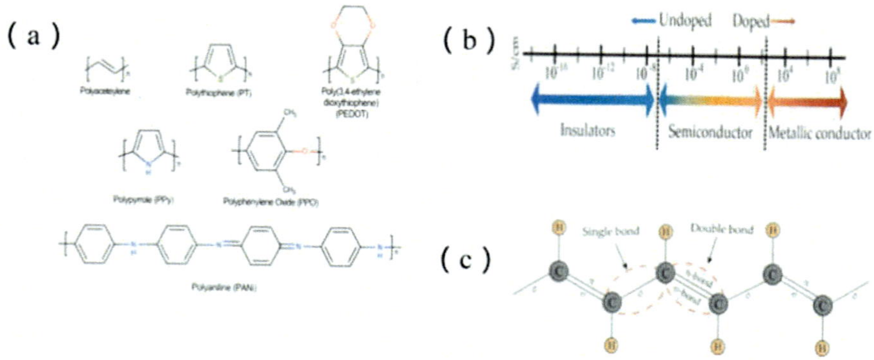

Fig. 1.1 **a** Chemical structures of π-conjugated polymers. Reprinted from ref. [16], copyright 2018, with permission from American Chemical Society. **b** The general conductivity range of conductive polymers (CPs). **c** The structure of polyacetylene: The backbone contains conjugated double bonds. Reprinted from ref. [12], copyright 2017, with permission from MDPI

Table 1.1 Conjugated conductive polymers

Conjugated π conductive polymer	Abbreviation	Formula	Electrical conductivity (S/cm)	Applications
Polyacetylene	PA	$[C_2H_2]_n$	10^5	Biosensors, colchicine detection, bioelectrodes
Polythiophene	PTh	$[C_4H_4S]_n$	10^0–10^3	Biosensors, enzyme immobilization, voltammetry of roxithromycin,conducting biomaterials
Polypyrrole	PPy	$[C_4H_2NH]_n$	2–100	Modulate cellular activities, nerve regeneration, biomedicine, biosensors, bacterial detection
Poly(p-phenylene)	PPP	$[C_6H_4]_n$	10^{-3}–10^2	Dental application, cell alignment
Polyaniline	PANI	$[C_6H_4NH]_n$	10^{-2}–10^0	Neural application, bioelectrodes

Reprinted from ref. [16], copyright 2018, with permission from American Chemical Society

arrangement, a hole is created. Then, a neighboring electron fills in this hole, resulting in a new hole. Thus, the charge flows through the polymer chain.

1.1.1 Doping of Conductive Polymers

The conductivity of conductive polymers depends on the nature and concentration of the dopants and the doping time. According to the molecular size, the dopants can be categorized into large polymeric species and small cations/anions. Typical examples of the dopants used for CPs and the corresponding conductivity are listed in Table 1.2.

Large dopants are difficult to leach from the polymer chain because they are strongly bound to it. Moreover, they influence the density, surface topography and physical properties of CPs. In contrast, small dopants readily dope the CPs in shorter time and easily dedope to result in the poor stability. The conductivity of CPs increases with the increasing doping level, as is strongly influenced by the dopant concentration and doping time. The increase of the dopant concentration improves the conductivity. Meanwhile, the dopant ions slowly diffuse into the conductive polymers due to their dense, ordered structure. The conductivity stepwise became saturated after long doping time, even several hours. Furthermore, the doping/dedoping process is reversible.

Various available doping methods in Table 1.3 consist of chemical doping, electrochemical doping, photodoping, non-redox doping, and charge-injection doping. Due to their low cost and convenience, chemical doping and electrochemical doping are most widely used. Vapor-phase doping and solution doping belong to chemical doping. For vapor-phase doping with vapors of iodine, bromine, AsF_5, and SbF_5 as dopant, the doping level is dependent on the vapor pressure and reaction time. For solution doping, the dopant and the products formed during doping are soluble in a solvent. For easy-to-control and more readily reversible electrochemcial doping, the doping level is controlled by the current passed when the DC power is applied on the CP-coated positive electrode.

1.1.2 Temperature Dependence

The conductivity of CPs can be comparable to that of the metals. Based on the investigation on the conductivity of CPs such as PPy, PPV, and PANI, their characteristic temperature dependence was explained using Zabrodski plots:

$$W(T) = -\frac{T[\mathrm{d}\ln\rho(T)]}{\mathrm{d}T} = \frac{\mathrm{d}(\ln\sigma)}{\mathrm{d}(\ln T)} \tag{1.1}$$

According to their reduced activation energy (W), CPs are divided into three regimes.

In the insulating regime, the resistivity is activated and σ follows a Mott variable range hopping mode:

Table 1.2 Typical examples of dopants for CPs and the conductivities obtained therefrom

CP type	Dopant	Chemical source	Doping method	Conductivity
Trans-polyacetylene	Na^+	$(C_{10}H_8)Na$	Solution doping	80
Poly(p-phenylene)	AsF_5	AsF_5	Vapor-phase doping	1.5×10^4
Poly(p-phenylene vinylene)	CH_3SO_3H	CH_3SO_3H,	Non-redox doping	10.7
	AsF_5	AsF_5	Vapor-phase doping	57
Poly(3-vinylperylene)	Cl_4^-	$(C_4H_9)_4\ N(ClO_4)$	Electrochmical doping	10^{-5}
PPy	AsF_6^-, PF_6^-, BF_4^-	$C_{16}H_{36}AsF_6N$, $(CH_3)_4\ N(PF_6)$, $(C_2H_5)_4\ N(BF_4)$	Electrochemical doping	30–100
	NSA	naphthalene sulfonic acid (NSA)	Electrochemical doping	1–50
	Cl_4^-	$LiClO_4$	Electrochemical doping	65
	Cl^-	NaCl	Electrochemical doping	10
	PSS/Cl^-	PSS/$FeCl_3$	Solution doping	4
	MeOH	MeOH	Vapor-phase doping	0.74
	HSO_4^-	$(C_4H_9)_4\ N(HSO_4)$	Electrochemical doping	0.3
	$C_{20}H_{37}O_4SO_3^-$	$C_{20}H_{37}O_4SO_3Na$	Solution doping	4.5
PANI	$C_{10}H_{15}OSO_3^-$	$C_{10}H_{16}O_4S$	Solution doping	300
	HCl	HCl	Non-redox doping	10
	I_3^-	I_2	Vapor-phase doping	9.3
	BF_4^-	HBF_4	Solution doping	(2.3×10^{-1})
PBTTT[1]	FTS[2]	$C_8H_4F_{13}SiCl_3$	Vapor-phase doping	$604–1.1 \times 10^3$
Poly(2-(3-thienyloxy)ethanesulfonate	$NaSO_3$	Na_2SO_3	Solution doping	5
PTh	Cl^-	$FeCl_3$	Vapor-phase doping	10–25
PANI-PPy	AsPB	Anionic spherical polyelectrolyte brushes (ASPB)	Electrochemical doping	8.3

Reprinted from ref. [12], copyright 2017, with permission from MDPI

[1]Poly(2,5-bis(3-tetradecylthiophen-2-yl)thieno[3,2-*b*]thiophene;[2]Trid-cafluoro-(1,1,2,2-tetrahydroocty1)-trichlorosilane

Table 1.3 Different methods used for doping CPs

Doping method	Controlled variables	Advantages	Disadvantages
Chemical doping	Vapor pressure, Exposure time to dopant	Simple way to obtain doping upon exposure of the sample to a vapor of the dopant or immersion into a solution with the dopant	Performed as slowly as possible to avoid inhomogeneous doping
			The doping levels obtained are not stable with respect to time
			Unexpected structural distortion may cause electrical conductivity decay
			Doping/dedoping shows low reversibility
Electrochemical doping	Amount of current passed	Doping level can be easily controlled by using an electrochemical cell with a controlled amount of current passed	Unexpected structural distortion may cause electrical conductivity decay
		Doping/dedoping is highly reversible and clean polymer can be retrieved	
		Can be achieved with many dopant species	
Photo doping	Radiation energy of light beam	Charge carrier is formed without chemical compound (dopants)	The electrical conductivity disappears rapidly when irradiation is discontinued due to recombination of electrons and holes
		No distortion of the material structure	
Non-redox doping	Protonic acid strength	Number of electrons generally does not change	Depends on the degree of oxidation of CPs and degree of protonation of the material
			Low conductivities are observed for some CPs

(continued)

Table 1.3 (continued)

Doping method	Controlled variables	Advantages	Disadvantages
Charge-injection doping	Applying an appropriate potential on the polymer structure	Does not generate counter ions Minimized distortion	Coulombic interaction between charge and dopant ion is very strong and can lead to change in the energetics of the system

$$\sigma = \sigma_0 \exp\left(\frac{T_0}{T}\right)^{1/(n+1)} \tag{1.2}$$

W(T) has a negative temperature coefficient and can be determined as

$$\log_{10} W(T) = A - x \log_{10} T \tag{1.3}$$

where $A = x \log_{10} T_0 - \log_{10} x$. From a plot of $\log_{10} W$ against $\log_{10} T$, the slope x and dimensionality of the sample can be obtained.

In the case of the critical regime, $W(T)$ is independent of temperature, and the slope of the $W(T)$ plot is zero. The resistivity is not activated; however, the conductivity is given by a power law

$$\sigma(T) = aT^{\beta} \tag{1.4}$$

In the metallic regime, $W(T)$ has a positive temperature coefficient and resistivity (ρ) is finite ($T \to 0$). Therefore, the electronic conductivity in the metallic regime is calculated by

$$\sigma = \sigma_0 + mT^{1/2} + BT^{p/2}, m = \alpha\left[\frac{4}{3} - \gamma\left(\frac{3F_\sigma}{2}\right)\right] \tag{1.5}$$

where σ_0 is the zero-temperature conductivity, B is a constant depending on the localization effects, α is a parameter depending on the diffusion coefficient, γF_σ is the interaction parameter, and p is determined by the scattering rate (for electron–phonon scattering, $p = 3$; for inelastic electron–electron scattering, $p = 2$ in the clean (weakly disordered) limit or 3/2 in the dirty (strongly disordered) limit). The term "$mT^{1/2} + BT^{p/2}$" is mainly determined by the interaction and localization contributions to the conductivity. In disordered materials, electron–electron interactions play an important role in transportation at low temperatures.

1.1.3 Charge Carrier Transport Models

By doping, the undoped polymer can be changed from insulator to conductor. In solid-state physics terminology, the oxidation and reduction processes correspond to *p*-type and *n*-type doping, respectively. During the *p*-type doping, the electron is extracted from the highest occupied molecular orbital (HOMO) of the polymer to the dopant, creating a hole. Conversely, the electron transfers from the doping species to the lowest unoccupied molecular orbital (LUMO) of the polymer for the *n*-type doping. The redox processes lead to the charge carrier in the form of polarons, bipolarons, or solitions moving along the polymer chains, producing the electrical conductivity. As to PTh in Fig. 1.2, positive polarons and bipolarons are generated by *p*-type doping and negative polarons and bipolarons are produced by *n*-type doping.

Both polarons and solitons act as charge carriers and facilitate the electrical conductivity of trans-polyacetylene. For trans-polyacetylene with odder number of carbons, two geometric structures (A and B phases) with the same energy form because the single and double bonds exchange electrons (Fig. 1.3a). Neutral soliton refers to the radical form with unpaired π electron, as is produced between the two phases (Fig. 1.3d). Positive or negative soliton is generated when the neutral soliton is oxidized or reduced by doping (Fig. 1.3c, e). The interaction between the charged soliton causes soliton band located in the middle of HOMO and LUMO of the polymer. It becomes larger when the doping level increases (Fig. 1.3b).

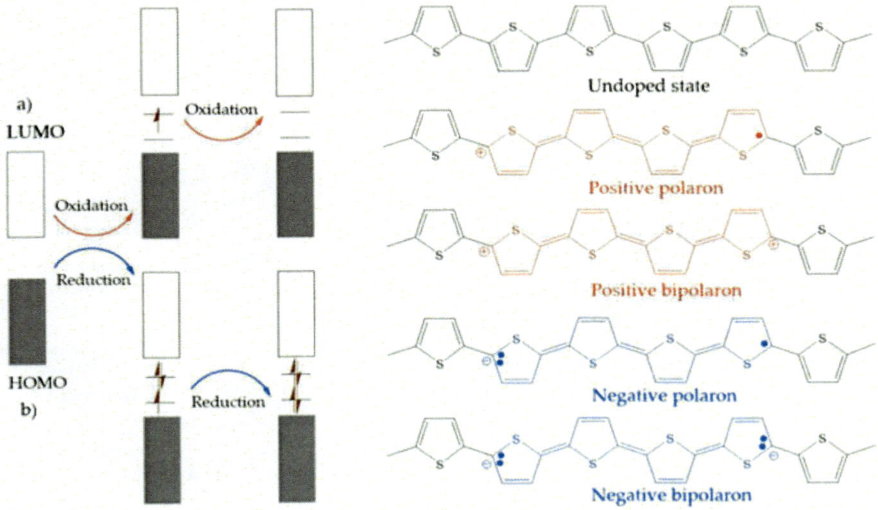

Fig. 1.2 Electronic band and chemical structures of polythiophene (PTh) with **a** *p*-type doping and **b** *n*-type doping. Reprinted from ref. [12], copyright 2017, with permission from MDPI

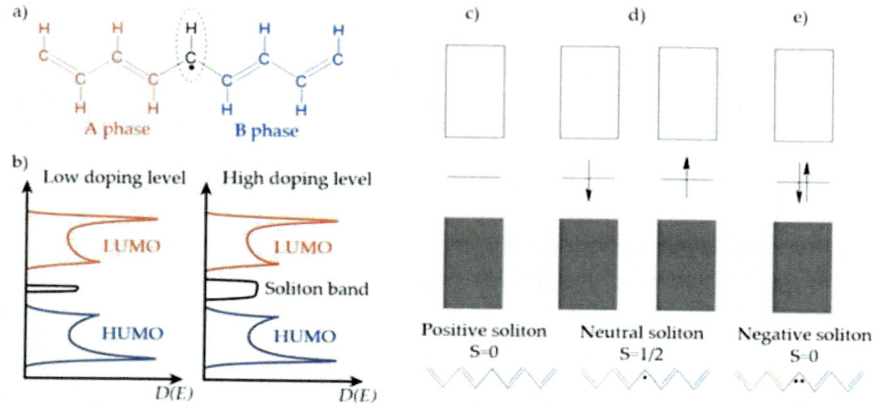

Fig. 1.3 **a** Schematic illustration of the geometric structure of a neutral soliton on a trans-polyacetylene chain; **b** a soliton band with light doping (left) and heavy doping (right); the band structure of trans-polyacetylene containing **c** a positively charged soliton; **d** a neutral soliton; and **e** a negatively charged soliton. Reprinted from ref. [12], copyright 2017, with permission from MDPI

1.1.4 Multiscale Charge Transport

Figure 1.4 shows the cartoon representation of typical microstructure of conductive polymers. Obviously, there are amorphous region and ordered region comprised of either aggregates or crystallites. The charge transport in the conductive polymers depends on the electron delocalization and π-orbital overlap. In the crystallites, the charge transport can be highly anisotropic. The rate is the fastest along the polymer

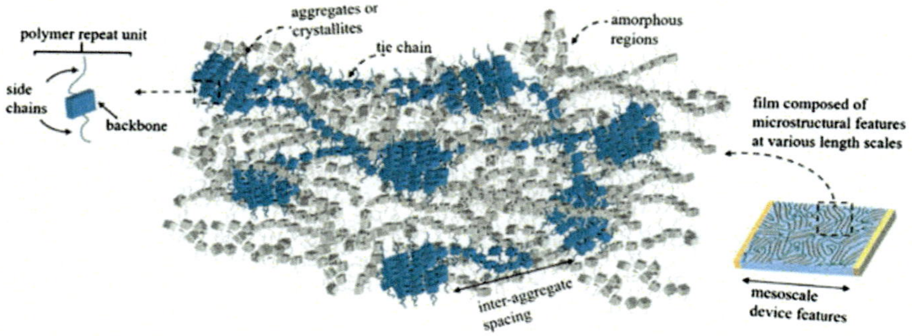

Fig. 1.4 Cartoon representation of typical microstructure of conductive polymers. The polymer thin film comprises both ordered and amorphous regions, which are typically much smaller than the device dimension. The ordered domain can be made up of either aggregates or crystallites. Polymer tie chains provide the connectivity between the ordered domains. Reprinted from ref. [10], copyright 2019, with permission from Wiley Periodicals, Inc

backbone and next fastest along the π-stacking direction. Nearly, no charge transports in the direction of side-chain stacking. Moreover, the charge transport is also relative to the crystalline texture in that the in-plane orientation of crystallites is more important than their out-of-plane orientation.

The polymer structure plays a critical role in the charge transport of the conductive polymers. Macroscopic charge transport of the conductive polymers requires the interconnected ordered domains. The intrinsic properties and the microstructural features of the conductive polymers relevant to the charge transport are summarized in Fig. 1.5.

The long-range order and high crystallinity increase the charge transport. Meanwhile, the tie chains, which bridge the adjacent crystallites, are important for the charge transport. Therefore, the polymer molecular weight (*MW*) of the conductive polymers is a primary parameter impacting the charge transport. The macroscopic charge transport requires a critical tie-chain fraction of 10^{-3}. The intracrystallite paracrystalline disorder is the bottleneck above this value, and the intercrystallite connectivity below it limits the charge transport.

For instance, the size, fraction, and orientation of ordered regions and the interconnection decide the charge propagation processes in regioregular P3HT. As shown in Fig. 1.6a, the intrachain and interchain charge transport at different rates occur in both the ordered and disordered regions [15]. Figure 1.6b shows the morphology of P3HT measured by scanning tunneling microscopy (STM) [9]. Crystalline domains appear lighter than the amorphous regions connected by tie chains. The morphology for macroscopic charge transport can be optimized by tuning the microstructural parameters via processing conditions.

Owing to the hierarchical structure of the conductive polymers, the charge transport represents a multiscale process. At the persistence length scale of a few nanometers, charge transport originates from the electronic coupling between the connected mers and is efficient along the polymer backbone. When the length is about the size of polymer aggregates or crystallites (typically 10 s of nanometer), multiple interchain hopping is necessary for the charge transport caused by the interchain electronic coupling. And, the rate is two or three orders of magnitude slower than intrachain one. At longer length scale above 100 s of nanometer, the charge transport is limited by the disordered regions or the interchain hopping within the ordered regions.

The planarity of the conductive polymer chain is crucial for the charge transport, especially when π–π interactions between neighboring chains are not sufficient to promote the mesoscale ordering. For example, although the long-range order of IDTBT, an indacenodithiophene–benzothiadiazole copolymer, is very limited, the transistor based on this material demonstrates high mobilities just because of its planar conformation with a largely torsion-free backbone. The transport properties of IDTBT approaches the intrinsic disorder-free transport limits probably because the backbone planarity subsumes the mesoscale structural disorder.

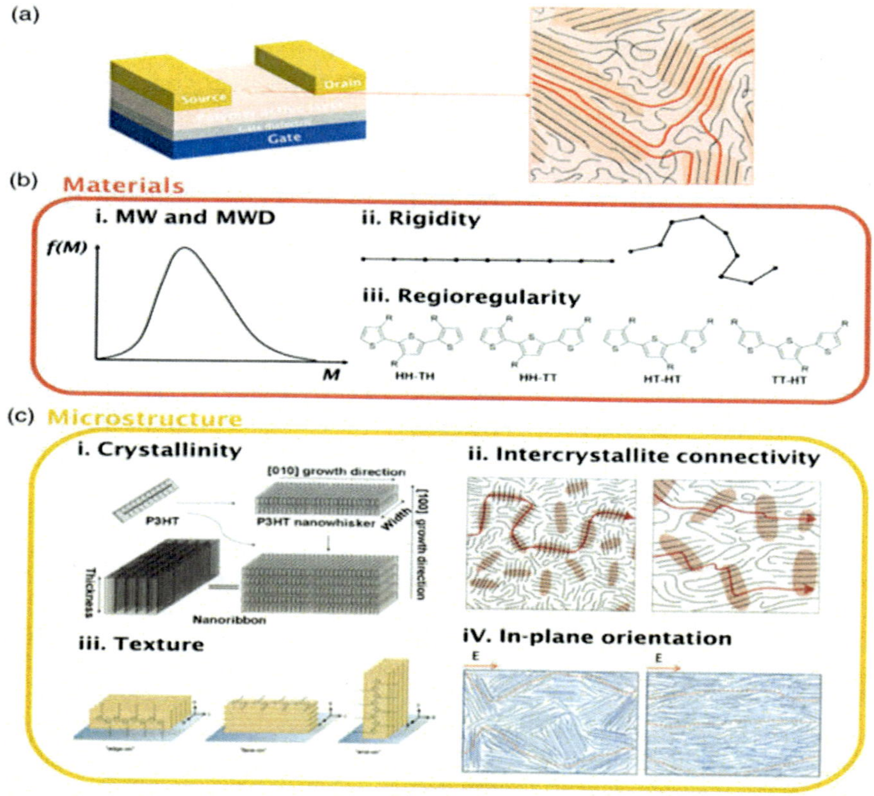

Fig. 1.5 **a** Field-effect transistor in a bottom-gate, top-contact configuration. The microstructure of a semicrystalline polymer active layer is schematically shown on the right. **b** The important materials parameters that impact the overall morphology include: (i) molecular weight (MW) and molecular weight distribution (MWD). (ii) Chain rigidity. Two chains with the same number of repeating units are schematically shown, one rod-like and the other semiflexible. (iii) regioregularity. For example, 3-alkylthiophene has four possible regioisomeric triad structures. H (alkyl group R at the head position) and T (alkyl group R at the tail position). **c** The important structural parameters that can be independently tuned by processing conditions include (i) crystallinity: nanoscale single-crystalline P3HT study has proposed the formation of P3HT nanowhiskers and nanoribbons, and (ii) intercrystallite connectivity: the two schemes show semicrystalline films with small isolated ordered domains and interconnected ordered domains, respectively; (iii) texture or the out-of-plane orientation of crystallites: the example shows three representative textures of P3HT crystallites: edge-on; face-on, also referred to as plane-on or flat-on; and end-on or chain-on; (iv) in-plane orientation: the semicrystalline films are normally isotropic in-plane, but the crystalline domains can be aligned with various special techniques. The two schemes illustrate semicrystalline films with non-orientated domains and oriented domains, respectively. Reprinted from ref. [10], copyright 2019, with permission from Wiley Periodicals, Inc

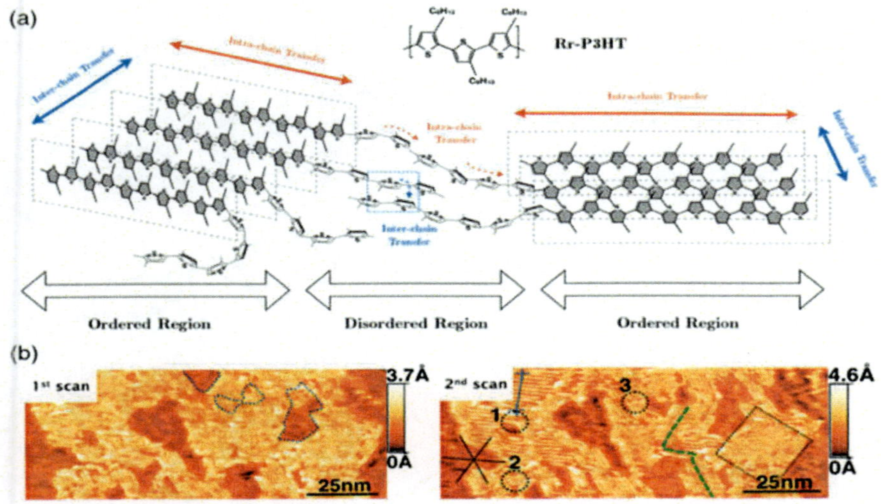

Fig. 1.6 **a** Charge transport processes at different length scales in a thin film comprising edge-on regioregular P3HT (Rr-P3HT). **b** Two scanning tunneling microscopy (STM) images of P3HT monolayer deposited on highly oriented pyrolytic graphite (HOPG), revealing its semicrystalline morphology. The contour of a chain connecting two neighboring ordered domains is underlined with a green dotted line in the second scan. Reprinted from ref. [10], copyright 2019, with permission from Wiley Periodicals, Inc

1.1.5 Multilevel Investigation of Charge Transport

In order to approach the intrinsic properties of conductive polymers, it is essential to comprehensively investigate their charge transport from different aspects. Figure 1.7 highlighted multilevel investigation on the charge transport in conductive polymer, from macroscale film, single crystal at micro-/nanoscale to polymer chain at molecular scale.

Conductive polymer films with optimal structure and order can enhance the charge transport and thus the performance of their devices. Ordered TA-PPE films were fabricated via a combination of "drop-casting self-assembly" and the friction transfer technique. The polymer chains oriented with side chains perpendicular to the substrate improve the charge transport in the two-dimensional direction in the polymer films. PEETs based on the ordered TA-PPE films exhibit charge carrier mobility of ~4.3 $\times$ 10^{-3} cm^2 V^{-1} s^{-1}, on/off ratio of 3.8 $\times$ 10^4, and threshold voltage of −28 V.

Improving the alignment of polymer chains in solid state to extreme degree is beneficial for fairly evaluating and studying the nature of charge transport. Especially single crystals with long-ranged molecular order, no grain boundaries, and low defect density are best candidates. Based on organic transistors of small single crystal, well-defined charge transport physics consists of transport-temperature dependence, structure–property relationship, and anisotropic transport property. Furthermore, the

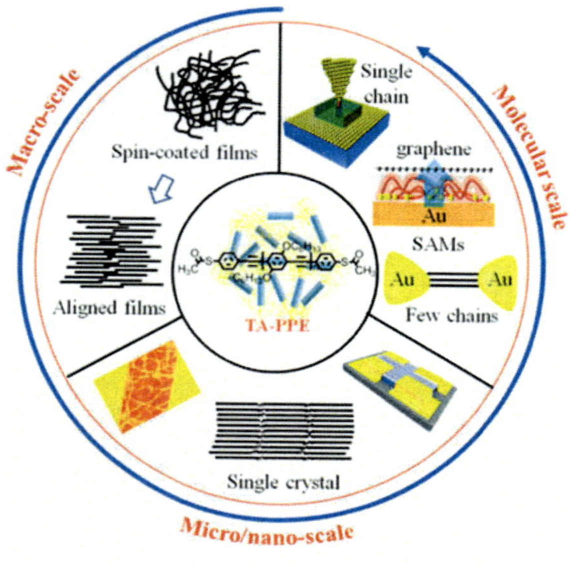

Fig. 1.7 Multilevel investigation of charge transport in conductive polymers. Reprinted from ref. [7], copyright 2016, with permission from American Chemical Society

devices had better be based on a few even a single polymer chain to investigate the mechanism of charge transport from the molecular level.

1.1.6 Effect of Film Morphology on the Charge Transport

Morphology of the conductive polymers plays a critical role in the charge transport and in turn the performance of their devices. Controlling the morphology can avoid the morphological defects such as misaligned crystalline grains and grain boundaries and thus improve the intrinsic charge transport characteristics.

The charge carrier mobility of the conductive polymer thin film is typically impeded by the lower intra and intermolecular interaction and the poor orientation of polymer chains. Therefore, it is an important challenge to optimize their morphology and alignment of the polymer thin films. Great progress has been achieved on controlling the morphology of the conductive polymer thin films via various processing techniques, including thermal annealing, polymer–dielectric interface modification, solvent-vapor annealing, solution treatments, and film deposition methods. Among them, solution treatments consist of tuning solubility, solution aging, adding nucleating agents, sonication, and UV irradiation. And, the film deposition methods include drop casting, spin coating, dip coating, and shear coating.

1.1.6.1 Solution Treatments Prior to Film Deposition

Solvent Solubility Tuning

The morphology of conductive polymer thin film such as shape and size of crystallites, crystallinity degree, and phase separation is influenced by the physical interaction between solvent and polymer. In binary solvent system, the major good solvent interacts with the incorporated lower boiling poor solvent via hydrogen bond, having positive influence on the molecular ordering and charge transport characteristics. Varying the solution temperature such as solution cooling results in different solvent solubility relative to conductive polymers. Besides these methods, other approaches to modulate the morphology consist of selecting marginal solvents and solution aging, etc.

Nucleation-Inducing Agent

Incorporation of heterogeneous additives, i.e., nucleation agents, can tune the formation of the polymer crystallites. Nucleation agents thermodynamically facilitate the polymer crystallization by providing the heterogeneous surface for the polymer solution. The commonly used nucleation agents consist of cerato ulminm (CU) and inert additives such as tris-tert-butyl-1,3,5-benzenetrisamide (BTA) and 1,3:2,4-bis(3,4-dimethylbenzylidene)sorbitol (DMDBS), etc.

Ultrasonic Treatment

It has been proved that the nucleation and crystallization of conductive polymers are initiated and promoted by ultrasonic treatment. Zhao et al. hypothesized that the disorder–order transformation within the individual polymer chains in solution was promoted by the ultrasonication and the cofacial stacking formed between the polymer chains. Namely, the polymer chain disentangled during the acoustic cavitation processes involved in ultrasonication, the chain conformation change was subsequently induced by a shear and finally well-ordered polymer aggregates formed through π-π interactions.

UV Irradiation

As aforementioned, ultrasonic treatment is favorable for nucleating the conductive polymer crystallites, as is a key factor impacting the charge transport. However, the ultrasound agitation results in the polymer nanofibers of less than ~200 nm in length so that a lot of grain boundaries impeded the efficient charge transport. In contrary, under the UV irradiation, the conductive polymer formed long fibers of above 1 μm, suppressing the grain boundaries. Reported by Chang et al., the mobility of the corresponding films remarkably improved using a synergistic combination of ultrasonication followed by UV irradiation.

1.1.6.2 Film Deposition Techniques

The solution treatment above remarkably enhanced the crystallinity of the conductive polymer film. However, the poor grain boundaries also impede the charge carrier mobility. To avoid this, alternative film deposition strategies were adopted, e.g., template-guided solution shearing, capillary force-assisted film deposition, centrifugal force driven film deposition, etc. Fortunately, by these methods, the polymer chains and their crystalline aggregates can be aligned so that high performance of the based devices was achieved for wide commercial application.

Template-Guided Solution-Shear Coating

Solution shearing techniques are scalable to industrial manufacturing process. By this means, locally ordered crystalline domains of conductive polymers were obtained due to the multiple degrees of conformational freedom. Highly aligned small-molecule single crystals can be effectively deposited by template-guided solution shearing coating, and the conductive polymer film represents high charge carrier mobility. Chang et al. reported the P3HT film with aligned crystalline nanowires (NWs) via a solution shear coating of P3HT solution containing the NWs, reducing the concentration of grain boundaries formed within the sheared polymer films.

Figure 1.8a, b showed more improved orientation and brighter birefringent textures of the sheared P3HT NW films than the spin-coated pristine, sheared pristine and spin-coated ones. The solution shearing enhanced the intramolecular and intermolecular ordering and coplanarization of the polymer chains. As shown in Fig. 1.8c, d, the charge carrier mobility of P3HT NW films (~0.32 $cm^2{\cdot}V^{-1}{\cdot}s^{-1}$) is about 53-fold greater than that of the spin-coated pristine film (~6 × 10^{-3} $cm^2{\cdot}V^{-1}{\cdot}s^{-1}$). It was confirmed that this approach can be applied to other conductive polymers [2].

Capillary Force-Assisted Film Deposition

Capillary force-assisted film deposition facilitates the assembly of the polymer chains and the unidirectional alignment of the polymer films, resulting in highly aligned conductive polymer thin films. By this method, unique optical and electrical anisotropies were acquired for the thiophene-based polymer films such as poly(2,5-bis(3-alkylthiophene-2-yl)thieno[2,3-b]thiophene) (PBTCT) and PBTTT.

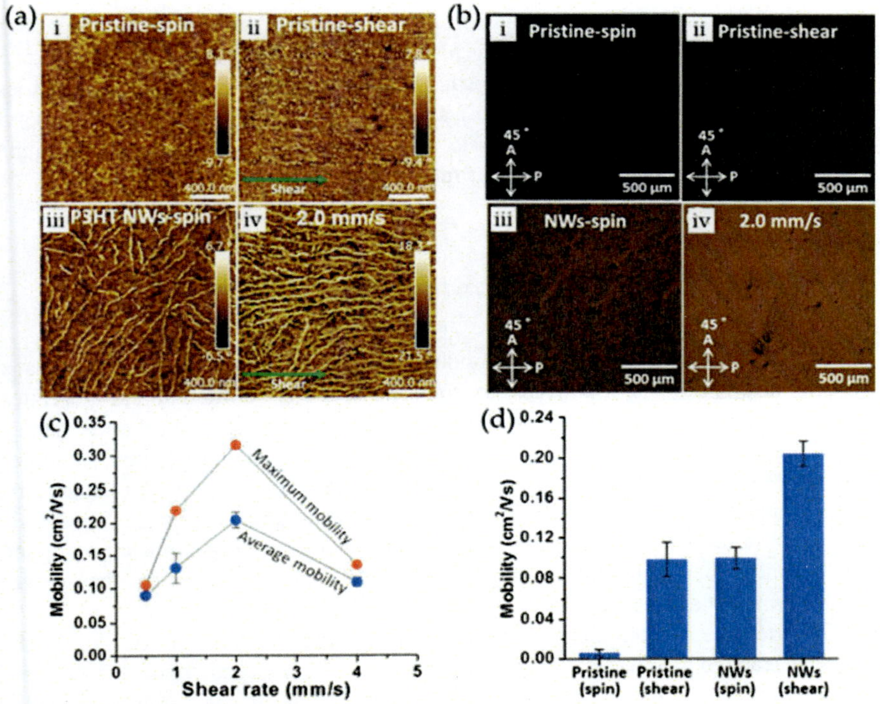

Fig. 1.8 **a** AFM and **b** POM images of pristine P3HT thin films: (i) spin-coated; and (ii) shear-coated at 2.0 mm/s from a pristine solution; and P3HT NW films: (iii) spin-coated; and (iv) shear-coated at 2.0 mm/s from a P3HT solution containing P3HT NWs. **c** Average and maximum field-effect mobilities of P3HT NW films shear-coated as a function of the shearing speed. **d** Mobility comparison of P3HT pristine films spin-coated and shear-coated at 2.0 mm/s and P3HT NW films spin-coated and shear-coated at 2.0 mm/s. Reprinted from ref. [3], copyright 2017, with permission from MDPI

Centrifugal Force-Driven Film Deposition

Besides capillary force, the alignment of conductive polymer films can also be achieved by centrifugal force. Yuan et al. first reported that 2,7-dioxtyl[1]benzothieno[3,2-b][1]benzothiophene (C8-BTBT) formed a highly aligned, meta-stable crystal packing structure within a PS matrix. During the deposition process, C8-BTBT/PS blend solution on the substrate far away from the spin center was fast rotated. The resultant film combined a slightly reduced in-plane intermolecular spacing and the aligned crystalline grains. Hence, the hole mobility was highly increased from 3 ~ 16 to 25 $cm^2 \cdot V^{-1} \cdot s^{-1}$.

The charge transport characteristics of the conductive polymer films were influenced by the morphology parameters, e.g., crystallinity, grain size, grain orientation, grain boundaries. During the solution processing, many processing parameters such as polarity, solubility, volatility and viscosity, pressure, temperature, fluid flow speeds, roughness, and surface tension of the substrate influence the morphological parameters. How the processing parameters affect the morphology is essential to be understood for optimizing the device with high performance. Besides, improving environmental stability, self-assembly mechanism, and process scale-up are also challenge. However, the flexible, stable, low-cost, high-performance organic electronics will be realized in the near future.

1.1.7 Opportunities for Improving Charge Transport

Although the microstructure of the conductive polymer influences the charge transport in various aspects, it is necessary to carry on the experimental and theoretical studies on the interconnected ordered domains and thus understand the mechanism due to the macromolecular nature. Upon the postsynthesis chain alignment strategies, the intrachain charge transport of conductive polymer is more efficient because the macromolecular nature of conductive polymers is leveraged. During the process, the intrachain defects are reduced and the polymer chains and crystallites are aligned. For instance, mechanical stretching can increase the effective persistence length, resulting in higher conductivity.

The inefficient interchain hopping in amorphous regions of the conductive polymer leads to the charge transport barriers. It is necessary to connect the ordered domains by tie chains. Such physical connections relate to the persistence length, contour length of the polymer chain and the characteristic spacing between neighboring ordered regions. So, the conductive polymers need high molecular weights and high crystallinity. In addition, due to higher persistence length, the polymers with rigid coplanar backbones facilitate the formation of tie chains.

Usually, the polymer chain length of the conductive polymer is much smaller than the typical device dimensions. Therefore, the macroscopic charge transport requires the interchain hoping within the ordered domains and can be enhanced by improving local interchain coupling. Moreover, because the backbone planarity results in more ready π-stacking of the less-twisted polymer chains and improvement of the interchain ordering, the interchain transport is enhanced. To improve the local packing, the π–π stacking distance can be modulated by side chains.

1.2 Electrochemical Properties

1.2.1 Reversible Oxidation/Reduction

The sensors and capacitors based on the conductive polymer originate from the mechanism that their doping/dedoping correspond to the charge/discharge process. By doping, the charge carriers form and the geometric structure changes. And, dedoping recovers the original geometric structure. Doping is categorized into *n*-doping (electrooxidation) and *p*-doping (electroreduction). For n-doping of the conductive polymers, the electrons transfer to the polymer backbone, into which the countercations intercalate from the electrolyte solution to balance the overall charge. In contrast, the electrons are extracted from the polymer backbone and then the counteranions insert into the polymer backbone to leverage the electronic charge.

By measuring the current obtained from an applied potential and at a fixed scan rate, cyclic voltammetry (CV) is considered as the most powerful electrochemical technique to study the redox process of conductive polymer. Negatively and positively charged polymer chains are produced by the reduction and oxidation, respectively. The dopant ions play a critical role in the function of current versus potential scan rate (v) and the voltammogram. For instance, the cyclic voltammogram of PPy doped with BF_4^- is symmetrical among the applied potential voltage of $-0.4 \sim +0.3$ V and the current is proportional with v. However, when very large or sluggish dopant ions are used for thick films, diffusion controls the charging/discharging process. Thus, the current increases proportionally with $v^{1/2}$ and the voltammogram turns asymmetrical.

The electrochemical properties and the electron transfer capability of the conductive polymers, especially at the nanometer scale, highly depend on the morphology and the structural characteristics. For example, CV measurement was carried on PANI with three different nanostructures of nanospheres, nanorods, and nanofibers in the sulfuric acid solutions. In spite of the similar shape, the integrated area in the CV voltammogram of PANI nanofiber is the largest and that of PANI nanosphere is the smallest, as shown in Fig. 1.9a [11]. Figure 1.9b indicates the impact of the scan rate on the peak current. Figure 1.9c shows that the anodic (I_{pa}) and cathodic (I_{pc}) peak currents increase linearly with the scan rate because the redox process is surface-controlled. The function of the anodic (E_{pa}) and cathodic (E_{pc}) peak potentials versus logarithmic scan rate is shown in Fig. 1.9d [13]. Based on the Laviron theory, the electron transfer coefficient (α) is calculated to be 3.6×10^{-1}–3.7×10^{-1} and the electron transfer rate constant (k_s) increases in the order of nanospheres < nanorods < nanofibers (nanospheres: 2.6×10^{-1} s^{-1}, nanorods: 3.1×10^{-1} s^{-1}, nanofibers:4.3×10^{-1} s^{-1}).

In a certain potential range, the redox process of the conductive polymers is stable and reversible. But overoxidation in accompany with irreversible structural change occurs at higher potentials. Overoxidation of the conductive polymers results in the rapid structural degradation and electroactivity loss. Besides, on the basis of the studies, the overoxidation is found to highly depend on the electrolyte pH and the presence of nucleophiles.

1.2.2 Pseudocapacitance

Electrochemical capacitors are categorized into electrochemical double-layer capacitors (EDLCs) and pseudocapacitors. Pseudocapacitors usually possess higher energy

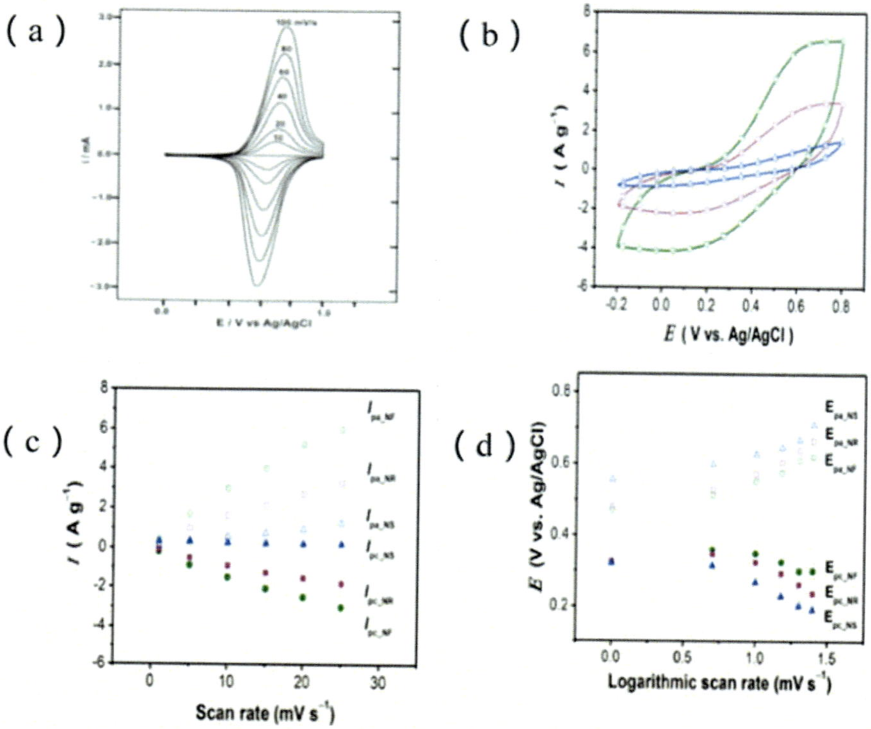

Fig. 1.9 **a** Voltammograms of a poly(*N*-phenyl-1-naphthylamine) film on Pt in 1 M $LiClO_4/CH_3CN$ solution at different scanning rates. CV analysis of PANI nanostructures with three different shapes (nanosphere, NS; nanorods, NR; and nanofibers, NF) performed in a 1 M sulfuric acid solution. **b** Cyclic voltammograms of electrodes consisting of PANI nanostructures at the same scan rate (25 $mV{\cdot}s^{-1}$); **c** plots of the peak current (the anodic peak current, I_{pa}; the cathodic peak current, I_{pc}) versus the scan rate; and **d** plots of the peak potential (the anodic peak potential, E_{pa}; the cathodic peak current, E_{pc}) versus the log of the scan rate. Reprinted from ref. [12], copyright 2017, with permission from MDPI

density and specific capacitance than EDLCs. During the charge/discharge process, the performance of the pseudocapacitors is similar to the battery because, based on the Faradic mechanism, the charge stores in the electrode of pseudocapacitors by redox process. The charging of the conductive polymers corresponds to an oxidation process. In this process, the electrons are extracted from the conductive polymers and the anions from the electrolyte enter the polymer backbone. And, the discharging process is on the contrary. For example, the redox reaction of PANI in sulfuric acid (*p*-doping) can be expressed as follows:

Charging process (oxidation): $PANI + nSO_4{}^{2-} \rightarrow PANI^{2n+}: nSO_4{}^{2-} + 2ne$.

Discharging process (reduction): $PANI^{2n+}: nSO_4{}^{2-} + 2ne \rightarrow PANI + nSO_4{}^{2-}$.

The capacitance of the pseudocapacitors based on the conductive polymers can be acquired by conducting CV on a three-electrode cell. The specific capacitance can be calculated from the redox cycles in the CV curves according to the following equation:

$$C = \frac{1}{mv(V_a - V_b)} \int_b^a i dV \tag{1.6}$$

where m is the mass of deposited material, v is the potential sweep rate, V_a and V_b are the high and low potential limits of the CV tests and i is the discharge current corresponding to the reduction peak in the case of *p*-doping (and the oxidation peak in the case of *n*-doping).

Mechanical stress in CP films has been linked to the cycle life of pseudocapacitors. Hu et al. reported cycle stability of an HCl-doped PANI film in 1 MH_2SO_4 at a rate of 20 $mV \cdot s^{-1}$. By the 500th cycle, the reduction peak virtually vanished, suggesting that the capacitor electrode was unstable. The insertion/extraction of dopants into and out of the polymer chain causes volume change during redox switching. During charge/discharge (doping/de-doping) cycles, substantial volumetric swelling, shrinkage, and cracking of CPs frequently result in mechanical damage of the polymer structure, degeneration of electrochemical performance, and fast decay of capacitance. As a result, attaining long-term cycle stability is a big problem for CP-based high-performance pseudocapacitors.

As their redox status changes, CPs swell and deswell, resulting in the volumetric change. This swelling/deswelling may be divided into two components: intrinsic swelling caused by variations in bond lengths and polymer backbone conformation and osmotic expansion of the polymer phase. This phenomenon has been offered as a potential use for a new generation of actuators. Chemical and electrochemical control of CP actuation is possible. Figure 1.10 schematically depicts the mechanism of electrochemical actuation in CPs. When electrons are removed from polymer chains by oxidation, they become positively charged. To ensure overall charge neutrality, small anions are injected into the polymer matrix. Ionic crosslinks develop between polymer chains and anions, causing the overall volume to grow. Negative voltages can be used to eject dopant anions during electrochemical reduction (Fig. 1.10a–c). This anion-driven actuation essentially induces swelling upon oxidation and deswelling upon reduction.

In other circumstances, when large anions are included during polymerization, they become immobilized and hence trapped inside the polymer structure. The CPs then enlarge further when cations from the solution are added for charge equalization during reduction (Fig. 1.10c–e). This sort of actuation is also known as cation-driven actuation. According to electrochemical quartz crystal microbalance research, the expansion reduces as the electrolyte concentration increases. The osmotic effect can explain the overall volume change generated by solvent molecules and ions. The effects of solvent molecules and electrolyte ions were studied on free-standing films of PPy doped with dodecylbenzene sulfonate.

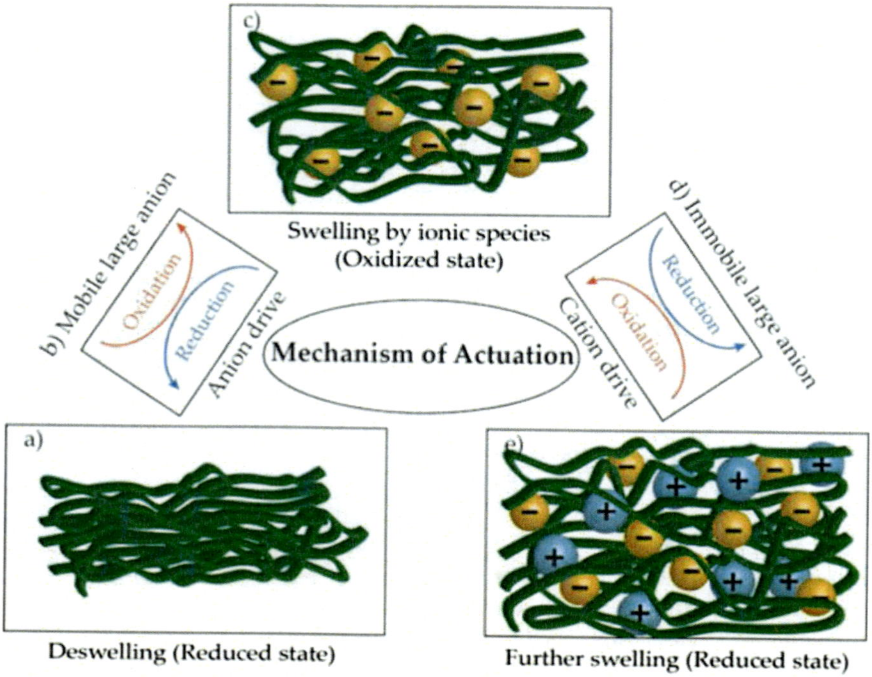

Fig. 1.10 Mechanism of electrochemo-mechanical actuation in CPs(**a**, **c**, **e**). Volume changes in CP via two different redox pathways (**b**, **d**). Reprinted from ref. [12], copyright 2017, with permission from MDPI

This finding is consistent with the results of Maw et al. and Aydemir et al., who discovered that solvent molecules migrate in and out of the electrolyte-associated layer. Many important aspects influence this actuation, including the chemical/physical characteristics of the CP and the size/type of dopant and electrolyte ions. In addition to adjusting the characteristics of the primary components, such as the CP and electrolytes, efficient actuation systems are critical for significantly improving the performance of CP-based soft actuators. Several noteworthy actuation systems, including linear actuators, bilayer–trilayer actuators, and out-of-plane actuators, have been developed.

1.2.3 Electrochromism

Electrochromism of CPs refers to noticeable color change affected by reversible redox processes. Utilizing the complexation of CPs, the absorption spectra can be enlarged from the visible to the infrared regions. For example, electrochromic properties of PANI (p-doping) have been broadly examined in acid, salt, and organic

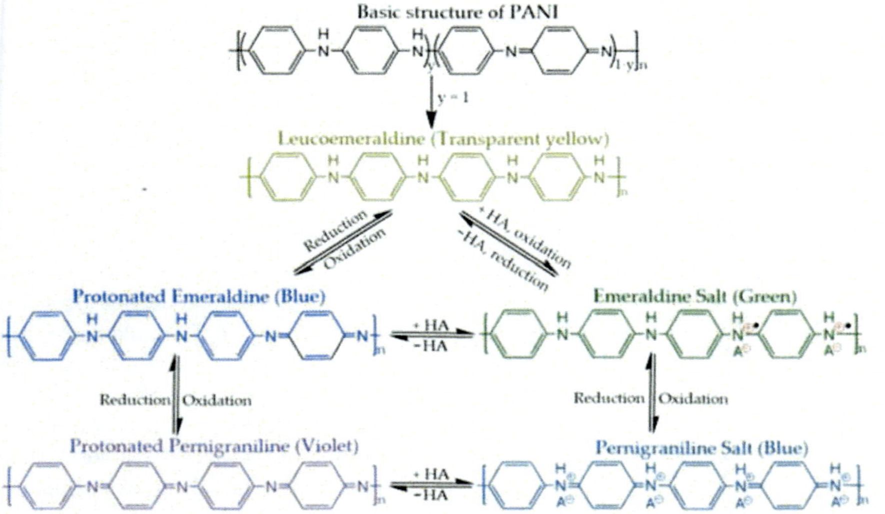

Fig. 1.11 Different redox/protonation states and colors of PANI. Reprinted from ref. [12], copyright 2017, with permission from MDPI

media by CV. Different redox/protonation forms of PANI shown in Fig. 1.11 relate to the pronation/deprotonation and insertion/extraction of anions. PANI displays color changes from transparent yellow to green, blue, and violet. Radical cations are formed by the injection of protons or anions to the nitrogen atoms in PANI chains.

The mechanism of electrochromism results from the injection/deinfusion of dopant ions through doping/dedoping processes. The energy gaps for the π-π^* transition is reduced because the doping initiates reorganizing the electronic structure of the polymer. The generation of sub-bands by charge carriers like polarons and bipolarons alters the CP absorbance, resulting in color shifts. The films in the undoped condition are usually colorless and transparent when the energy gap for pristine CP thin films is greater than 3.0 eV. In the visible spectrum, thin films in the doped state can also show the greatest absorption spectra. In the undoped condition, pure CPs with an energy gap of about 1.5 eV strongly absorb visible light and generate high-contrast colors. The absorption wavelength shifts to the near-infrared region upon doping. In different redox conditions, the colors of CPs may differ.

Chemical structures, redox capability, temperature, and the pH of the electrolyte solution all influence the electrochromic characteristics of CPs. The speed at which protons/dopant ions migrate into and out of the polymer matrix has a significant impact on the color change switching time. To improve the rate of color change, the morphology and microstructural features of CPs, such as their diameters and porosity, should be adjusted.

CPs are appealing candidates for electrochromic applications such as electrochromic displays, rearview mirrors, and smart windows because of this intriguing characteristic. To provide a long life cycle duration, rapid color change switching, and strong color contrast, it is required to improve the features of CPs in order to extend their use in electrochromic devices.

1.3 Wettability of Conductive Polymers

To accurately characterize the surface wettability, apparent and dynamic contact angle measurements are frequently required. The Young–Dupre equation, as stated by Marmur, yields the stable apparent contact angle on a smooth, homogeneous, stiff, insoluble, and non-reactive solid surface. This contact angle is calculated using the formula $\cos\theta^{Y} = (\gamma_{SV} - \gamma_{SL})/\gamma_{LV}$, where γ_{SV}, γ_{SL} and γ_{LV} are the solid–vapor, solid–liquid, and liquid–vapor surface tensions, respectively. The surface is intrinsically hydrophilic if the contact angle is less than 90° and intrinsically hydrophobic if the contact angle is greater than 90°. Superhydrophilic surfaces have very narrow apparent contact angles (between 5° and 10°) and water spreads quickly. Superhydrophobic surfaces have an apparent contact angle greater than 150°, whereas the hysteresis (H) and sliding angles (α) provide information on the adhesion between the droplet and the surface.

Conductive polymers offer special features that make surface hydrophobicity and roughness geometry easier to adjust. Because of their many doping states, conductive polymers may include doping agents into their structures, allowing them to modify or switch the surface wettability from superhydrophobic to superhydrophilic. Various conductive polymers can be employed, with different optical and electrical properties, and different substituents can be added to conductive polymers to increase or decrease surface hydrophobicity or add other capabilities.

1.3.1 Formation of Soluble Polymers

Conductive polymers can be prepared in solution by the chemical oxidative polymerization using appropriate oxidizing agents. The soluble conductive polymer in the used solvent can be deposited by spin-coating, dip-coating, or spraying, and the film surface is always smooth. A low-voltage electrochemical approach was used to reversibly switch poly(3-alkylthiophene)s between hydrophobic and hydrophilic states, changing the polymer state from a dedoped to a doped state (Fig. 1.12a). Various anions were introduced during the process, with SO_4^{2-}anions having the greatest impact: water ≈105.9° for the dedoped state and water ≈76.7° for the doped state. The effect on the doping anions was greatly enhanced by deposition of this polymer on a substrate patterned with micrometer-sized posts: water ≈147.4° for the dedoped state and water ≈62.2° for the doped state.

For a PEDOT derivative containing an imidazolium substituent, multiresponsive surfaces of with reversible switchable wettability were created (Fig. 1.12b) [6]. While the properties of these polymers were investigated in terms of their

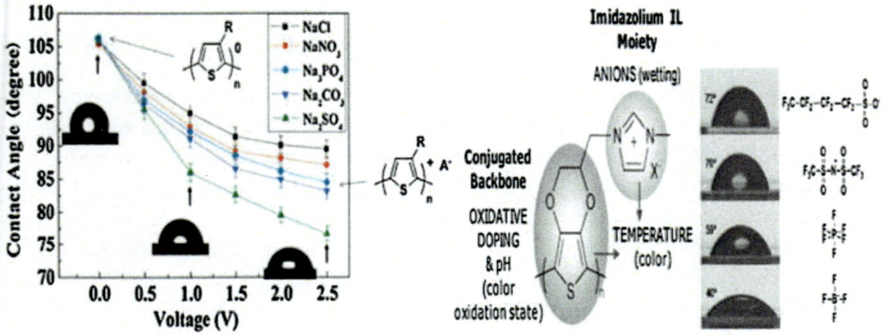

Fig. 1.12 **a** Contact angles of smooth poly(3-hexylthiophene) films at different oxidation potential and using different electrolytes. **b** Switchable surface wettability by exchange of anions in a PEDOT polymer containing imidazolium moieties. Reprinted from ref. [5], copyright 2014, with permission from Elsevier

sensitivity to pH, temperature, and oxidizing agents, the wettability of spin-coated films was reported in terms of anion exchanges due to the imidazolium moiety. Exchanging the counteranion of the imidazolium moiety with fluorinated bis(trifluoromethane)sulfonamide or nonafluoro-1-butanesulfonate anions increases the surface hydrophobicity from 40° to 70–72°. To improve wetting, the polymer was deposited as rough surfaces on ZnO nanowire arrays. After anion exchange, the polymer wettability increased from 24° to 107°.

1.3.2 Formation of Micro-/Nanostructures in Solution by Self-Assembly

If the synthesized polymer is not soluble in the used solvent but is stable in solution, different polymer micro- and/or nanostructures can be produced in solution depending on the conditions and monomer used. Polyaniline is known to form nanostructures in solution, specifically nanofibers, due to the presence of hydrogen bonds. The self-assembly of these 1D structures can result in 3D microstructures, which can then be deposited on substrates via methods such as dip-coating, spin-coating, or spraying. When the polymer cannot form nanostructures by itself, surfactants can be applied to activate the formation of these structures. Otherwise, by the combination with micro- or nanoparticles or other chemicals, the composite materials of conductive polymers can be produced.

1.3.3 Use of Hydrophobic Doping Ions

The surface morphology at micro- or nanoscale can be easily controlled via the polymerization methods, such as electrochemical deposition on the substrate, the chemical oxidative polymerization in solution, or the vapor-phase polymerization. Thus, the surface wettability changes from superhydrophilicity to superoleophobicity and can be adjusted by varying the nature of the polymerizable core, the functionalization with different hydrophobic/hydrophilic substituents or introducing diverse hydrophobic/hydrophilic doping ions.

The counterions of the dopants enter the backbone of the conductive polymers to neutralize the charges. Polyaniline with rough surfaces exhibits water of 102°, 65°, and 86° for leucoemeraldine, emeraldine, and pernigraniline forms, respectively. If the surfaces are adequately structured, using hydrophobic doping ions produces superhydrophobic surfaces. The surface wettability can change from superhydrophobic to superhydrophilic by the removal of these doping ions. Due to the dedoped state of the conductive polymers as the natural form, the shortcoming of this strategy is that the doping anions were released from the surfaces upon aging and the stability highly depends on the polymer.

As a dopant, sodium dodecyl sulfate (SDS) was used to control the roughness geometry and wettability of polyaniline structures. A multi-scale roughness (microtwigs at nanoscale) able to repel water formed for PANI with water = 138.9° by precisely controlling the SDS concentration. A minor decrease or increase of the SDS concentration yielded superhydrophilic surfaces. Owing to SDS as the hydrophobic function, reduction reaction can switch the surface wettability from hydrophobic to hydrophilic. By this method, the surface hydrophobicity can also be tuned utilizing perfluorinated salts as doping agents. Moreover, if the doping agents are electrochemically removed, the surface wettability shifted from superhydrophobic to superhydrophilic.

The sub-micron polyaniline fibers with helical structure were prepared via the electropolymerization of aniline in the presence of perfluorooctanesulfonic acid and exhibited switchable superhydrophobic properties with low hysteresis (water = 153°, $H = 8°$). Using perfluorooctanesulfonate anions, superhydrophobic micro-/nanostructured polypyrrole, and PEDOT films with switchable wettability were also produced.

Incorporating poly(allylaminehydrochloride) and dextran into poly(3-hexylthiophene)s films can also modify the surface wettability and roughness. The hydrophilic or superhydrophilic surfaces are produced using poly(3,4-ethylenedioxythiophene)/poly(styrenesulfonate) (PEDOT:PSS) as starting materials. The polymers are soluble in solvent after the protons of PEDOT:PSS are exchanged by hydrophobic cations: 1-dodecyl-3-methylimidazolium(DoMIm), hexadecyltrimethylammonium (HDTMAm), and 1H,1H,2H,2H-perfluoro-1-decyl-3-methylimidazolium (PFDMIm). After exchange, the film hydrophobicity increases from 48° to 100°.

1.3.4 Grafting of Substituent on the Monomer Before Polymerization

The sulfonated group provides greater surface wettability. By electrodeposition at constant voltage of functionalized and unfunctionalized PEDOT and ProDOT in dichloromethane and with tetrabu-tylammonium perchlorate, various nanostructures such as nanofibers, nanodots, nanonetworks, and nanotubes were observed. Unfunctionalized EDOT, ProDOT, and EDOT derivatives functionalized with nonpolar functional groups such as alkyl chains generated nanofibers and nanoporous structures preferentially. EDOT derivatives functionalized with polar groups, such as hydroxyl, carboxylic acid, triethylene glycol, and perfluorocarbon and, on the other hand, resulted in nanodots when electropolymerized at 25 °C and tubular structures when electropolymerized at 0 °C. The fluorinated derivatives produced superhydrophobic surfaces, whereas the other derivatives produced superhydrophilic surfaces.

Fluorinated poly(3,4-ethylenedioxypyrrole) derivatives (PEDOP) were shown to be the best candidates for producing superoleophobic characteristics (Fig. 1.13a). They enabled the fabrication of polymer films with microstructures, particularly nanoporosities (50 nm) that impeded the penetration of oil droplets (hexadecane > 140), as shown with re-entrant structures. Indeed, increasing the length of the bridge by one methylene unit (Fig. 1.13b) or replacing the oxygen atoms in the bridge with sulfur atoms (Fig. 1.13c) resulted in the absence of surface nanoporosities and, as a result, significantly lower surface oleophobicity [1, 4]. The introduction of short *F*-butyl tails, on the other hand, resulted in substantially decreased surface oleophobicity (hexadecane $\approx 110°$).

Indeed, because short *F*-butyl tails have fewer interactions than long perfluoroalkyl tails, their high mobility frequently impedes the creation of surface nanostructures during electrodeposition, resulting in substantially lower surface oleophobicity than with longer fluorinated tails. Also investigated were EDOP compounds with two fluorinated tails (*F*-butyl and *F*-hexyl). The presence of the two fluorinated tails resulted in extremely significant steric hindrances during electrodeposition and rather low surface oleophobicity.

1.3.5 Vapor-Phase Polymerization

By vaporizing a monomer and an oxidizing agent, conductive polymers may be produced in the vapor phase. The procedure is referred to as oxidative chemical vapor deposition. Gleason's group created superhydrophobic nanoporous PEDOT films on a variety of substrates (silicon wafers, quartz plates, and paper mats) utilizing EDOT as a monomer and $CuCl_2$ as an oxidizing agent and under well-defined conditions. The size of the nanoporosities may be readily changed by changing the temperature of the substrate. The surfaces were superhydrophobic after chemical vapor deposition of a perfluorinated polyacrylate. A solution comprising pyridine and iron(III) p-tosylate

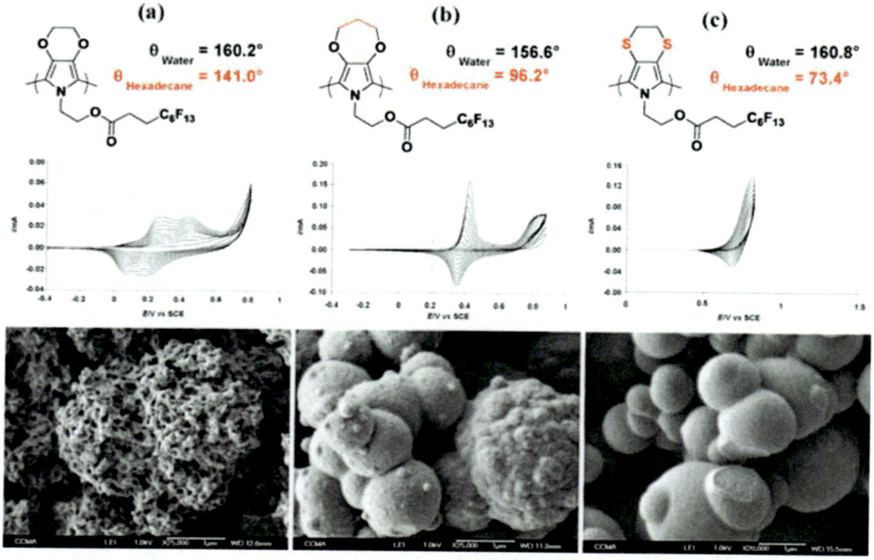

Fig. 1.13 Contact angles with water and hexadecane, cyclic voltammogram and surface morphology of fluorinated **a** poly(3,4-ethylenedioxypyrrole); **b** poly(3,4-propylenedioxypyrrole); and **c** poly(3,4-ethylenedithiopyrrole) obtained by electrodeposition. Reprinted from ref. [5], copyright 2014, with permission from Elsevier

was spin-coated on substrates before exposure to EDOT vapor to make PEDOT films doped with tosylate. The surface contact angle was 52.5°, while that of the reduced state was 30.1°.

1.4 Polyacetylene

1.4.1 Properties and Structure

For all intents and purposes, PA is regarded as a Nobel Prize-winning macromolecule. PA is a conductive polymer with several functional variants that have been widely studied in the literature. Electrical conductivity, photoconductivity, gas permeability, supramolecular assemblages, chiral recognition, helical graphitic nanofiber production, and liquid crystal are some of its important properties. The early finding of electrical conductivity in the doped form sparked a lot of interest in CPs, spawning an intriguing field of study on synthetic metals. PA has a linear polyene chain $[-(HC=CH)_n-]$ as its chemical structure. Because of the existence of repetitive units of two hydrogen atoms, its backbone offers a significant potential for decoration with pendants.

1.4.2 *Polyacetylene Synthesis*

A number of methods can be adopted to synthesize polyacetylene. Among them, Ziegler–Natta catalysis uses titanium and aluminum in the presence of gaseous acetylene. By this method, tuning the temperature and catalyst amount efficiently develops polyacetylene with fine structure. But metal probably exists in the monomer triple bonds. Substituting the catalyst with $CoNO_3/NaBH_4$ results in polyacetylene with stabilities to oxygen and water. The radiation polymerization methods using glow discharge, γ-radiation, ultraviolet, which do not need catalyst and solvents, could also be used in developing polyacetylene.

1.5 Polythiophene

1.5.1 *Properties and Structure*

Polythiophene is intriguing because of its stable conductivity and strong electrical conductivity (10^3 S cm^{-1}), and its conductivity changes depending on the dopant and polymerization. Because CPs are built on conjugated systems, they are nontransparent and refractory by nature, with polythiophene being a notable example. Previous research has studied how the length of conjugated sequences in polythiophene affects conductivity. It was discovered that oligomers composed of 11 thiophene units had conductivity comparable to larger molecular weight polythiophene. This is supported by the discovery that short oligomers of thiophene show polymer characteristics, with conductivity and carrier mobility rising as conjugation length increases up to the thiophene hexamer. Transparency is one of the most significant features dependent on the application, such as photographic films coated with antistatic coatings, which should be greater than 90%. Dilution can improve CP transparency, which impacts conductivity. Polythiophene dilution can be accomplished using a variety of ways, including block copolymerization, alkyl side chain grafting onto the conjugated backbone, blending with a transparent polymer, and generating composites via thiophene polymerization absorbed in an insulating polymer.

Plasma polymerization, electrochemical techniques, and thin layer polythiophene deposition might also be carried out. Plasma polymerization allows for the creation of very thin pinhole-free layers that adhere securely to practically any substrate without the need of any solvents. Electrochemical polymerization of PEDOT, a conductive polymer with high conductivity, on a neural electrode might be used to create semiconducting films. PEDOT can electrically connect with neurons, but it lacks the chemical capability to covalently attach biological molecules. Three functionalized carboxylic acids were also investigated to polymerize 3,4(ethylenedioxy)thiophenes (EDOTs); however, the lengthy linkers to the acid group reduce their water solubility (the preferred polymerization solvent).

PEDOT, being a highly conductive polymer, might be fabricated using electrochemical or oxidative chemical polymerization. Counterion inclusion affects the electrical and morphological characteristics of PEDOT polymer film. As a result, counterion techniques for incorporation might be employed to better understand its essential features. Biological buffers, such as phosphate-buffered saline (PBS) solution, can be employed, particularly in applications requiring biological incorporation, such as implant coatings and biosensors. PEDOT might potentially be doped with different counter ions for electrochemical polymerization.

1.5.2 Compound

By employing appropriate catalysts, there is improved monomer selection and synthesis in chemical synthesis and development. To improve conductivity, various materials were hybridized with polythiophene, including tosylate anion, poly(styrenesulfonate), sodium chloride, lithium perchlorate, sodium phosphate monobasic monohydrate, Br, poly(sodium 4-styrenesulfonate), poly(styrenesulfonic acid)/Au, methyl- or benzyl-capped diethylene glycol, tetraethylene glycol (ethyleneoxide). Furthermore, PEDOT performed better in the presence of EDOT-OH, C_2-EDOT-COOH, C_4-EDOT-COOH, C_2-EDOT-NHS, and EDOT-N_3. Counterions have been widely studied in addition to polyanion poly(sodium 4-styrenesulfonate), sodium chloride (NaCl), lithium perchlorate ($LiClO_4$), sodium phosphate monobasic monohydrate ($NaH_2PO_4H_2O$), and ions in PBS (i.e., KH_2PO_4, NaCl, and Na_2HPO_4).

The cell proliferation and adhesion of PEDOT thin films interact with physiochemical properties and doping for stent application. PEDOT film fabricated by spin coating and vapor polymerization was doped by tosylate anion (TOS) or poly(styrenesulfonate) (PSS). Cell proliferation was conducted by PEDOT:TOS PEGylation with RGD peptides for biofunctionalization and reducing immunogenicity. Owing to its hydrophilicity and nanotopography, the high adhesion of PEDOT:TOS positively influences cytocompatibility. Furthermore, the hydrophilicity and conductivity are enhanced by PEDOT:TOS PEGylation.

Surface modification with a protein-resistant polymer can efficiently improves the biocompatibility and mechanical properties in biometallic implants. The initiation of PEDOTs containing −Br group by electropolymerization has also been investigated. By surface-initiated electropolymerization, [2(methacryloyloxy)ethyl]dimethyl-(3-sulfopropyl)-ammonium hydroxide (PSBMA), poly((oligoethylene glycol methacrylate), 2,2′-Bipyridine, polyethylene glycol methacrylate (POEGMA), and zwitterionic poly([2-(methacryloyloxy)ethyl]dimethyl-(3-sulfopropyl)ammonium hydroxide) were grafted to PEDOT deposited. By quartz crystal microbalance, (QCM) PSBMA-grafted PEDOTs and POEGMA protein resistance were confirmed. The density of brushes and the protein-binding characteristics displayed by the surface can be adjusted by the feed content of the PEDOT-Br containing the initiator in the monomer mixture solution for electropolymerization. In addition, the cell surface adherence is also impeded by the polymer-grafted PEDOTs.

The counterions studied consist of small anions such as lithium perchlorate, sodium chloride, and sodium phosphate monobasic monohydrate as well as polyanion poly(sodium 4-styrenesulfonate) (PSSNa). When PEDOT was electrochemcially polymerized with an ion mixture in phosphate-buffered saline (PBS) solution, the chlorine anion from sodium chloride fulfilled the role of the counterion. Such performance is significant for the potential use of PEDOT in biomedical application. As different mixtures of PSSNa, lithium perchlorate, sodium p-toluenesulfonate (TosNa), and PBS counterions were used, the polyanion PSS^- was generally integrated by PEDOT rather than $ClO_4{}^-$ and Cl^- anions.

High-quality polythiophene films with various alkyl side chains were successfully deposited in the presence of sodium dodecylbenzenesulonate (SDBS) and a N_2 atmosphere on the poly(tetrafluorethylene) (PTFE) substrate. The protein adsorption in succession and, hence, film hydrophobicity were modulated by tuning the length of alkyl side chains. The hydrophobic polythiophene films with longer alkyl side chains demonstrated the capacity for protein adsorption and $PC1_2$ cell proliferation. The prepared films lack cytotoxicity of the prepared films toward the two employed cell lines and provide positive support for cell adhesion and proliferation. Polythiophene films deposited have potential use for biomedicine.

Layers with high transparency (>80%) and conductivity (10^{-6} Scm^{-1}) were produced by pulsed plasma polymerization of thiophene. How the conductivity of these plasma-polymerized thiophene (PPT) layers was affected by power, pressure, pulse time, duty cycle, and location in the reactor was estimated. Pressure most markedly impacts conductivity within the employed ranges. There results could relate to the influence of deposition factors on thiophene monomer fragmentation. Thiophene did not fragment significantly at high pressure, resulting in more conductive layers. A pulsed plasma with the greatest efficiency facilitates the decrease of fragmentation, permitting the refilling new monomer to the reactor. A minimum fragmentation during deposition is necessary to obtain the plasma polymerized (PP) layers maintaining the conjugated monomer structure. During plasma polymerization, various methylated and halogenated thiophenes serve as monomers to study the response of fragmentation to the substitute(s). Halogenated thiophenes demonstrated higher fragmentation during deposition than methylated thiophenes. Additionally, for a particular substitute, fragmentation is higher in the case of substitution on the 3-position than the 2-position. Actually, disubstituted thiophenes exhibit more limited fragmentation than monosubstituted thiophenes. In addition, the limited fragmentation leads to the presence of more conjugated structures in the PP layers of methylated thiophenes. Hence, these PP layers presented higher conductivity after iodine doping.

Thin (<100 nm), ultrasmooth (roughness R_{rms} < 5 nm), and functionalized PEDOT films was electropolymerized with aqueous microemulsions to produce nanobiointerfaces. The decrease of acid concentration boosts the catalysis of microemulsion polymerization, resulting in a film from different functionalized (ethylenedioxy)thiophenes (EDOTs).

PEDOT-coated cell culture surfaces are prepared by solution-casting polymerization (SCP). The absorption of PEDOT thin films in the NIR region induced the cell harvesting. The proliferation and harvesting of human mesenchymal stem cells (hMSCs) may be quantitatively modulated by regulating NIR absorption of the PEDOT thin films or by developing PEDOT film with the thickness in the range of 70–300 nm. This light-triggered cell detachment technique underpinned by PEDOT films can temporally and spatially adjust both cell harvesting and patterning. Furthermore, in spite of the temperature increase caused by NIR, the survival and proliferation of the harvested stem cells, their inherent properties and ability of multilineage differentiation keep constant. No photodeterioration occurs when the PEDOT films were exposed to NIR. Therefore, when different tissue-derived cells are cultured, the PEDOT surface may be applied for recurrent hMSC culture and disengagement or for effectively extracting or excluding a special subset from a nonuniform population. The coating of PEDOT films 70–300 nm thick was promoted by SCP on Petri dishes made of polystyrene or ITO glass. Pure PEDOT films (SP) doped with tosylate demonstrate transparent blue. Through cyclic voltammetry-based SP surface electrochemical dedoping, the partially doped PEDOT (P1P and P2P) and dedoped PEDOT (neutral PEDOT, NP) coatings were fabricated by terminating the potential at +0.4, −0.2, and −1 V, respectively.

1.6 Polypyrrole

In 1963, Weiss et al. first achieved polypyrrole (PPy) from pyrolysis of tetraiodopyrrole. PPy is synthesized by oxidation of pyrrole in methanol. As one of the biocompatible electrically conductive polymers, PPy possesses easy synthesis, impressive conductivity, biocompatibility, outstanding redox properties, and environmental stability. Therefore, it can be potentially applied in biomedicine such as neural implants, biosensors, and molecular memory devices. PPy with an amorphous structure is insoluble. Cellular activities such as deoxyribonucleic acid (DNA) synthesis by ES, migration, and cell proliferation in biological environments can be adjusted by PPy.

1.7 Polyaniline

PANI is promising because of the advantages including facile preparation, great environmental stability, and high electrical conductivity, enabling it to be employed on different sensor applications, such as pH switching electrical conducting biomaterials, electrically active redox biopolymers, and matrixes for nanocomposite CP preparation. Only the electrical properties of PANI can be adjusted by protonation or charge transfer doping. PANI exists in three idealized oxidation states affecting its conductivity. After doping with acid, the emeraldine PANI with the partially oxidized structure is conductive and very stable at room temperature. However, pernigraniline and leucoemeraldine are oxidized and reduced forms, respectively, and demonstrate poor conductivity, even after doping. PANI can be obtained from a three-component-state physical mixture of leucoemeraldine $[(C_6H_4NH)_n]$, emeraldine $[([C_6H_4NH]_2[C_6H_4N]_2)_n]$, and per-nigraniline $[(C_6H_4NH)_n]$.

References

1. Bellanger H, Darmanin T, Elisabeth TDG, Guittard F (2012) Superhydrophobic hollow spheres by electrodeposition of fluorinated Poly(3,4-ethylenedithiopyrrole). RSC Adv 2(29):10899–10906
2. Chang M, Choi D, Egap E (2016) Macroscopic alignment of one-dimensional conjugated polymer nanocrystallites for high-mobility organic field-effect transistors. ACS Appl Mater Interfaces 8(21):13484–13491
3. Chang M, Lim GT, Park B, Reichmanis E (2017) Control of molecular ordering, alignment, and charge transport in solution-processed conjugated polymer thin films. Polymers 9(6)
4. Darmanin T, Guittard F (2009) Molecular design of conductive polymers to modulate superoleophobic properties. J Am Chem Soc 131(22):7928–7933
5. Darmanin T, Guittar DF (2014) Wettability of conducting polymers: from superhydrophilicity to superoleophobicity. Prog Polym Sci 39(4):656–682
6. Döbbelin M, Tena-Zaera R, Marcilla R, Iturri J, Moya S, Pomposo JA, et al (2009) Multiresponsive PEDOT–Ionic liquid materials for the design of surfaces with switchable wettability. Adv Func Mater 19(20):3326–3333
7. Dong H, Hu W (2016) Multilevel investigation of charge transport in conjugated polymers. Acc Chem Res 49(11):2435–2443
8. Goel M, Heinrich CD, Krauss G, Thelakkat M (2021) Principles of structural design of conjugated polymers showing excellent charge transport toward thermoelectrics and bioelectronics applications. Macromol Rapid Commun 40(10):1800915
9. Grévin B, Rannou P, Payerne R, Pron A, Travers J-P (2003) Scanning tunneling microscopy investigations of self-organized Poly(3-hexylthiophene) two-dimensional polycrystals. Adv Mater 15(11):881–884
10. Gu K, Loo Y-L (2019) The polymer physics of multiscale charge transport in conjugated systems. J Polym Sci Part B Polym Phys 57(23):1559–1571
11. Guay J, Paynter R, Dao LH (1990) Synthesis and characterization of Poly(diarylamines): a new class of electrochromic conducting polymers. Macromolecules 23(15):3598–3605
12. Le TH, Yukyung K, Hyeonseok Y (2017) Electrical and electrochemical properties of conducting polymers. Polymers 9(12):150
13. Park HW, Kim T, Huh J, Kang M, Lee JE, Yoon H (2012) Anisotropic growth control of polyaniline nanostructures and their morphology-dependent electrochemical characteristics. Acs Nano 6(9):7624
14. Roth S (1984) Charge transport in conducting polymers. Springer Berlin Heidelberg
15. Segatta F, Lattanzi G, Faccioli P. (2018) Predicting charge mobility of organic semiconductors with complex morphology. Macromolecules 51(21):9060–9068
16. Toktam N, Amelia S, Tan A, Seifalian AM (2018) Conductive polymers: opportunities and challenges in biomedical applications. Chem Rev 118:6766

Chapter 2
Preparation of Conductive Polymers

In the 1970s, the discovery of conductive polymers (CPs) completely changed the traditional concept of insulating polymers and opened up a new research field of organic electronics. In comparison with silicon and inorganic semiconductors, the unique advantages of conductive polymers such as lightweight, low cost, and easy-solution processability make them promising in the applications of large area flexible devices and circuits. Over the past decades, significant progress has been made in this field due to the synthesis of abundant new semiconductive polymer materials and the improvement of their performances. For instance, the charge carrier mobility of conductive polymer field-effect transistors has been improved from the initial value of 10^{-5} cm^2 V^{-1} s^{-1} to current values of over 40 cm^2 V^{-1} s^{-1}. The powder conversion efficiency (PCE) of conductive polymer photovoltaics has also been increased to a new record value of over 14%. Moreover, conductive polymer light-emitting diodes are currently being industrialized. Even so, to meet the requirements for more fields of application, conductive polymer-based optoelectronic devices and circuits still urgently need more excellent performances with favorable stability, elasticity, and strength. The preparation methods of conductive polymers are various and play a critical role in the morphology and properties.

2.1 Electrochemical Polymerization

In comparison with other polymerization methods, the electrochemical polymerization (EP) has the characteristics of free catalyst, controllable film thickness, the reaction at room temperature, and rapid polymerization. Additionally, the morphologies of electrochemically polymerized conductive polymer film can be controlled by varying the electrochemical cycle numbers and EP polymerization rate. Different from spin-coating or other coating methods, the fast polymerization process and film-forming rate lead to the EP films usually exhibiting significant roughness. Currently, various conductive polymers with 1D- and 2D- conjugation have been obtained by

Y. Wang and W. Feng, *Conductive Polymers and Their Composites*,
https://doi.org/10.1007/978-981-19-5363-7_2

reasonably designing the monomer precursors and properly controlling the EP conditions, which presents great potential for applications in the organic optoelectronic field.

As shown in Fig. 2.1a, the device for EP process theoretically should comprise one compartment and three electrodes: the working electrode, counter electrode, and reference electrode. Cyclic voltammetry (CV) curves in Fig. 2.1b demonstrate the possible reversible or irreversible redox reactions and polymerization during the process of EP. To reduce the polarization of the working electrode, the size of the counter electrode should generally be larger or the same as that of the working electrode. The working electrodes include metals (such as gold and platinum, nickel), alloys (e.g., stainless steel), and glassy carbon and conductive oxide (such as indium tin oxide, ITO, and coated glass electrode). The electrochemical polymerization systems, the purposes of experiments, their surface properties, shapes, and costs decide the choice of working electrodes.

The potential for oxidation and reduction is determined in the first cycle of the curves which correspond to the processes of the monomer losing electrons and obtaining electrons, respectively. The two pairs of oxidation and reduction peaks illustrate the two electrochemically active parts of the monomer. In addition, if two different monomers are polymerized, the difficulty of losing and attaining electrons can be compared by the oxidation and reduction potentials. As both the oxidation potential and the reduction potential correspondingly change with the increase in cycle numbers, the conductive polymers form and the thin films are then deposited on the surface of the working electrode. Besides, by calculating the relationship between the film thickness and the cycle numbers, the thickness of the resulting polymer films can be controlled.

2.1.1 Principles of Electrochemical Polymerization of Aromatic Monomers

Aromatic compounds used for electrosynthesizing CPs as the monomers include benzene, thiophene, pyrrole, aniline, furan, etc. Relatively stable cation radicals can form after the oxidation of these compounds. Then, a dimer is produced by the coupling of two cation radicals or one radical and a monomer and subsequent removal of two protons. Figure 2.1c shows the widely accepted mechanism of thiophene dimer formation. This reaction comprises an electrochemical process of oxidation and a chemical process of coupling and excluding protons. Oligomers yielded by further oxidation of the dimer and coupling are soluble in electrolytes and electroactive for polymerization. Through the same procedures, elongation of oligomers leads to the final product. More than two electrochemically active positions of the usual aromatic monomers results in a crosslinked structure of most electrosynthesized CPs. For simplification, Fig. 2.1c only presents the α,α'-coupling of thiophene. CPs become insoluble and are deposited at the working electrode surface when the

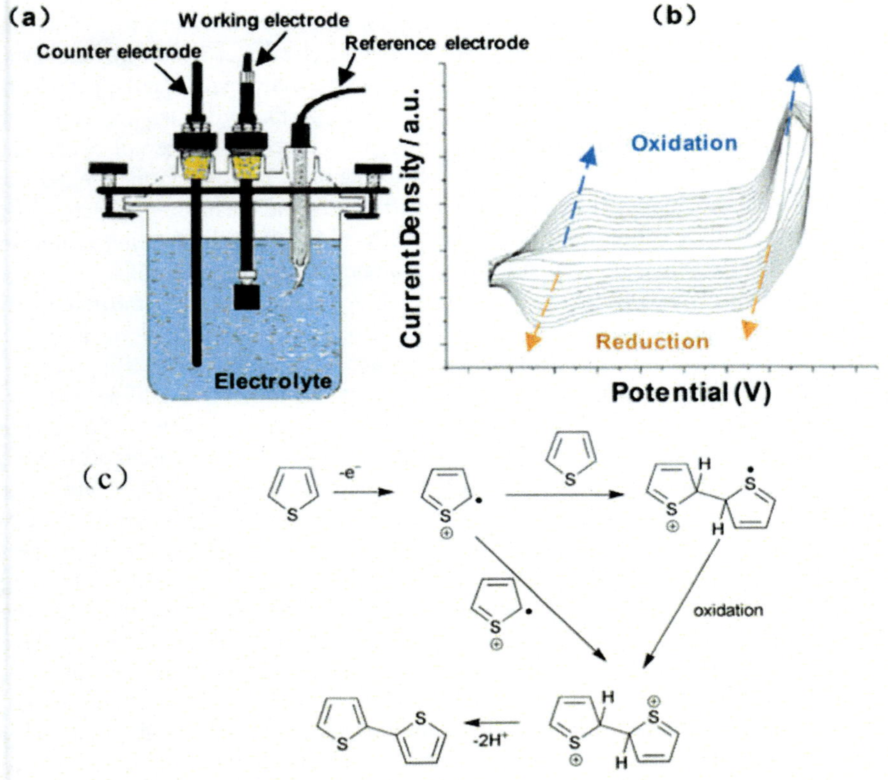

Fig. 2.1 **a** Schematic of the EP equipment showing one compartment and three electrodes, the counter electrode, working electrode and reference electrode, which are submerged in the electrolyte. **b** An example of the cyclic voltammetry curves during the process of EP. Current density changes with the potential change and broad peaks were observed when molecules are in redox. The directions of the arrows represent the shift of the redox peak along with the potential change. The onset potential of oxidation and reduction can be determined from the first cycle. The two pairs of redox peaks observed in this diagram correspond to two different active sites. **c** Electrochemical polymerization process of thiophene. Reprinted from ref. [15], copyright 2018, with permission from Royal Society of Chemistry

oligomer chains grow long enough or the crosslinked structure is formed. Although this mechanism has been widely accepted, several facets of the reaction are not yet completely understood.

CPs can also be prepared at constant current density or by cyclic voltammetric scanning. The total amount of the CP can be controlled simply by electrolysis time at constant current density. Under constant potential or current density conditions, the as-obtained polymers are in the conductive (doped or oxidized) state. However, the potential at which the polymerization stopped decides those prepared by cyclic voltammetric scanning.

A redox reaction consists of polymerization at the working electrode and the reduction of protons to hydrogen gas at the counter electrode. Therefore, small gas bubbles are released from the surface of the counter electrode during the process of electrochemical polymerization. Some gas bubbles probably diffuse and adhere to the surfaces of the working electrodes and can be utilized as soft templates for the electrosynthesis of CP nanospheres and microcontainers. The oxidation potentials of the monomers are mainly dependent on their structures and the nature of electrolytes. The corresponding cation radicals can be stabilized by electron donating groups or large dense aromatic rings so that the oxidation potentials are reduced.

Some electrolyte molecules with electrocatalytic properties activate the monomers and lower their oxidation potentials. If the oxidation potentials of aromatic monomers are lower than that of water decomposition, they can be electrochemically polymerized in aqueous media. The classic examples are aniline, pyrrole, and 3,4-ethylenedioxythiophene (EDOT). However, the poor solubility of aromatic compounds in water causes very low polymerization rates. Acids can improve both the solubility of pyrrole and aniline and the ionic conductivity of the electrolytes. The solubility of EDOT can also be improved by adding surfactants. Most electrochemical polymerizations are conducted in organic media, in which the monomers have good solubility. Aprotic solvents with poor nucleophilic character including acetonitrile, benzonitrile, dichloromethane, etc., are commonly chosen because the electropolymerization reaction progresses via cation radical intermediates. Diverse ionic liquids are also utilized as the solvents for electrochemical polymerizations.

The supporting electrolytes with high ionic conductivity are incorporated. According to its solubility, degree of dissociation, and nucleophilicity, the supporting electrolytes are chosen. Lithium perchlorate can be used as the supporting electrolyte because it is soluble in water and acetonitrile. Quaternary ammonium salts such as $(n\text{-}C_4H_9)_4NClO_4$ and $(n\text{-}C_4H_9)_4NPF_6$ are commonly used as supporting electrolytes for electrosynthesizing CPs in both aqueous and organic solvents. Although BFEE is a good electrolyte for the electropolymerization of CPs, its ionic conductivity is very low due to the lack of ionic charge carriers. Its conductivity is greatly improved, and the polymerization rate is accelerated by introducing a small amount of water or strong acid such as CF_3COOH [8].

2.1.2 2D Conductive Polymers Prepared by Electrochemical Polymerization

Based on the monomer precursors designed with three or more electrochemical active sites, 2D CPs can be obtained via EP. The electroactive molecules with three active sites in a coplanar structure tend to yield a planar hexagonal structure. However, holes exist in the middle of the hexagon and their size depends on designing the size of the monomers. The EP method offers a potential method to fabricate novel 2D conductive polymers. But many challenges still remain in this field. Designing appropriate monomer units and precisely controlling the electrochemical reaction of monomers on the electrode directly decides the crystallinity and the structure orders of the resulting 2D polymers.

The wide application of the conductive polymers obtained by EP consists of antistatic and anticorrosion materials, solar cells, catalyst support, polymer electrochromic devices, electrochemical sensors, and electrochemical supercapacitors. Furthermore, it is possible to produce high-performance polymer materials by fabricating copolymers via EP. EP may be further applied in preparing functional conductive polymers.

2.2 Photopolymerization

The chemical and electrochemical methods commonly used to synthesize CP also have disadvantages. For example, although chemical polymerization is cheaper and more efficient than other methods, it has impeded some applications such as microelectronic and multilayer coating. Via electropolymerization, CPs could not be coated on the insulating substrates. Compare with the mentioned methods, photopolymerization of CPs possesses rapid and facile processability, environmental friendliness due to avoiding volatile organic compounds, possibility at lower temperature, and ability to coat CPs on the insulating substrates. In addition, according to the application domain, the morphology and final properties of products in photopolymerization process are modulated by simply varying some process conditions such as excitation wavelength. On the other hand, CPs prepared by photopolymerization have more thermal stability and vantage counterion–counterion structural than those obtained from chemical polymerization.

2.2.1 Photopolymerizations Leading to CPs

According to the mechanism, photopolymerization approach to synthesizing CPs can be generally categorized into two types: (i) direct (self-synthesized) photopolymerization via photoexcitation of the monomer as the repeating unit of CP and (ii) photopolymerization by using photocatalytic systems. By these methods, the well-known CPs such as polypyrrole, polythiophene, polyaniline as well as their derivatives can all be prepared.

2.2.1.1 Direct Photoexcitation of Monomer to Synthesize CP

Self-synthesized photopolymerization via photoexcitation of the monomer as the repeating unit of CP is a facile and simple route. By this approach, the photoinduced synthesis of CPs involves the photoexcitation of monomer itself without using any photosensitizer. For instance, PPy was reported to be fabricated by self-synthesized photopolymerization of pyrrole. Through quenching by unexcited pyrrole molecules, pyrrole molecules as photosensitizer yield both cationic and anionic radicals after excitation. A pyrrole oligomer formed by ELT possibly absorbs light at longer wavelength, and new radical centers advance the polymerization process. In a similar fashion, using potassium dichromate as electron scavenger, a series of thiophene monomers grafted with acetic acid, boronic acid, and carboxylic acid were photopolymerized. The formation process of CP is very slow, particularly in deoxygenated solutions. Moreover, the low conductivity of CP obtained from direct photopolymerization originates from small concentration of dopant molecules and large-scale distance between polymer chains.

No sufficient investigation has been conducted on the preparation of other common CPs such as polyaniline and polyacetylene and their derivatives through direct photopolymerization. However, poly-1,2-azepines as an uncommon CP was synthesized by photoexcitation of phenyl azides. Furthermore, the CPs can also be synthesized by direct photopolymerization of oligomeric structures. For instance, direct photopolymerization of oligothiophene can yield polythiophene. The mechanism lies in the photoinduced ELT process from the triple excited state of oligothiophene to an electron-acceptor to form the cation radicals of the monomers. A series of highly conjugated thiophene derivatives such as 5,8-bis(2,3-dihydrothieno[3,4-b][1,4]dioxin-5-yl)-2,3-di(thiophen-2-yl)quinoxaline (DTDQ) and 4,7-di(2,3-dihydrothieno[3,4-b][1,4]dioxin-5-yl)benzo[1,2,5]thiadiazole (DTDT) in the presence of iodonium salt can achieve self-photosensitization.

2.2.1.2 Photopolymerization via Photocatalytic Systems

A photosensitizer in format of ligand metal complexes such as $[Ru(bpy)_2]Cl_2$ and $[Cu(dpp)]BF_4$ was usually irradiated by light in the photocatalytic polymerization of CPs. The corresponding excited state containing charge acts as a strong electron–donor in the presence of an electron-acceptor (such as cobalt complex, and p-nitrobenzyl bromide). Additionally, as organic–inorganic complexes such as ferrocene and its derivatives easily produce electron-acceptor in halogenated solvents, they can be used in photopolymerization. Correspondingly, photocatalytic polymerization of CPs can be achieved via two major ways: (i) photosensitization applying a coordination complex such as ruthenium complexes and (ii) photosensitization utilizing transition metal salts or metallic particles.

Photosensitization Applying a Coordination Complex

Photosensitization of CPs by different coordination complexes was significant. Ruthenium complexes such as ruthenium(II) trisbipyridyl, $[Ru(bpy)_3]_2$ as well-known catalysts for photopolymerization have attracted much attention because they absorb ultraviolet and visible lights very well. Depending on selection of reducer or oxidizer quenchers, ruthenium complexes induce photoreduction and photo-oxidation processes, respectively, to catalyze photopolymerization. Owing to its properties such as intense absorption at 452 nm and high stability to irradiation, $[Ru(bpy)_3]^{2+}$ complex was used as efficient photosensitizer in common photocatalytic systems to synthesize CPs. For instance, photopolymerization of pyrrole was completed in an aqueous monomer solution containing $[Ru(bpy)_3]Cl_2$ and $[CoCl(NH_3)_5]Cl_2$ as a sacrificial oxidant.

Because decreasing PH of the polymerization solution enhanced the propagation and hence formation rate of polymer, PH was considered as a key factor to improve the photopolymerization of CPs. Some structured materials such as clay layers and DNA can act not only as templates for photopolymerization of CPs but also as counter ion for doping of CPs. Photocatalytic polymerization leads to CPs with higher efficiency via light irradiation to a monomer-substituted ruthenium complex. Besides ruthenium complexes, other coordination and organometallic complexes can also be applied to photocatalytic synthesis of CPs. For instance, titanium dioxide (TiO_2) nanoparticles are usually utilized as a photosensitizer in photoinduced polymerization of pyrrole and thiophene.

Photosensitization Using Transition Metal Salts or Metallic Nanoparticles

Transition metal salts as well as metallic nanoparticles have photocatalytic action on synthesis. The role and fate of metal nanoparticles incorporation in photopolymerization of pyrrole was investigated by Sadik et al. de Azevedo et al. also investigated PANI photopolymerized using $AgNO_3$ salt. Cai et al. also studied the photopolymerization of 5-Mercapto-2,2′-bithiophene (BTSH) functionalized with metallic nanoparticles such as copper (Cu), gold (Au) and silver (Ag). In addition, based on intramolecular electron transfer from the ligand to the metallic nanocores, a mechanism was proposed for this process. Moreover, different conductive polymer nanostructures and their related hybrid materials can be easily synthesized by photocatalytic systems containing metallic particles [5].

2.3 Direct Arylation Polymerization

2.3.1 Direct Arylation Polycondensation (DArP)

Conductive polymers generated via (hetero)aryl-(hetero)aryl crosscoupling polycondensation possess various potential applications in the field of organic electronics, e.g., as active components of organic light-emitting diodes (OLEDs), organic photovoltaic devices (OPV devices), and organic field-effect transistors (OFETs).

However, toxic or environmentally risky by-products are usually produced by "conventional" (hetero)aryl-(hetero)aryl coupling schemes. Multistep procedures and expensive purification steps are required by the syntheses of these organometallic monomers. Here, so-called direct arylation schemes replaces standard coupling procedures because the process has no use for organometallic reagents or related anion equivalents and the products have high yields, low catalyst loadings, and high molecular weights. In 1999 direct arylation schemes for the aryl–aryl coupling of small molecules were first transferred into polymer synthesis by Lemaire et al. Using $Pd(OAc)_2$ as catalyst, equimolar amounts of nBu_4NBr and excess K_2CO_3 as base in DMF at 80 °C, the regioregular poly(3-alkylthiophene)s (P3AT, R=H, CH_3, $(C_2H_4)CH(CH_3)(C_2H_5)$, C_8H_{17}) was synthesized by polycondensation of 2-iodo-3-alkylthiophenes in a Heck-type reaction.

Research on the conditions suitable for polymerization was the initial key point for DArP. For this, Fagnou and Gorelsky contributed to critical mechanistic insight. Moreover, Ozawa, Kanbara, and Leclerc carried on the pioneering works as the starting point of DArP. Initially, the significant work improved the site selectivity of C–H activation so that disappearance of branching (β) defects eliminated the structural faults with impact on polymer solar cell device performance. Based on the groundwork of these and other studies, polymerization conditions that inhibit undesired couplings, such as homocoupling and branching defects, were determined, and advance was achieved from simple homopolymers to donor–acceptor copolymers, semirandom, and semi-alternating copolymers. Besides, under state-of-the-art conditions, polymers with high molecular weights (*Mn*) and yields were synthesized from certain monomers with ultralow loadings (ppm) of Pd-catalysts.

The well-known poly(4,4-di(2-ethylhexyl)-cyclopenta[2,1-b:3,4-b']dithiophene-alt-2,1,3-benzothiadiazole) (PCPDTBT) was initially synthesized using $Pd(OAc)_2$ (4 mol percent) as catalyst and 1.5 equiv. PCPDTBT with a molecular weight *Mn* of 40,300 was produced in 80% yield using K_2CO_3 as the base in DMAc without the introduction of carboxylic acids (based on the amount of high molecular weight dichloromethane and chloroform fractions after solvent extraction).

Nonetheless, it is a novel and not completely optimized approach of conductive polymer synthesis, and no standard procedure is available to create conductive polymers with high yields and molecular weights. Until date, several reaction conditions in terms of ligands, catalysts, and additives were employed. Furthermore, if hydrogens in other reactive places within the monomers are not safeguarded, the coupling of non-activated condensation monomers can result in structural unit misconnections and polymer chain branching. Direct arylation polycondensation techniques will not be able to completely replace traditional aryl–aryl couplings after Stille, Suzuki, Negishi, or Kumada, but they are a potent complementary strategy toward high molecular weight conjugated (co)polymers [6].

DArP, like other methods for conductive polymer synthesis, relies primarily on a limited number of solvents that have been demonstrated to be capable of solubilizing a wide range of conductive polymers, such as tetrahydrofuran (THF), toluene, chlorobenzene (CB), and dimethylformamide (DMF), as shown in Fig. 2.2A. Given that the solvent is present in the largest quantity compared to any other component involved in the reaction conditions, it poses the greatest risk to the process. Most of the aforementioned solvents are reproductive poisons, carcinogens, or, in the case of THF, organic peroxide formers. Furthermore, the production of these solvents can be quite energy demanding, making them unsustainable in the long run. Figure 2.2B shows sustainable solvents. With these concerns in mind, Fig. 2.2C depicts a list of solvents (anisole, CPME, and 2-MeTHF) that may permit more broad use with DArP. These solvents have been employed in the synthesis of conductive polymers by DArP, as illustrated in Fig. 2.2D. In terms of polymer examples created using DArP utilizing these solvents, PBDTBT (Fig. 2.2D) was made with anisole in high yield (96%) with excellent Mn (13.2 kDa). CPME has been demonstrated to be effective for a wide range of monomers since its first use in DArP by Ozawa et al. Future work with sustainable solvents will necessitate their usage in a range of procedures, in order to demonstrate good tolerance for polymerization conditions and monomer structural diversity. Many DArP reactions, in particular, need highly polar, coordinating solvents, such as dimethylacetamide (DMA). Identifying alternative sustainable solvents with this role and universal applicability for the synthesis of a wide range of conductive polymers would be a huge step forward. Monitoring the output and identifying new sustainable solvents that are usable in aryl–aryl cross-coupling processes should lead to the discovery of novel solvents compatible with DArP, as sustainable chemistry is a fast expanding discipline.

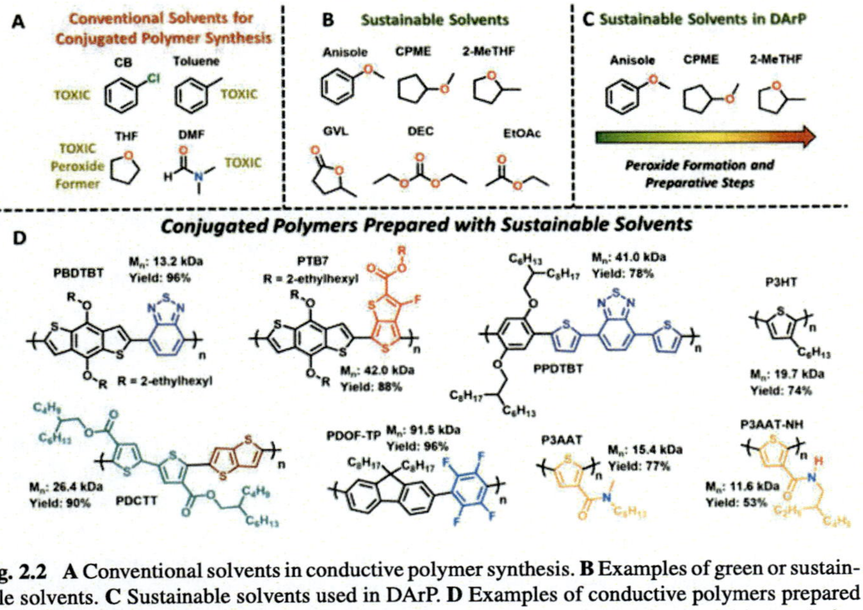

Fig. 2.2 **A** Conventional solvents in conductive polymer synthesis. **B** Examples of green or sustainable solvents. **C** Sustainable solvents used in DArP. **D** Examples of conductive polymers prepared using DArP with sustainable solvents. Reprinted from ref. [12], copyright 2020, with permission from Royal Society of Chemistry.

Besides the solvent, a first-row transition metal, such as a Cu-catalyst, replaces the Pd-catalyst, as also arouse the interest. Cu-catalyzed direct arylation conditions are remarkably different from those of Pd. Figure 2.3A shows many protocols employing an amine ligand, such as phenanthroline and phosphine ligands. Even though there is a stoichiometric excess of one reactant in all cases, the crosscoupled products in high yields (75–96%) were generated using each of the protocols. The reactivity of Cu-catalysts is lower than Pd-catalyzed protocols so that it requires higher temperatures (>120 °C). Octafluorobiphenyl were chosen as the ideal monomer for studying Cu-DArP under high temperature. Although carboxylic acid additives are usually used for Pd-catalyzed direct arylation, they are not suitable for Cu-DArP since these ligands can cause the disproportionation of Cu^{I}, leading to Cu^{0} and Cu^{II} (Fig. 2.3B). As shown in Fig. 2.3C, the polymerization conditions are optimized by increasing concentration and changing the base from K_2CO_3 to K_3PO_4 for PDOF-OD. Then as the loading of the Cu^{I}-phenanthroline catalyst is decreased from 50 to 5 mol%, a polymer of 16.4 kDa in 54% yield is produced. Because a C–H bond was functionalized with no use of a directing-group, conventional Pd-catalysts were replaced with Cu-DArP. Large amount of future work on Cu-DArP will be carried out, and it is necessary to optimize the polymerization conditions. The C–H activation and the stabilization of Cu^{III} species after oxidative addition should be promoted by incorporating more sustainable solvents, and optimizing Cu-catalysts via ligand design or through the use of additives.

2.3.2 *Oxidative Direct Arylation Polymerization (Oxi-DArP)*

There are still certain components of this technique that are unsustainable. Solvents utilized in many DArP procedures, for example, are toxic and dangerous, require multistep, energy-intensive manufacturing pathways, and are not supplied from sustainable or renewable sources. Another consideration is the source of the transition metal catalyst. DArP methods have virtually always relied on palladium, because first-row transition metals might provide a more sustainable alternative. Finally, because halogen functionalization of the monomers is not necessary, under-utilized polymerization techniques such as Oxi-DArP may provide more streamlined synthetic routes for monomer synthesis.

Oxidative direct arylation is also known as dehydrogenative direct arylation and is efficient in the synthesis of biaryl compounds. The substrates or monomers used for conventional direct arylation require functionalization. For example, C–H/C–X crosscoupling reactions necessitate installing a halogen. However, since oxidative direct arylation advances through an oxidative C–H/C–H coupling pathway, the simplified synthesis of monomers acknowledges rapid access and shorter synthetic pathways so that the overall sustainability is improved.

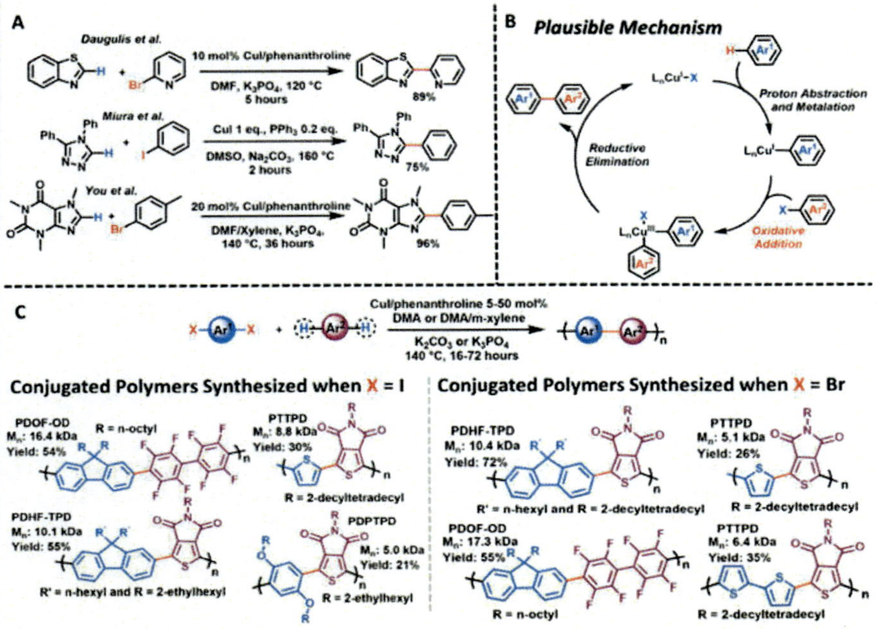

Fig. 2.3 **A** Examples of small molecules synthesized via Cu-catalyzed direct arylation. **B** A plausible mechanism for Cu-catalyzed direct arylation. **C** Examples of conductive polymers prepared using Cu-DArP. Reprinted from ref. [12], copyright 2020, with permission from Royal Society of Chemistry

The initial oxidative direct arylation studies on small molecules used Pd-catalyst with Ag-oxidant. The same reaction conditions were applied for synthesizing Pd-catalyst with Ag-oxidant but very limited in scope. Furthermore, specific structural functionalities of the monomers required additional steps to the syntheses so that this methodology to synthesize conductive polymer is not simplified. To conquer this limitation, the conditions were developed to polymerize an unsymmetrical monomer.

The random copolymers P3HET-TPD-5% (13.9 kDa and 54% yield) and P3HET-BTz-5% (11.7 kDa and 68% yield) were fabricated by Oxi-DArP. Under the polymerization conditions for polymerizing an unsymmetrical monomer, homopolymers demonstrate high levels of regioregularity and copolymers exhibit minimized homocoupling defects. The aforementioned work indicates that Oxi-DArP simplifies the synthesis of conductive polymers, reduces the number of synthetic steps and the associated waste for monomer preparation to enhance the sustainability.

In spite of this, several aspects of Oxi-DArP need to be improved. The monomer scope retains comparatively narrow due to unclear relations between the monomer structure, catalyst and oxidant. With respect to sustainability, many Oxi-DArP protocols employ stoichiometric amounts of silver-oxidants such as Ag_2CO_3. Thus, it is critical to find more sustainable oxidants or identify more general conditions for

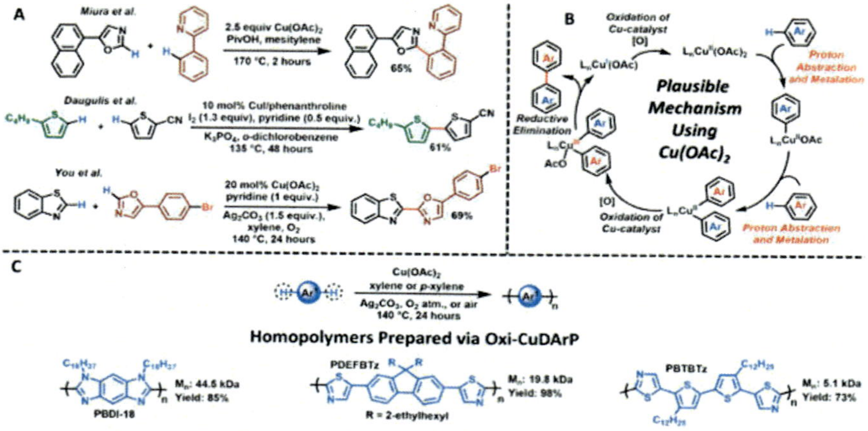

Fig. 2.4 **A** Examples of small molecules synthesized via Cu-catalyzed oxidative direct arylation. **B** A plausible mechanism for Oxi-Cu-DArP. **C** Examples of conductive polymers prepared using Oxi-Cu-DArP. Reprinted from ref. [12], copyright 2020, with permission from Royal Society of Chemistry

known oxidants, such as molecular oxygen and $Cu(OAc)_2$. Practically, this methodology is still inferior to the conventional polymerization methods for synthesizing functional materials with equal application in organic electronics.

Figure 2.4 depicted that Cu-catalyzed Oxi-DArP combined the sustainable aspects of Oxi-DArP and Cu-DArP. In this synthetic method, using a copper catalyst, a C–H/C–H crosscoupling occurs through dehydrogenative or oxidative direct arylation. No need of monomer functionalization reduces the number of synthetic steps and related workup and purification. Copper catalyst affords stainability and low cost for this synthetic method. Different from the transition metal catalyst for Oxi-DArP, Fig. 2.4B illustrates the mechanism of Cu-catalyzed dehydrogenative crosscoupling. The proposed Cu^{II}–Cu^{III}–Cu^{I} cycle comprises two separate oxidations of the Cu-catalyst before and after the reductive elimination. Aza-heterocycles such as imidazole, thiazole, and oxazole are commonly applied because the nitrogen in the heterocycle is capable of integrating to a metal center such as copper so that further improving the acidity and reactivity of the adjacent C–H bond decreases the energy needed for proton abstraction and functionalization. Figure 2.4C shows this structural feature of the representative polymers. In comparison with the other methodologies described, the number of polymers prepared via Cu-catalyzed Oxi-DArP is limited.

2.3.3 *Mechanisms of Direct (Hetero)arylation Reactions*

Well-defined, high molecular weight materials require reactions with both high selectivity and yield. It is critical to understand the mechanism of the direct (hetero)arylation polymerization reaction and potential competitive pathways for improving its efficiency. Figure 2.5a depicts direct (hetero)arylation reactions evolving from simple intramolecular reactions conducted at high temperature into highly selective intermolecular couplings at room temperature.

Aforementioned five-membered heteroarenes such as pyrroles, thiophenes, and indoles possess an intrinsic electronic bias in favor of α-substitution according to a CMD mechanism. However, another competing mechanism with lower energy than the CMD process leads to different selectivity. As shown in Fig. 2.5b, using $P(OCH(CF_3)_2)_3$ along with PdX_2 (X=Cl, Br, or I) or Pd_2dba_3 as the precatalyst, the direct (hetero)arylation reaction selectivity (>99%) may be adjusted to the β C–H

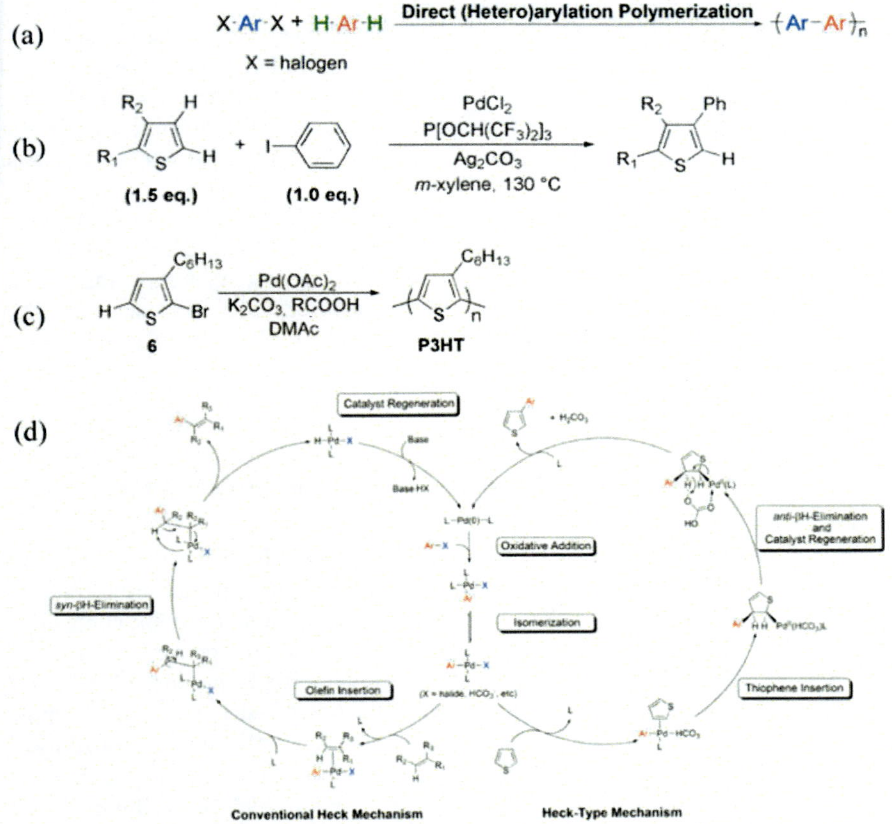

Fig. 2.5 **a** Direct (hetero)arylation polymerization reaction. **b** Reaction conditions leading to β-selective direct (hetero)-arylation products, as proposed by Itami and colleagues. **c** Heck and Heck-type reaction mechanisms described by Fu et al. **d** Analysis of β-branching of phosphine-free DHAP conditions for P3HT. Reprinted from ref. [13], copyright 2016, with permission from American Chemical Society

bonds for thiophene, thienothiophene, and benzothiophene derivatives. This adaption from α- to β-selectivity is completely catalyst-driven.

Fu et al. well explained this. Based on DFT calculations, a Heck-type mechanism was promoted by applying $P(OCH(CF_3)_2)_3$ as the phosphine. Figure 2.5c indicates that this special case is distinct from a standard Heck mechanism and an anti-β proton elimination follows the insertion of thiophene on the catalyst because the cyclic C–C bonds of thiophene cannot rotate into a syn configuration. In spite of this, the energy barrier of the Heck-type pathway is lower than that of the CMD process. By in-depth DFT experiments, the electronic or steric influences responsible for this mechanistic change could not be isolated. Nevertheless, the mildly acidic nature of the C–H bond present on the phosphine ligand resulted in H-bonding of $P[OCH(CF_3)_2]_3$ with the carbonate base on the catalyst. This strong bond (11.2 $kcal{\cdot}mol^{-1}$) can elevate the energy barrier of a CMD transition state over that of a Heck-type mechanism.

In 2010, poly(3-alkylthiophene)s (P3HT) with high molecular weight were first synthesized by Ozawa et al. using direct (hetero)arylation polymerization (DHAP). A notable improvement lies in that P3HT has high number-average molecular weight ($\overline{M}n$) of 30.6 kg/mol and regioregularity (RR > 98%) comparable to Kumada, Negishi, Suzuki, and Stille-prepared P3HT.

Regioregular P3HT was fabricated by Thompson et al. via direct (hetero)arylation reactions using $Pd(OAc)_2$, K_2CO_3, a catalytic amount of pivalic acid (PivOH), and the polar aprotic solvent DMAc (Fig. 2.5d). By studying reaction temperature, time, and catalyst concentration, it was found that the regioregularity increases from 82.6 to 89.3% as the reaction temperature decreases from 120 to 20 °C but higher catalytic loading leads to lower regioregularity.

2.4 Acyclic Diene Metathesis (ADMET) Polymerization

Acyclic diene metathesis (ADMET) polymerization is an efficient way to synthesize π-conjugated materials. The promising characteristics of this approach are as follows: (1) the resultant polymers are defect-free because either termination of the conjugated units or any negative impurities such as halogen, sulfur are not present, (2) because the reaction proceeds via metallacycle intermediate initially proposed by Thorn-Csányi et al., the resultant polymers (oligomers) possessed highly trans olefinic double bonds; (3) the resultant polymers prepared by Ru catalyst demonstrated well-defined polymer chain ends (as vinyl group), as modified the conjugated materials.

The results for ADMET polymerization of 2,7-divinyl-9,9-dialkylfluorenes by ruthenium-carbene catalysts were summarized in Table 2.1 (Fig. 2.6a). Although the repeating units cannot be perfectly controlled by this condensation (step growth) polymerization, the optimization of the reaction conditions such as catalyst, monomer/catalyst molar ratios, and initial monomer concentration could achieve PFVs with high molecular weight and uniform distributions. Due to an equilibrium shown in Fig. 2.6b, it is necessary to continuously remove ethylene by-product from the reaction medium for attaining high molecular weight polymers. The all-trans internal olefinic double bonds of the resultant polymers possess defect-free nature (without termination of conjugation).

Table 2.1 Acyclic diene metathesis (ADMET) polymerization of 2,7-divinyl-9,9-dialkylfluorene using various Ru-carbene complex catalysts

Alkyl in C_9R	Conc.[b]	Ru cat. (equiv)[c]	Solvent (mL)	Temp/°C	Time/h	M_n [d] × 10^{-4}	M_w/M_n [d]	Yield/%
n-C_8H_{17} (1)	90	A (20)	toluene (1.0)	50	8	–	–	–
n-C_8H_{17} (1)	90	B (40)	toluene (1.0)	50	7.5	1.84	1.8	75
n-C_8H_{17} (1)	180	B (40)	toluene (1.0)	50	8	2.75	2.0	90
n-C_8H_{17} (1)	150	B (40)	CH_2Cl_2 (1.2)	40	5	2.58	2.2	89
n-C_8H_{17} (1)	180	D (80)	CH_2Cl_2 (1.0)	40	5	0.45	1.6	>99
n-C_8H_{17} (1)	180	D (40)	C_6H_5Br (1.0)	90	3.5	0.33	1.8	75
2′-ethylhexyl (2)	180	B (40)	toluene (1.0)	50	3	2.10	2.1	75
2′-ethylhexyl (2)	270	B (40)	toluene (1.0)	50	3	3.30	2.2	82
2′-ethylhexyl (2)	180	B (40)	toluene (1.0)	50	8	3.00	2.6	79
n-C_6H_{13} (3)	100	B (40)	toluene (1.0)	50	3	2.30	2.5	93
n-C_6H_{13} (3)	200	C (40)	toluene (1.0)	50	5	2.10	2.1	82
n-C_6H_{13} (3)	103	B (35)	CH_2Cl_2 (2.0)	40	8	3.20	2.0	>99

Reprinted from ref. [14], copyright 2015, with permission from MDPI
[a]Conditions: solvent 1.0–3.0 mL, $RuCl_2(CHPh)(PCy_3)_2$ (Ru(**A**)), $RuCl_2(CHPh)(IMesH_2)(PCy_3)$ (Ru(**B**), $IMesH_2$ = 1,3-bis(2,4,6-trimethylphenyl)-2-imidazolidinylidene), $RuCl_2$(CH-2-O^{i}Pr-C_6H_4)($IMesH_2$) (Ru(**C**)), $RuCl_2(CHPh)(IMesH_2)(3\text{-}BrC_5H_4N)_2$ (Ru(**D**)); [b]Initial monomer concentration in μmol/mL; [c]Initial molar ratio based on monomer/Ru; [d]GPC data in THF versus polystyrene standards

(a) (b)

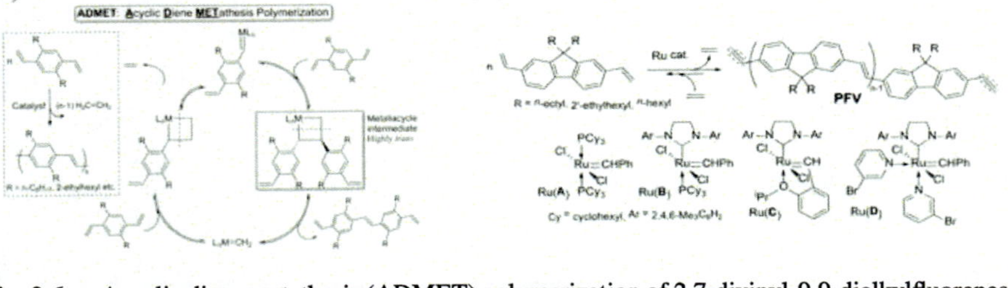

Fig. 2.6 **a** Acyclic diene metathesis (ADMET) polymerization of 2,7-divinyl-9,9-dialkylfluorenes by ruthenium-carbene catalysts. **b** Proposed mechanism for ADMET polymerization of 2,5-dialkyl-1,4-divinylbenzene. Reprinted from ref. [14], copyright 2015, with permission from MDPI

2.5 Biocatalytic Synthesis

Chemical and electrochemical polymerizations are the most common methods for producing conductive polymers (CP). However, chemical synthesis is quite unfavorable in terms of environmental impact since it is done in highly acidic medium and necessitates huge volumes of an oxidant, as well as the usage of its reduction products. Furthermore, the chemical oxidation of monomers is frequently connected with the production of hazardous by-products. Chemical polymerization takes place through an autocatalytic process with a long induction time. Electrochemical polymerization is also carried out in highly acidic solution, and it necessitates the use of a conducting electrode with geometrically constrained dimensions. As a result, employing natural biocatalysts for in vitro polymer synthesis to reduce environmental load (by minimizing by-products) appears to be highly appealing and promising.

The biocatalytic synthesis takes place under "soft" circumstances (moderately acidic pH values of the reaction mixture at room temperature), its kinetics may be regulated, and the resultant polymer is free of oxidant disintegration products. Furthermore, hazardous by-products of the process are not present. The creation of conductive polymers by catalysis with enzymes has piqued the interest of academics from around the world. Enzymes can give not only the ecological purity of the process but also the production of a finished product with peculiar qualities when synthesizing chiral conductive polymers. Horseradish peroxidase (HP) is commonly used in the synthesis of conductive polymers, particularly PANI, because it is a commercially available and well-known enzyme. Using HP as a catalyst, the template technique yielded a pseudo-soluble chiral PANI. Chiral PANI was also produced on a short-chain DNA template utilizing HP and myeloperoxidase-11 as oxidative polymerization of aniline biocatalysts. PH-stable peroxidases from soybeans and palm leaves were employed to avoid biocatalyst deactivation at acidic pH levels during PANI production.

For the production of conducting PANI, a fungal laccase with a high redox potential was suggested. In contrast to peroxidase-catalyzed polymerization, ambient oxygen acts as the oxidant during laccase-catalyzed polymerization of monomers, simplifying the synthesis. Laccases from basidial fungi are also active and stable at acidic pH levels. A conducting PANI was also created on PAMPS, polystyrene sulfonate, and SDBS templates utilizing a high redox potential fungal laccase from Trametes hirsuta (=0.78 V with respect to the NHE).

To study specific features of the template polymerization of aniline using laccase from T. hirsuta and to elucidate differences in the mechanisms of chemical and enzymatic PANI synthesis, the reaction medium redox potential was measured during the synthesis under conditions of disconnected chain with concurrent recording of UV–visible spectra of the resulting products. The redox potential ($pH = 3.5$) of the solution containing aniline and PAMPS fell by ~1520 mV upon the start of the chemical polymerization of aniline with APS and reached a minimum in a few minutes. This was linked to the emergence in the electronic spectra of the reaction medium of an absorption band at wavelength 320–330 nm, which is indicative of the π-π* transition in aromatic rings during aniline oxidation. The medium redox potential then significantly rose without any changes in optical density in the range of 500–1100 nm.

Fifty minutes later, the optical density in the electronic spectra of the reaction medium rose in the 600 nm region, which corresponded to the oxidized chains of aniline oligomers. The growth rate of medium redox potential remarkably reduced and the reaction medium potential quickly became maximum and subsequently declined. Simultaneously, for the interpolymer complex PANI/PAMPS, the shift of absorption band to longer wavelength from 600 to ~730 nm suggested the production of PANI emeraldine salt.

Immediately after the laccase was added in the enzymatic polymerization of aniline, the redox potential of the reaction medium obviously increased and subsequently reduced for ~20 min. Upon the addition of the enzyme, the absorption in the solution electronic spectra rose in the region of 700 nm, as suggested that the conducting interpolymer complex PANI/PAMPS was produced.

During the polymerization of aniline with APS as an oxidizer, APS progressively decomposes in acidic medium. Since the oxidation potential of aniline oligomers is greatly lower than that of the monomer, the growing PANI chains are invariably in the oxidized state and a fraction of unoxidized aniline exist in the reaction media. After APS is decomposed completely, oxidized chains of PANI are reduced to the conducting emeraldine salt by the unoxidized aniline. Therefore, the induction period happened in the chemical polymerization of the PANI emeraldine salt. In contrast, there is no induction period in the enzymatic synthesis of aniline. MALDI TOF spectrometry indicates that the PANI chains grow because halfoxidized forms of aniline oligomers are produced.

Figure 2.7 illustrates the formal mechanism of the polymer chain growth at the initial stages during the chemical and laccase-catalyzed polymerization of aniline. In spite of the polymer chain growth within the reaction solution, the oxidation of aniline or its oligomers with low molecular weight relates to the laccases and peroxidases for synthesis of PANI. As to PANI, the reaction rate of the enzymatic oxidative polymerization is comparatively low and high concentrations of the enzymes are essential to increase it, whereas the enzymatic polymerization has controlled kinetics and no induction period. Redox mediators of the enzymes appear promising for fabrication of conductive polymers with high molecular weight and conductivity because they can greatly accelerate the monomer polymerization. Different from aniline, EDOT with high ionization potential cannot be oxidized with enzymes alone. PEDOT synthesis on a polysulfostyrene template using HP was only once reported. The corresponding redox mediator of the enzyme enables the enzymatic oxidation of EDOT to yield the radical and subsequent growth of the polymer chain. Synthesis of conducting polypyrrole (PPy) utilized the enzyme–mediator systems based on laccase and HP and ABTS as a mediator.

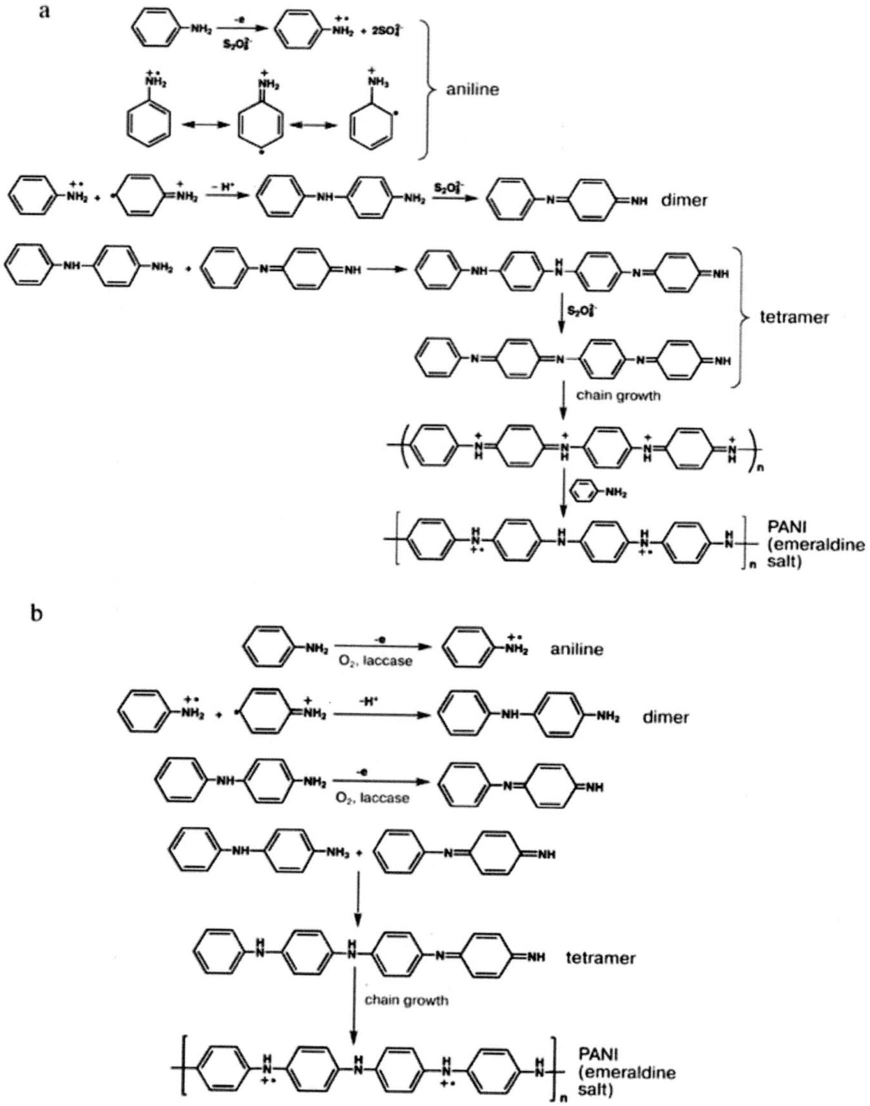

Fig. 2.7 Formal scheme of initial stages of aniline polymerization in chemical (**a**) and laccase-catalyzed (**b**) syntheses. Reprinted from ref. [11], copyright 2014, with permission from Springer Nature

2.6 Vapor Phase Oxidative Synthesis

2.6.1 *Vapor Phase Polymerization (VPP)*

Several polymerization approaches using the vapor phase to deposit CPs include pulsed laser deposition, plasma-enhanced chemical vapor deposition, oxidative chemical vapor deposition, and vapor phase polymerization (VPP). VPP is a facile method to produce thin films of both soluble and insoluble CPs, and nanocomposites thereof. In 1986, conductive PPy was deposited when Ojio et al. exposed pyrrole monomer vapor to a poly(vinyl alcohol) (PVA) film containing iron(III) chloride. The conductivity and transparency of the PVA/PPy composites are greatly dependent on the temperature, pyrrole polymerization time, and $FeCl_3$ concentration [2].

For the first attempt, a vapor phase component was introduced to the oxidative polymerization by a stepwise deposition protocol. A substrate preloaded with an oxidant was passively exposed to monomer vapors in a closed setting. Upon the passive diffusion of the volatile monomer to the oxidant-coated substrate surface, the polymerization is initiated. Over the past decade, the VPP process has quickly become a popular process to prepare "state-of-the-art" CPs with potential applications.

While VPP has undergone numerous changes over the years, including oxidant variation, vacuum incorporation, and use of additives, the sequential nature of the technique has remained consistent. Because solution processing step necessitates compatibility between solvent and substrates, it is great challenge to upscale VPP. Figure 2.8 illustrates the process of VPP occurring at the liquid–vapor interface. In the simple procedure, a variety of both substrate and monomer are used. Both soluble and insoluble polymers can be synthesized on various substrates by VPP only if the precursor monomer can be vaporized and the substrate itself can be initially over-coated by an oxidant solution. Both monomer and oxidant can more easily be enhanced to manipulate polymer properties than other polymerization methods. For instance, additives can be introduced into the oxidant solution to control the produced polymer. Great improvements in CP films including remarkably enhanced electrical conductivity have been achieved by Fabretto et al. using such additives.

During the VPP process, the oxidative polymerization occurs when the oxidant (in solution) and monomer (as a vapor) intimately contact at the liquid–vapor interface. Various factors such as the concentration of the oxidant and monomer at the interface, their replenishment rate, and physical constraints with respect to the physical mixing of the two chemical species determine the polymerization rate at this interface. The conditions such as temperature, pressure, and parameters (chamber, oxidant, etc.) used in VPP decide these aforementioned factors and straightly influence the properties of the resulting CP thin film such as optical properties and conductivity. These conditions and parameters have impact on the inherent nature of the CP (doping level, morphology, conjugation length, etc.) and then define its properties. According to this, certain CPs for particular applications can be engineered. In order to regulate the properties of the resultant CP, the inputs to VPP (chemistry, engineering, or

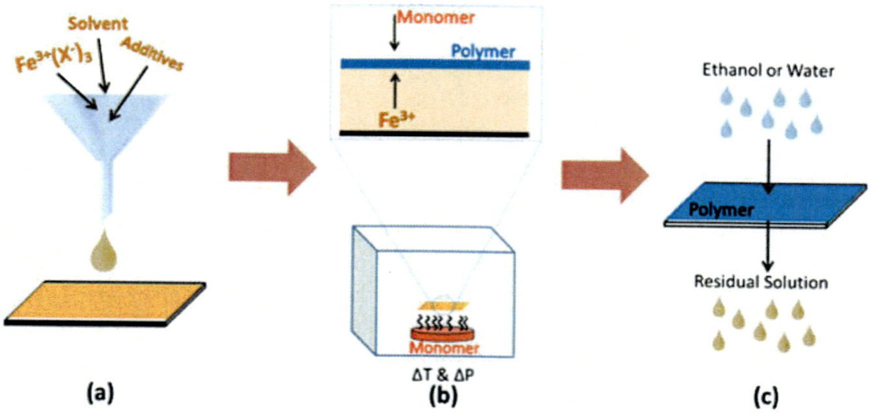

Fig. 2.8 VPP process involving **a** deposition of the oxidant solution (onto the substrate, black) commonly containing an Fe^{3+} salt in a solvent (typically an alcohol such as ethanol) with possible additives, **b** exposure of the oxidant to monomer vapor at a given temperature (*T*) and pressure (*P*), where oxidant/monomer is transported/condensed at the interface to initiate polymerization, and **c** washing away excess oxidant and monomer to yield a CP thin film. Reprinted from ref. [4], copyright 2017, with permission from Elsevier

otherwise) were controlled. The choice of the oxidant (a salt formed by combing the oxidizing agent and the doping anion) is the most paramount parameter.

2.6.1.1 Oxidants

Oxidants with varying oxidation strengths have been introduced into VPP and lead to interesting results. The strength of the oxidizer is associated with the standard electrode potential of the cation. As the most commonly used cation, Fe^{3+} possesses a standard electrode potential of 0.77 V to transform to Fe^{2+}. Its typically paired anions include Cl^- and Tos^-, and various sulfonates are used to dope the CPs. They are necessary for stabilizing the polaron and bipolaron states along the conjugated backbone of the CP, as become the origin of the CPs conductivity. It is well known that various dopants change the properties of the final polymer.

Many monomers can be polymerized by $FeCl_3$, while $Fe(Tos)_3$ is limited to selected monomers, because of its lower effective oxidation strength. Although the standard electrode potential of the cation reduction (Fe^{3+} to Fe^{2+}) is consistent, different anions vary the polymerization rate, as results in diverse effective oxidation strength. However, in spite of certain additives included, $Fe(Tos)_3$ is chosen for many applications because its effective oxidation strength is lower, and thus, the polymerization rate is slower. Via VPP using this oxidant, the most electrically conductive and topographically smooth CPs are yielded and are systematically utilized in optoelectronic devices.

It is assumed that slower rate of polymer formation forms long polymer chains with long conjugation lengths. As a result, charge transfer through the polymer chain and between ordered polymer grains can proceed with minimal delay owing to structural or chemical flaws. Subramanian et al. investigated the use of several sulfonate-based anions to dope polypyrrole (PPy) through VPP by varying the length of the alkyl chain connected to the parent sulfonic acid anion. Benzenesulphonic acid, p-ethylbenzenesulphonic acid, and p-dodecylbenzenesulphonic acid were among the variants. When the alkyl chain was stretched, the conductivity decreased by an order of magnitude. Surprisingly, doping level (ratio of anions per monomer, S:N ratio) originating from the X-ray photoelectron spectroscopy (XPS) remained rather steady at roughly 0.3. The lower conductivity was not caused by a decrease in the quantity of charge carriers; however, their mobility is likely to be reduced as a result of the anion impact on the structure/morphology of the resulting polymer.

Other oxidant research has looked on using Fe^{3+} salts to make PPy and PEDOT. Winther-Jensen et al. studied the anions of camphor-sulfonate, chloride, 4-ethylebenzenesulfonate, tetradecysulfonate, and para-toluenesulfonate in these studies. Although the conductivity of the resultant PEDOT films utilizing the atypical sulfonates was lower than that of $Fe(Tos)_3$, the comparison allowed for an in-depth investigation of the source of the conductivity. X-ray diffraction (XRD) tests revealed that PEDOT synthesized by VPP utilizing $Fe(Tos)_3$ had a bigger interchain spacing than the other variations, which was determined to be the source of its higher conductivity. The interchain distance improves the intermolecular forces operating between neighboring polymer chains. This alteration was proposed to decouple charge transport on neighboring chains, allowing for better electrical conductivity. These findings suggest that the electrical characteristics of a CP are regulated not only by individual chain attributes (conjugation length, doping level, etc.), but also by the structural features of the CP films (chain spacing, crystallinity, morphology, etc.).

Winther-Jensen's study also featured an important discovery about oxidant selection and conductivity. Ion exchange was performed on PEDOT synthesized using $FeCl_3$ or $Fe(Tos)_3$ to study if the dopant anion affected the CP characteristics or whether the properties were defined during the polymerization process. There was no substantial change in interchain spacing or conductivity after ion exchange on the synthesized CP films. This study showed that while anion specificity was necessary during polymerization, once the CP was produced, the sheer presence of any anion (regardless of type) was enough to make the CP conductive. These observations led to the conclusion that electrostatic interactions between the anion and the CP regulate charge transfer, regardless of the anions chemical composition. For example, following ion exchange to form PEDOT-Cl, synthesized PEDOT-Tos remained extremely conductive, but synthesized PEDOT-Cl kept a reduced conductivity after ion exchange with PEDOT-Tos.

The absolute concentration of oxidant used has an effect on the characteristics of the CP film as well. EDOT, pyrrole, and 3-hexylthiophene are some of the monomers that have been reported in terms of oxidant concentration (wt% of oxidant in solution). For these monomers, there was a tendency in that increasing the oxidant concentration increased the thickness of the resultant CP film. This is not a surprising outcome, as a larger concentration should allow for more monomer molecules to be oxidized, only on the basis of concentration. Furthermore, it is anticipated that an excess of oxidant is desired, assuring the continued availability of the oxidant to the presenting monomers after polymerization has commenced.

However, as Jang et al. demonstrated the manufacture of VPP poly(3-hexylthiophene) (P3HT), increasing the oxidant concentration may have a detrimental influence on some attributes such as surface roughness. They found that the surface harshness of the films was increased from 9.5 nm to 83.2 nm by an increment in $FeCl_3$ from 5 wt% to 20 wt%. A comparative report by Ali et al. on VPP PPy showed that remarkable decrease in conductivity of about 300 S cm^{-1} when the $FeCl_3$ concentration rose from 3 wt% to 7 wt%. This study likewise indicated that different oxidant concentrations impacted CP film morphology in that smooth islands produced at lower $FeCl_3$ concentration were connected after $FeCl_3$ concentration was increased.

All these studies recommend that the kinetics of the polymerization is an important factor in determining the CP properties. In general, at higher oxidation polymerization, the faster polymerization rate makes the CP films thicker, rougher, and less conductive. In contrast, thin, smooth, highly conductive films can be yielded by slowing the polymerization kinetics by some means. In order to produce slow kinetics during the VPP process, it is necessary to appreciate the monomer vapor pressure and the temperature of the polymerization.

2.6.1.2 Polymerization Film Formation

Many groups in the literature suggested the CP thin films during VPP were formed with both a "top-down" and a "bottom-up" method. "Top down" signifies that the growth of freshly produced polymer becomes descending into the oxidant layer and "base up" inferred that it goes upwards and away from the oxidant layer. Nair et al. proposed that in a "top-down" approach, polymerization happens either at the surface of the oxidant or in the oxidant bulk and the monomer diffuses through the polymer layer, as is the decisive step in the reaction. Fabretto et al. clearly described the mechanism as follows: "bottom up" represents the upward diffusion of the oxidant mixture through the produced polymer layer, while "top down" demonstrates the downward diffusion through the forming polymer layer to approach supplementary oxidant. It was proved by X-ray photoelectron spectroscopy (XPS) and time of flight secondary ion mass spectroscopy (ToF–SIMS) depth profiling that the "bottom-up" mechanism is the most plausible. Figure 2.9 schematically shows that Fe species increase with the depth and there is nonzero value for Fe^{3+} at the top surface even after long polymerization time, as confirmed the "bottom-up" approach.

In support of "bottom-up" approach, Brooke et al. successively synthesized two different CPs including EDOT and pyrrole using the same oxidant solution. Their identification determined using XPS and ToF–SIMS analysis uncovered the growth mechanism present in this system. If the oxidant diffuses upwards through the initially produced CP layer, the second polymerized monomer would occur above the first

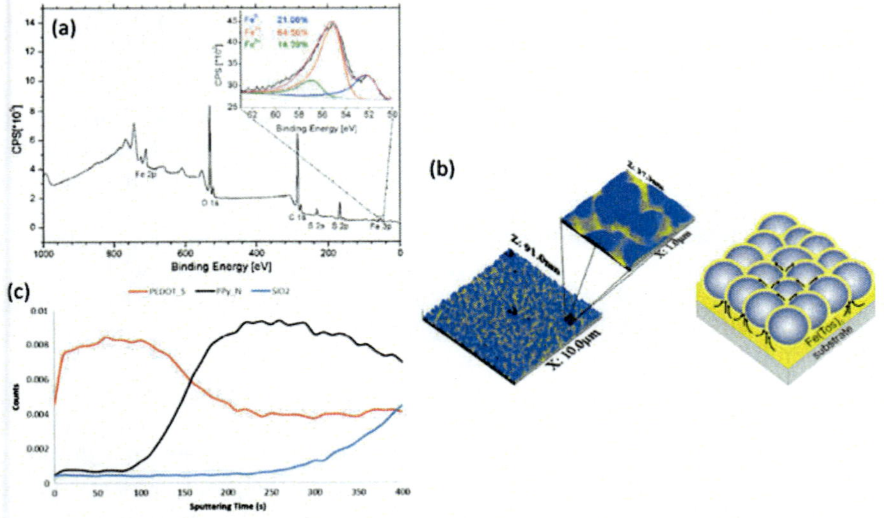

Fig. 2.9 Evidence of the "bottom-up" process in VPP. **a** XPS spectra of unwashed PEDOT after VPP demonstrating a finite concentration of Fe^{3+} at the top surface of the film, with **b** a schematic representation of the process. **c** ToF–SIMS depth profiling of a PEDOT/PPy multilayer showing the initially synthesized PPy resides underneath the later fabricated PEDOT. Reprinted from ref. [4], copyright 2017, with permission from Elsevier

for "bottom-up" approach. XPS spectra indicate the initial observation of only sulfur and the subsequent presence of only nitrogen, as was confirmed by ToF–SIMS. Therefore, since the polymerization of EDOT was followed by pyrrole, a "bottom-up" approach during the VPP process was proved in their system. In order to confirm the existence of this growth mechanism within the VPP on porous substrates such as fibers and CP foams, it is essential to conduct the in-depth elemental investigation on their mechanistic behavior.

2.6.1.3 Advances in Vapor Phase Polymerization

Since the polymerization of the monomers happens at the liquid–vapor interface and is independent on the substrate, the polymer film structures can be patterned and created via VPP. VPP has the ability to pattern and create complex film structures as the monomer arrives and polymerizes in situ (substrate independent and polymerization occurring at the liquid–vapor interface). Especially, the following three main aspects of the VPP process can be modified: (i) the substrate material and geometry, (ii) additives in the oxidant solution and (iii) synchronous or consecutive delivery of various monomers. In spite of the modified process, the overall point of these variations is to improve the resultant polymeric film in some way. The changes with such process are viewed as the advances to the basic VPP method.

Winther-Jensen and West carried out pioneering work on VPP PEDOT and proposed that the apparent acidity of the oxidant solution caused an unwanted acidic side reaction in the absence of any performance improving additives. Due to the acidic side reaction, the electrical properties of the resultant polymer films are poor and the chemical structure of PEDOT are changed; for instance, the dioxane ring was cleaved. Then, a basic inhibitor was introduced to increase the pH and hence avoid the acidic side reaction by applying pyridine in a molar ratio of 0.5–1 with the oxidant ($Fe(Tos)_3$). The final base-inhibited PEDOT-Tos possessed electrical conductivity of 1000 S cm^{-1} and a visually smoother morphology after a post-ethanol wash. Because the pH of a non-polar liquid cannot be measured by a stringent technique, whether introducing the base inhibitor alters the pH or simply slows the polymerization kinetics is undetermined.

Other research groups widely utilized base inhibitor additives to manufacture PEDOT via VPP. For instance, when the molar ratio of pyridine to $Fe(Tos)_3$ increases from 0.25 up to 0.5, the electrical conductivity arises from 180 S cm^{-1} to 500 S cm^{-1} and optical transmission increases from 60 to 80% at 450 nm, while the RMS surface roughness decreases from 3.8 nm to 2.1 nm. According to the mechanism proposed by Le Truong et al., the pyridine as a substitute ligand coordinate with Fe^{3+} ions and impedes the rapid reduction of Fe^{3+} to Fe^{2+}. Namely, pyridine reduces the noticeable reactivity of the oxidant and thus the thickness of the resultant PEDOT via VPP, as shown in Fig. 2.10.

Meanwhile, pyridine may benefit the stabilization of the EDOT radical in the monomer and oligomers during the polymerization process. This assumption adjusts well to the previously mentioned remarks in regards to the polymerization kinetics essential to accomplishing great quality CP films. Critically, an upper limit on the pyridine concentration with a molar proportion of pyridine to oxidant of 0.75 reduced the conductivity supposedly because the impurity caused by extra pyridine discontinues the charge within the PEDOT layers. The scheme in Fig. 2.10 proposes that the introduction of an alcohol "actually" dislodges Tos particles from the coordination shell of the Fe^{3+}. Figure 2.10c indicates that pyridine displaces the alcohol because of the solitary pair of electrons present on the pyridine particle without consideration of likely existence of water in the system. As shown by Subramanian and Winter-Jensen in separate studies, water straightforwardly coordinates with Fe^{3+}, even in the presence of Tos^-. Even though trace amounts of water may greatly coordinate to the Fe^{3+} center, assuredly the critical effect of water on the VPP of PEDOT cannot be excluded from this debate, as supposed by Mueller et al.

Regarding other oxidant additives, the utilization of poly(ethylene glycol) (PEG) containing molecules as another common additive for the VPP of EDOT was initiated by Fabretto et al. using a triblock copolymer, PEG-PPG-PEG (2700 g mol^{-1}) in their vacuum VPP chamber. This class of additives has demonstrated the ability to yield uniform, highly conducting PEDOT with the properties comparable to the transparent inorganic materials.

Zuber et al. reported another novel function of the PEG derivatives in the VPP process. A series of VPP experiments at different humidity levels without additives demonstrated that the crystallization of $Fe(Tos)_3$ occurs over a relative humidity (RH)

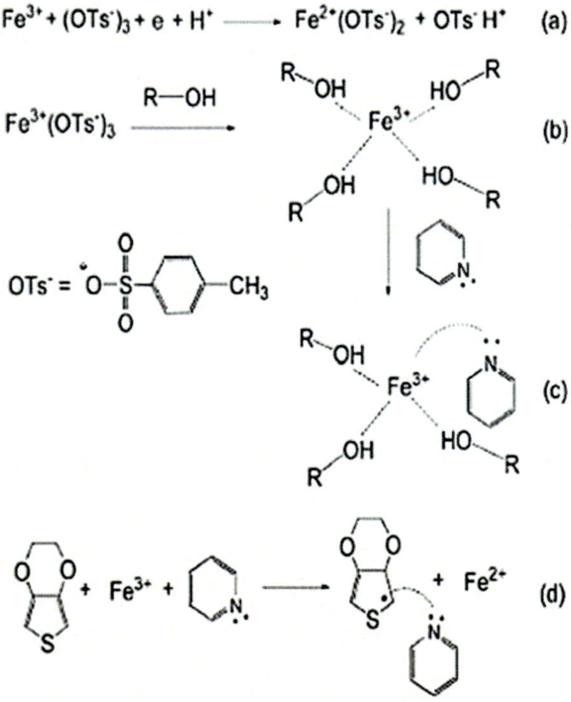

Fig. 2.10 Proposed mechanism of the EDOT polymerization with the aid of pyridine. **a** Fe(Tos)3 interaction with EDOT hydrogen's, **b/c** coordination of pyridine with Fe(Tos)3 through unpaired electrons on the nitrogen and **d** stabilization of the radical cation of EDOT by pyridine. Reprinted from ref. [7], copyright 2007, with permission from Springer Nature

of 48% and nearly half the substrate (a glass microscope slide) significantly crystallizes by 59% RH. It was assumed that the oxidant crystal forms because the initial monomer polymerization serves as nucleation sites for water condensation. Incorporating a random copolymer, PEG-ran-PPG at a concentration of 15 wt% thoroughly suppressed the crystallization up to 62% RH. However, higher levels of PEG-ran-PPG restrict the crystal growth but reduce the conductivity so that reconciliation needs to be established. 5 wt% PEG-ran-PPG led to the highest conductivity of 528 S cm^{-1} and apparent crystal formation because the PEG-ran-PPG inhibited both the crystal formation and polymerization itself. However, suppression of crystal formation weights the proposed polymerization mechanism by direct interaction between the additive and the oxidant yielding, which in this case results in a decrease in polymerization kinetics and suppression of oxidant crystallization.

Reported by Evans et al., another PEG derivative, a PEG20-PPG70-PEG20 (5800 g mol^{-1}) triblock copolymer significantly enhanced the conductivity of the PEDOT thin films to a maximum value of 3400 S cm^{-1}. The exceptionally high conductivity originated from the increased molecular ordering of the VPP PEDOT film. Comparing no PEG-PPG-PEG to 15 wt% of a 2900 g mol^{-1} PEG-PPG-PEG (Pluronic L64) and 15 wt% of a 5800 g mol^{-1} PEG-PPG-PEG (Pluronic P123), the ordering of the [100] stacking in the XRD analysis of the film ascended. The highly conductive VPP PEDOT possessed sheet-like structures, while the morphology of lower conductive PEDOT exhibited a "cauliflower" structure. With regard to the sheet-like morphology, the increase of [200] and [300] peaks in XRD indicate a lamellar structure at the molecular scale, as supports that the CP properties depend on the molecular structure and morphology. As the PEG-PPG-PEG additive produces hydrophobic and hydrophilic domains in the oxidant solution, the tosylate anion interacts with the PPG and Fe atom coordinates within the PEG in the hydrophilic regions. The polymerization kinetics of the VPP was decreased, and the structure/morphology of the resulting polymer was controlled by such portioning of the constituents within the solution. Vucaj et al. first demonstrated that incorporating graphene and MoS_2 into PEDOT-Tos created the composite films with 2D nanosheets dispersed within. The introduction of surfactant exfoliated nanosheets to the oxidant solution was followed by being put through a classical VPP process.

2.6.2 Oxidative Chemical Vapor Deposition

Mohammadi et al. exposed pyrrole monomer vapor to iron chloride ($FeCl_3$) vapor under vacuum and a polymer thin film formed by the condensation of monomer and oxidant. This alternate polymerization approach was referred to as chemical vapor deposition, as actually eradicated troubles with CP solubility and solvent/substrate compatibility.

Oxidative chemical vapor deposition (oCVD) is different from VPPs, especially in the way oxidants are supplied. In oCVD, both the monomer and the oxidizer are supplied to the substrate surface in a vapor phase in a single step. In a standard oCVD process, the substrate in a vacuum chamber is placed on an inverted, temperature-controlled stage. At room temperature, iron (III) chloride, a solid oxidant, is placed directly into a crucible in a vacuum chamber facing upward in the reverse stage and heated to a high temperature for sublimation. The monomer, often in the form of a liquid, is placed in a heated vacuum jar and delivered to the system via a heated line at a controlled flow rate. The operating pressure during deposition is usually manipulated by butterfly valves.

During the deposition of a PEDOT thin film by oCVD, iron(III) chloride in a crucible within a vacuum reactor was heated to vapor, as reacted with EDOT monomer vapor. Produced by this solventless deposition technique, PEDOT films demonstrated conductivities as high as 10^5 S cm^{-1} and transparency of 84%. Afterward, the modification of oCVD reactor configuration remarkably promoted the surface roughness of the deposited PEDOT. The substrate was placed upside down on the oxidant crucible, helping to avoid the buildup of iron (III) chloride particles likely to fall onto the substrate.

Gleason et al. employed the oCVD protocol with $FeCl_3$ to prepare crystalline PEDOT samples with the conductivities as high as 6259 S cm^{-1} when their substrates were heated to 300 °C during deposition and polymer film thickness (10 nm) was accurately controlled. Taking these parameters into account, X-ray diffraction studies have elucidated that the crystal transition is induced from "edge-on" to "face-on" orientation as the substrate temperature rises and the P1-Cl film thickness decreases. Such a feat enhances the influence of vapor deposition protocol and chamber parameters on the properties of conductive polymer thin films. P1-Cl film was also achieved by Parsons et al. by experimenting with chamber parameters and deposition protocols.

Based on the investigations of Andrew, the film properties of P1-Cl (Fig. 2.11) samples processed using the VPP protocol have the fundamental differences from those via oCVD. Cyclic voltammetry and electrogravimetry measurements show that repeated doping/dedoping cycles rarely trap ion mass in the P1-Cl film coated on the gold electrode using oCVD. In comparison, films produced by the VPP approach indicate significant ion mass trapping at each doping/dedoping cycle, as suggests the relative stability of oCVD-generated P1-Cl to redox cycling. It was hypothesized by Andrew etc., that the superior ion transport of oCVD-P1-Cl originated from the microscopically smooth surface including uniformly sized nanopores, as was measured by scanning electron microscope (SEM). By contrast, P1-Cl constructed using VPP showed a wide distribution of pore size and a rough surface that could interfere with ion transport.

In the follow-up of this study, the vapor-deposited P3 was incorporated into a two-layer solar cell and the physically vapor-deposited C60 was used as the acceptor layer. Solar cells containing a P3 layer deposited by oCVD showed similar power conversion efficiencies to solution-treated bilayer devices consisting of poly(3-alkylthiophene). In particular, thiophene-based polymers in lack of 3,4-dialkoxy substituents on repeating units were dedoped during the post-deposition solvent rinse step. UV–vis absorption spectra of P4 film recorded immediately after deposition and after various rinsing times obviously indicate the dedoping. In contrast, P2-Cl

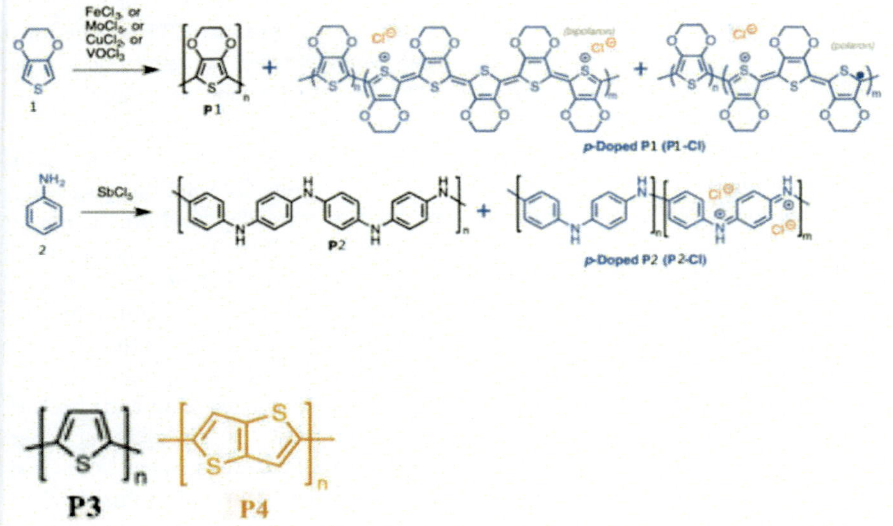

Fig. 2.11 Conductive polymers prepared by oxidative chemical vapor deposition. Reprinted from ref. [3], copyright 2019, with permission from Royal Society of Chemistry

vapor deposition films made with 3,4-dialkoxythiophene monomers (Fig. 2.11) are constantly *p*-doped after rinsing with a solvent and long exposure to the surroundings. It was confirmed by X-ray photoelectron spectroscopy and energy dispersive X-ray spectroscopy on P2-Cl and P4 (Fig. 2.11) films, respectively, that the metal salts were completely washed away from both samples. Therefore, such different behavior cannot be explained by calling for metal impurities. Alternately, whether the polymer film is dedoped during rinsing is dependent on the inherent ability of each polymer repeating unit structure to stabilize polaron or bipolaron charge carriers. The nature of the linkages between the repeating units of the vapor-deposited polythiophene is also taken into account. Since oxidative polymerization constructs all known vapor deposition recipes, the thiophene radical cation is responsible for chain extension, as it exhibits significant electron densities at both the α- and β-positions of the thiophene ring. Therefore, a statistical mixture of both α-linkages and unwanted β-linkages and crosslinks can be created by the intramolecular coupling of these radical cations.

2.7 Catalyst-Transfer Polycondensation

Conductive polymers are typically prepared by transition metal catalyzed crosscoupling polycondensation reactions usually proceeding by a gradual growth mechanism. Nevertheless, under appropriate reaction conditions, these polycondensation reactions can follow a controlled chain-growth mechanism. For example, the catalyst "moves" from monomer to monomer as the polymerization progresses, rather than diffusing from the growing chain. Figure 2.12 illustrates various methods of chain growth polymerization for conductive polymers. To convert uncontrolled polycondensation reactions into "catalytic transfer polymerization" (CTP) is the most common strategy in developing new chain growth syntheses. As regard with these reactions, a system-specific catalyst-monomer affinity prevents the catalyst from detaching the growing chain after reductive elimination.

The majority of semiconducting macromolecules produced are made up of repeated aromatic rings with pendant alkyl chains to allow for solution processing. Transition metal catalysts (e.g., Ni, Pd) are used to enhance C@C crosscoupling reactions of bifunctional monomers carrying complementary reactive groups (X-Ar-X, where X is a halide, and Y-Ar′-Y, where Y=H, $B(OR)_2$, SnR_3). These reactions normally occur via a step-growth condensation process, with tiny molecule by-products being eliminated as the polymerization develops. To synthesize high molecular weight materials (at high conversion) with dispersities of around 2, pure monomers, strong stoichiometric control over the bifunctional reagents, and effective coupling are required. The molecular weight and dispersity of conductive polymer are essential characteristics for determining its physical properties. When compared to a chain-growth reaction, step-growth reactions do not allow the same level of control over these parameters.

When difunctional monomers (X-Ar-Y) are polymerized, polycondensation may occur via a chain-growth process. In this situation, (i) molecular weights are

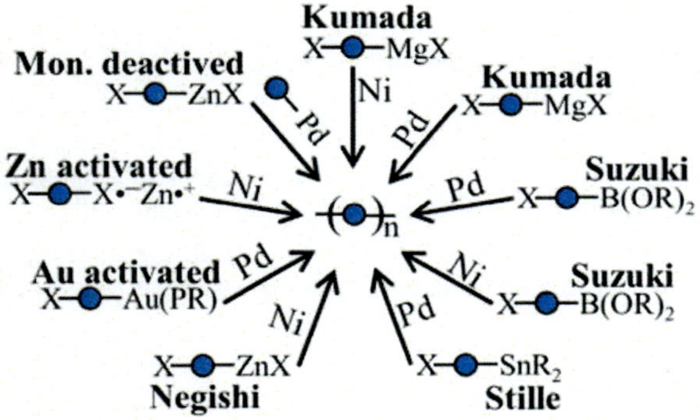

Fig. 2.12 Reported methods for chain-growth polymerizations of conductive polymers. For all reactions, the success of a chain-growth mechanism is highly dependent on the monomer used. In most cases, the affinity of the monomer to the transition metal catalyst is critical. Reprinted from ref. [1], copyright 2017, with permission from American Chemical Society

predictable based on catalyst loading, (ii) molar mass distributions (dispersities) are quite narrow, (iii) end groups are controllable, and (iv) backbone composition is tailorable (e.g., block, graft, and gradient copolymers). The crosscoupling process for the chain-growth mechanism goes as predicted, with the basic stages being oxidative addition (OA), transmetalation, and reductive elimination (RE). It is suggested that after each RE the metal catalyst coordinates with the developing polymer through back donation from a metal d-orbital to the π^* orbital of the π-electron system. This bonding contact causes selective activation of the halogen at the polymer chain end, ensuring that the chain grows. McCullough and Yokozawa separately discovered this for poly(3-hexylthiophene) and named it catalyst-transfer polycondensation (CTP).

2.7.1 *Choice of Functional Groups and Crosscoupling*

Bromine is the most often used halogen (X group) for monomers because Ar-Br bonds offer an ideal balance between monomer stability and reactivity in oxidative addition (relative bond strengths: Ar-I < Ar-Br < Ar-Cl). Aryl chlorides have been utilized with nickel catalysts in a few occasions, and aryl iodides have been used with various palladium-based catalysts. Many of the monomers employed in CTP have two distinct halogen substituents, which allows for the selective activation of one spot on the arene to generate a single regioisomer (X-Ar-Y) for further polymerization. Transmetalation reagents (Y groups) used in these polycondensations are frequently extremely nucleophilic metalcarbon bonds (X-Ar-Li, X-Ar-MgX, X-Ar-ZnX). These molecules result in excellent yields and quick polymerization kinetics. In situ insertion, exchange, or deprotonation processes on the arene produce active monomers. The monomers, however, are not stable in air or water and cannot be separated. These basicity and nucleophilicity of these species (especially X-Ar-Li and X-Ar-MgX) restrict their functional group tolerance.

CTP may also be carried out with monomers containing milder transmetalating agents (X-Ar-B(OR)$_2$, X-Ar-SnR$_3$). These isolable entities are not nucleophilic, and crosscoupling can be used in the presence of a variety of functional groups. These monomers often require activation (typically a base or fluoride source), which might change reaction kinetics, yield, and molecular weight distribution. These couplings are frequently slower, and the transmetallation process is more intricate, which might confound reaction analysis. Owing to the functional group tolerance of this reaction and the ecologically benign boronic acid by-products, organoboron reagents (Suzuki–Miyaura coupling) have become a reagent of choice. Stille–Migita coupling (X-Ar-SnR$_3$) is possibly the gentlest coupling approach used to date, but it is also the least reactive for CTP. The poisonous nature of the tin species utilized in the synthesis of these monomers and the tin by-products of the coupling is a restriction of this polymerization method. Aurylated monomers (X-Ar-AuPtBu$_3$) are a relatively new addition to the controlled polymerization family.

2.7.2 Choice of Transition Metal Catalyst

As previously stated, M^0 donation into the aromatic system's π^* orbitals is crucial for metal binding to the growing polymer and achieving chain growth. It has been proposed that the catalyst travels along the π-system to promote oxidative addition of C-X chain end after each crosscoupling cycle. This ring-walking process is retrospective of the intramolecular migration, as was observed with Ni complexes and polycyclic aromatic hydrocarbons. Optimizing the π-bonds of metal polymers and achieving controlled polymerization necessitates proper selection of metals and ancillary ligand(s).

Only two metals used in CTP reactions include nickel and palladium. Although both metals work well in the polymerization of various monomers, their key characteristics are different. Generally, in regard with Ni, the oxidative addition is much easier and the π-complexes formed with alkenes tend to be stronger than Pd ((PR$_3$)$_2$ M(CR$_2$=CR$_2$)), perhaps due to the lower ionization energy of nickel. Initially, these features may indicate that nickel catalysts are more efficient at polymerization. However, the situation is more complex if the conjugated building block has certain features. For instance, electron-deficient monomers or monomers with fused aromatic rings can enhance π-complexity with metals, especially nickel, and promote the activation barrier for oxidative additions. Finally, both group metals have their advantages. More work is needed to compare the two under the same conditions.

The ancillary ligand is definitely important for CTP and is used to regulate the electronic and steric environment around the metal center. Based on great progress on ligand design, in particular for Pd, a wide scope of catalysts have been yielded, as enhances efficient small molecule crosscoupling. These advances are also highly associated with CTP. Both monodentate and bidentate ligands are used in these polycondensations. Adjusting the σ-donation and steric environment of L-ligands will have influence on the rates of oxidative addition, transmetallation, and reductive elimination. The increase of the σ-donation increments the electron density of M^0 species and promotes oxidative addition. In addition, this function requires a balance between binding and easy oxidative addition, which may improve back donation and possibly reinforce metal-π-complexes. In order to guarantee expedient transmetalation and reductive elimination, it is necessary to balance the strong σ-donation with the proper steric bulk of the ligand.

Both monodentate and bidentate ligands provide different advantages. The monodentate ligand (L_1M^0) can create exceptionally reactive and sterically available active catalyst for polymerization. Bidentate systems (L_2M^0) apply cis geometry around the metal and provides the bite angle as an adjustable parameter. Owing to the chelate effect, bite catalysts may also show enhanced catalytic stability. In contrast, both have their drawbacks in that monodentate catalysts are usually less steady, while bidentate systems tend to be less reactive. Most precatalysts used in CTP are metal dihalides. The precatalyst undergoes two transmetalations, and then, the two arene ligands are reduced to create an active M^0 catalyst. A considerable amount of research has sought to convert one halide in the precatalyst to arene, as improves the solubility of the catalyst and the rate at which the polymer starts (because only one metal exchange reaction is required). Overall, the conductive polymers produced possess narrower molar mass distribution and well-defined end groups.

2.7.2.1 Nickel Catalysts in CTP

Kumada Catalyst Transfer Polymerization (KCTP)

In 1999, McCullough et al. utilized Kumada-type coupling with a [1,3-bis(diphenylphosphino)propane]-nickel(II) chloride ($Ni(dppp)Cl_2$) catalyst to prepare regioregular poly(3-hexylthiophene-2,5-diyl) (P3HT). Shortly after the advancement of KCTP for P3HT, this method was employed for the controlled synthesis of other conductive polymers. The polymerization progresses by a chain-growth mechanism for the poly(9,9-dioctylfluorene) (PF8) synthesized by Kumada coupling with a $Ni(dppp)Cl_2$ catalyst.

Other electron-rich monomers that can be polymerized via KCTP reactions by a controlled chain-growth mechanism consist of furan, selenophene, tellurophene, pyrrole, thiazole, cyclopentadithiophene, and paraphenylene. Only limited electron-deficient monomers were polymerized by a controlled chain-growth KCTP. Challenges related to KCTP of electron-deficient monomers start in the initial step of the response. Many electron-deficient monomers are irreconcilable with the metal-halogen exchange steps essential to form organometallic monomers. The challenge of the polymerization step lies in that most electron-deficient monomers have only a weak π-donation to the metal catalyst and do not form a stable associated complex.

The KCTP reaction generally incorporates a nickel catalyst. Nevertheless, there is no less than one illustration of palladium-catalyzed KCTP.

Phosphines

Nickel is the most widely used metal in CTP and is most often used in the Kumada Tamao and Negishi couplings. Bidentate phosphine is the most widely used coligand, which is partly due to the commercial availability of a wide range of diphosphane nickel precatalysts. The strong σ-donating properties of phosphine ligands make them very suitable for this polymerization. The first catalyst used for CTP was (1,3-bis (diphenylphosphino) propane) nickel dichloride (Ni (dppp) Cl_2). Along with Ni (dppe) Cl_2 (dppe, bis (diphenylphosphino) ethane), it is still the precatalyst of choice. Much research has been done to understand the steric and electronic contributions of these ligand frameworks during polymerization. Using 31P NMR spectroscopy, in situ probing of polymerization reactions is possible to gain a better understanding of catalyst activation, dormancy, and degradation.

Polymerization test cases are often 3-alkylthiophene monomers and have been extensively studied for both Ni(dppp)Cl_2 and Ni(dppe)Cl_2. Interestingly, the rate-determining step was found to be different for these two catalysts. In the case of dppp, the metal exchange reaction is rate determining, but in the case of dppe, the reductive elimination is rate determining. This is rationalized by subtle changes in the steric parameters of these ligands, with a larger bite angle with the propyl linker (918 vs. 868) accelerating reductive elimination. This small change presents a surprising possibility for catalyzing CTP through ligand design.

Both dppp and dppe-based catalysts have been used to polymerize a wide range of monomers. These include phenylene, thiophene, pyrrole, furan, selenophene, tellurophenes, thiazole, pyridine, fluorene, thienopyrazine, dithienosylol, carbazole, and rerangeimide. In each of these cases, there is a dramatic difference when exploring these different conjugated building blocks. Recently, it was reported by Bielawski et al. that Ni(dppp)Cl_2 can polymerize a donor–acceptor thiophene-benzotriazole alternating polymer. Ni(dppp)Cl_2 and Ni(dppe)Cl_2 also allow access to more complex polymer structures, especially diblock copolymers.

Several research groups have investigated the modification of diphosphane substituents for polymerization catalysis. Controlled Negishi polymerization of 3-hexylthiophene was performed using dicyclohexylphosphinoetan (Ni(dcype)Cl_2). The Ni(depe)Cl_2 catalyst was found to perform controlled Kumada polymerization of pyrrole in the presence of LiCl, although it was inferior to Ni(dppe)Cl_2 in the preparation of polyphenylene.

A heteroreptic phosphine-pyridine catalyst was used by Sommer et al. to polymerize thiophene with bulky side chains. The electronic properties of monodentate phosphines are similar to those bidentate phosphines. However, the steric properties of monodentate phosphines are more dynamic because the ligands can adopt transgeometry or dissociate and reassociate during polymerization. They are often less stable in solution and thus use less than Ni-CTP bidentate phosphines. Solitary phosphines have so far catalyzed CTP only for thiophenes.

Ring-walking and chain growth for bithiophene and terthiophene monomers were investigated by Kiriy et al. using $Ni(PPh_3)_2Cl_2$ catalysts. Noonan et al. investigated $Ni(PCy_3)_2$(1-Nap)Br pioneered by Percec's group to polymerize (3-hexylester)-thiophenes using the Suzuki–Miyaura coupling. The catalyst worked relatively well, but only at high loads (5 mol%). Polymerization with a single-dental nickel phosphine catalyst appears to result in more chain transfer and termination.

NHC ligands have recently been investigated as an alternative to phosphine in the catalyst of CTP. Due to their strong σ-donation and adjustable steric parameters, these ligands are very attractive for this controlled polymerization process. To date, only one NHC ligand has been used in CTP, with a 2,6-$(iPr)_2C_6H_3$ substituent on the nitrogen atom of the ring (abbreviated as IPr). Two Ni-IPr precursors are used to polymerize the thiophene monomer.

Diimine is a stronger π-acceptor ligand than phosphine and NHC and is commonly used for the polymerization of ethylene. A highly modular framework of these ligand structures is ascribed to the aryl group on the N atom of the imine, as allows the substituents to be adjusted to improve the polymerization behavior. The linker can also be adjusted, but so far only the acenaphthene skeleton has been used for the CTP reaction. Thiophene and benzotriazole can be polymerized by this class of catalyst in a chain growth manner. Excitingly, an externally launched nickel diimine has recently been developed. Due to the modularity of the ligands and the ease of synthesis of these compounds, this catalytic framework offers great opportunities. Several chain transfer problems have been pointed out, and further development of these catalysts is needed.

2.7.2.2 Palladium Catalysts in CTP

Suzuki Catalyst Transfer Polymerization (SCTP)

Palladium-catalyzed Suzuki polycondensation is one of the most common methods used in the synthesis of conductive polymers. In 2007, it was shown that Suzuki coupling can likewise continue by a controlled chain growth mechanism. Different from KCTP, it is not necessary to generate monomers in situ, as it eliminates problems associated with incomplete metal-halogen exchanges and unreacted Grignard reagents. Generally, in comparison with KCTP, SCTP is relatively less controlled and is more effective with monomers composed solely of carbon and hydrogen. The reason for these trends remains unclear. Controlled chain-growth SCTP of fluorene, paraphenylene, and thiophene monomers has been reported. Excitingly, Kirii group succeeded in chain-growth polymerization of an alternating fluorene-benzothiadiazole monomer.

Stille Catalyst Transfer Polymerization (StCTP)

Similar to Suzuki polycondensation, palladium-catalyzed Stille polycondensation is a common method used in the production of conductive polymers. Until recently, this method was only employed to manufacture conductive polymers by an uncontrolled step-by-step growth mechanism. By a controlled growth method confirmed by kinetic studies, P3HT was prepared from the asymmetric monomer 2-bromo-5-trimethylstannylthiophene. For two consecutive monomer loadings, the molecular weight increased linearly with the monomer transformation and dispersity remained genuinely consistent at ~1.2. These outcomes definitively show controlled chain growth behavior with no termination reaction. Like SCTP, StCTP uses asymmetrically functionalized monomers to avoid the challenges of in situ monomer synthesis.

Phosphines

Palladium catalysts are more common in polymerization using a milder crosscoupling strategy (e.g., Suzuki–Miyaura, Still-Migita). The ligands used for palladium tend to be larger and more electron donating because palladium is larger and has a higher electronegativity than nickel. It has been suggested that the π-complex of palladium is weak as an advantage of the monomer that nickel may tie too strongly. Interestingly, only monodentate, bulky, electron-rich phosphines succeed with Pd-CTP. By Pd-CTP using tri(tert-butyl)-phosphine ($PtBu_3$), the monomers polymerized consist of thiophene, phenylene, fluorene, phenylacetylene, phenanthrene, rerangeimide, and alternating copolymer of fluorene and benzothiadiazole is performed to prepare. These palladium complexes are more resistant to polycyclic frameworks compared to the corresponding nickel ones. The first example of sp^2-sp coupling in CTP was also related to this catalyst. Several other sterically demanding phosphines, tris(1-adamantyl) phosphine, tricyclohexylphosphine and (o-tolyl)-phosphine have been sought for polyfluorene synthesis.

Pd-NHC is an important class of catalyst for CTP. They are effective catalysts in a variety of crosscouplings due to strong σ-donor properties, bench stability, commercial availability, and solubility. CTPs of thiophene, fluorene, and phenylene used palladium NHC. The PEPPSI-IPr (pyridine fortified precatalyst preparation, stabilization, initiation) complex is the most common Pd-NHC used in CTP. This particular catalyst is interesting in that its 3-chloropyridine ligand may influence the polymerization on basis of computation. The search for various NHCs with Pd in the chain-growth polymerization is as limited as the Ni-NHC complex. There is no doubt that catalyst design will significantly affect improving the overall polymerization process as the range of crosscouplings and monomers increases for controlled synthesis of conjugate polymers. There are many opportunities and even more exciting discoveries, especially in the context of ancillary ligands.

2.7.3 New Monomers and Catalysts for CTP

The most evident area for advancement in this field is an increase in the number of conjugated building blocks that will be subjected to CTP. While numerous aromatic monomers operate well (e.g., thiophene, selenophene, phenylene, fluorene), modifications in the aromatic ring structure (especially when going to more complex polycyclic aromatics) can have a significant influence on polymerization behavior. Catalyst design and a better knowledge of catalyst–monomer interactions will be critical in improving this polymerization process for conjugated materials. In a number of investigations, computation has previously been used to give improved mechanistic knowledge. The community would benefit greatly from the development of a comprehensive model for analyzing catalyst–monomer interactions with a wide range of monomers. New catalysts will be critical in the extension of CTP to additional monomers.

Because of their strong σ-donating properties, carbenes have received a lot of interest as ligands in small molecule crosscoupling. Despite these advancements, only the IPr ligand has been employed in controlled polymerization up to this point. Organ et al. created PEPPSI-IPr, a strong palladium precatalyst that has been used in Kumada–Corriu, Suzuki–Miyaura, and Stille–Migita CTP. Given the variety of carbene ligands that have been created in recent years, these structures appear to be attractive candidates for developing novel metal catalysts for CTP. Furthermore, these ligands are frequently easily produced using low-cost reagents [9].

2.8 Controlled Polymerization

One of the most essential core technologies is the ability to achieve well-controlled polymerization, which results in polymers with limited polydispersities and defined molecular weights. Many approaches for synthesizing conductive polymers are comprised of chemical oxidation polymerization, electrochemical polymerization, and organometallic polycondensation. Dehydrogenation is required in chemical and electrochemical polymerizations: (n H–Ar–H (Ar)n; H–Ar–H= aromatic or heterocyclic molecule). Conductive polymers may also be made in general by employing dihalo-organic chemicals and a variety of organometallics in catalytic processes such as Suzuki–Miyaura coupling, Kumada–Tamao–Corriu coupling, and Stille coupling. One possible disadvantage of the presented approaches is that it is difficult to manage the polydispersity, molecular weight, and end groups of the polymer. To achieve the aim of controlled polymerization and produce advanced polymer structures, a novel polymerization process must be used.

2.8.1 Polyphenylenes

As previously stated, PPP syntheses are frequently hampered by the quick production of quasi-crystalline particles of low molecular weight PPP segments in solution. Various methods to prepare PPP are described in the literature; nonetheless, the resultant polymers exhibit large molecular weight dispersion, high 1,2-isomer concentrations, many structural flaws, and high impurity levels.

To create well-controlled and specified polymers, derivatives that are soluble in common organic solvents were generated. The dehydrogenation of poly(1,3-cyclohexadiene) (PCHD), which is easier to handle than PPP, is one of the dependable pathways to PPP. PCHD is made up of a primary chain created by a 1,2-addition (1,2-CHD unit) and a 1,4-addition (1,4-CHD unit). The base initiator, which is commonly amine or alkyllithium reagent, can influence the unit ratio of 1,2-CHD/1,4-CHD in the polymer chain.

2.8.2 Polythiophenes

In contrast to PPP, PTh is thought to have a near to planar shape, which boosts its charge mobility. Unsubstituted PThs are insoluble and only soluble in solutions containing arsenic trifluoride and arsenic pentafluoride. Using the 1,3-bis(diphenylphosphino)propanenickel(II) chloride [Ni(dppp) Cl_2]-catalyzed polymerization of Grignard-type monomers, the Yokozawa and McCullough groups independently developed methods for the synthesis of highly rr poly(3-hexylthiophene) (P3HT) with controlled molecular weights and narrow molecular weight distribution. Externally started rr-P3HT was synthesized with a regulated molecular weight and a restricted polydispersity.

The started polymerization of 3-hexylthiophene monomers was formed from an externally introduced cis-chloro(aryl) (dppp)nickel complex. Throughout the polymerization, the molecular weight of the resulting polymer grew linearly in proportion to the conversion of monomer with low polydispersity, reaching a Mn of 11.2 kDa and a PDI of 1.1 versus polystyrene standards.

Yokozawa et al. described polymerization for the synthesis of polyfluorenes using Suzuki–Miyaura coupling reaction, as opposed to the GRIM technique employing the Ni-catalyzed polycondensation system. They concentrated on arylpalladium (II) halide complexes featuring a bulky phosphine ligand as the initiator complex in their investigation. The polymerization procedure was carried out in the presence of an externally introduced initiator, $tBu_3Pd(Ph)Br$. The Pd(II) complex's phenyl ligand functions as an initiating unit. The 2-(7-bromo-9,9-dioctyl-9H-fluorene-2-yl)-4,4,5,5-tetramethyl-1,3,2-dioxaborolane polymerization occurred by chain-growth polymerization from the initiator. The molecular weight of the resulting polymers was regulated by a rather limited polydispersity [10].

References

1. Aplan MP, Gomez ED (2017) Recent developments in chain-growth polymerizations of conjugated polymers. Ind Eng Chem Res 56(28):7888–7901
2. Bhattacharyya D, Howden RM, Borrelli DC, Gleason KK (2012) Vapor phase oxidative synthesis of conjugated polymers and applications. J Polym Sci Part B Polym Phys 50(19):1329–1351
3. Bilger D, Homayounfar SZ, Andrew TL (2019) A critical review of reactive vapor deposition for conjugated polymer synthesis. J Mater Chem C 7:7159–7174
4. Brooke R, Cottis P, Talemi P, Fabretto M, Murphy P, Evans D (2017) Recent advances in the synthesis of conducting polymers from the vapour phase. Prog Mater Sci 86:127–146
5. Heydarnezhad HR, Pourabbas B, Tayefi M (2018) Conducting electroactive polymers via photopolymerization: a review on synthesis and applications. Polym-Plast Technol Eng 57(11):1093–1109
6. Kowalski S, Allard S, Zilberberg K, Riedl T, Scher FU (2013) Direct arylation polycondensation as simplified alternative for the synthesis of conjugated (co)polymers. Prog Polym Sci 38(12):1805–1814
7. Le Truong T, Luong ND, Nam JD (2007) Poly(3,4-ethylenedioxythiophene) vapor-phase polymerization on glass substrate for enhanced surface smoothness and electrical conductivity. Macromol Res 15:465–468
8. Li C, Bai H, Shi G (2009) Conducting polymer nanomaterials: electrosynthesis and applications. Chem Soc Rev 38:2397–2409
9. Matthew A, Baker, Chia-Hua, Tsai, Taaffe KJ, Noonan (2018) Diversifying cross-coupling strategies, catalysts and monomers for the controlled synthesis of conjugated polymers. Chemistry 25(50):13078–13088
10. Okamoto K, Luscombe CK (2011) Controlled polymerizations for the synthesis of semiconducting conjugated polymers. Polym Chem 2(11):2424–2434
11. Otrokhov GV, Morozova OV, Vasil'Eva IS, Shumakovich GP, Zaitseva EA, Khlupova ME et al. (2013) Biocatalytic synthesis of conducting polymers and prospects for its application. Biochem Mosc 78(13):1539–1553
12. Pankow RM, Thompson BC (2020) Approaches for improving the sustainability of conjugated polymer synthesis using direct arylation polymerization (DArP). Polym Chem 11(3):630–640
13. Pouliot J-R, Grenier FO, Blaskovits JT, Beaupré S, Leclerc M (2016) Direct (hetero)arylation polymerization: simplicity for conjugated polymer synthesis. Chem Rev 14225
14. Tahmina H, Kotohiro N (2015) Acyclic diene metathesis (ADMET) polymerization for precise synthesis of defect-free conjugated polymers with well-defined chain ends. Catalysts 5(2):509–517
15. Zhang Q, Dong H, Hu W (2018) Electrochemical polymerization for two-dimensional conjugated polymers. J Mater Chem C 6:10672–10686

Chapter 3
Nanostructured Conductive Polymers

Nanostructured materials have been widely studied in both basic research and potential applications due to their unique properties at the nanolevel. The development of new multifunctional materials is driven primarily by sustainable technologies and new ways to easily realize nanostructures. In comparison with bulk polymers, nanostructured conductive polymers (NCPs) have easy synthesis, high surface area, excellent conductivity, high carrier mobility, improved electrochemical activity, particular optical properties, and biocompatibility so that they have attracted much attention.

3.1 Synthesis of Nanostructured Conductive Polymers

Multidimensional nanostructured conductive polymers (NCPs) synthesized with various morphologies and structures consist of 0D (nanospheres, nanoparticles), 1D (nanorods, nanowires, nanofibers, nanobelts, nanoribbons, hollow structure nanotubes), 2D (nanosheets, nanodisks, nanoclips). Various physical and chemical strategies have been applied to the production of NCPs, including physical template synthesis (i.e., hard and soft methods), and template-free approaches (i.e., self-assembling or interfacial polymerization, electrospinning, seed approaches). Table 3.1 shows typical examples of the manufacturing methods of the developed CP nanomaterials.

3.1.1 Fabrication of Conductive Polymer Nanoparticles

As a rule, nanoparticles of conductive polymers are attainable by post-polymerization dispersion of separately prepared polymers or directly by polymerization in a dispersed heterophase system. Both approaches have their advantages and limitations. The production of nanoparticles by post-polymerization can rely on commercially available polymers with predetermined specifications and does not require

Y. Wang and W. Feng, *Conductive Polymers and Their Composites*,
https://doi.org/10.1007/978-981-19-5363-7_3

Table 3.1 Representative examples of fabrication methods for CP nanomaterials developed

CP	Polymerization method	Details
PANI	Amyloid nanofiber template polymerization	Amyloid nanofibers were successfully used as templates for the formation of conductive core–shell nanowires
	Planar DNA template	Production of CPs with controlled shapes on 2D polyelectrolyte templates was investigated for the first time
	Electrospinning using poly(amic acid) fiber as a template	Hollow nanofibers with controllable wall thicknesses were successfully obtained
	Dedoped chemical polymerization	Water-dispersed CP nanofibers with high capacitance were achieved by double doping
	Biphase interfacial polymerization	The mechanism for self-assembly in crystalline 1D nanostructures was investigated
	Surface-initiated polymerization	A new approach for multimodal core–shell nanoparticles with a stable doping state was reported
	Interfacial polymerization	A novel hollow PANI nanocapsule with holes in the wall was synthesized
PPy	Time-dependent template-assisted polymerization	A new synthesis approach for the precise control of wall morphologies of colloidal microparticles was studied
	Modified pulse potentiostatic method	A good method to control the shape of micelles at the substrate/electrolyte interface and control the morphology of CPs was proposed
PEDOT	Galvanostatic electrodeposition	Good result combining a carboxylated polystyrene template made by nanosphere lithography with SDS as a molecular template was achieved
	Non-spontaneous emulsification	A novel method using colloidal chemistry to fabricate multifunctional CPs was developed
	Electron pulse-enabled in situ polymerization	The mechanism of CP growth was investigated experimentally and via modeling

organic and polymer synthesis equipment or expertise. If desired, the polymer can be extensively purified after polymerization. Direct polymerization, on the other hand, is not confined to polymers dissolvable in natural solvents and on a basic level can provide a wider range of nanoparticles in terms of size control and particle structure.

3.1.1.1 Post-Polymerization

The post-polymerization formation of the polymer particle dispersion is also called the secondary dispersion. Note that such processes also apply to fabricate aqueous dispersions of polyurethanes in a larger scale or, more recently, polyolefin by different methods. Of the various techniques available for post-polymerization dispersion, polymer solution in an organic solvent was utilized as a starting point for the dispersion of the conductive polymer. Nanoparticles are most commonly formed by the following two methods: (1) the removal of solvent from emulsion droplets requires a solvent immiscible with continuous phase of the final particle dispersion, (2) precipitation of the polymer upon rapidly adding the polymer solution to an excess of the continuous phase requires a solvent miscible with the continuous phase.

Emulsion/Miniemulsion Technique

The most often utilized approach in the manufacture of conductive polymer nanoparticles is the emulsion/miniemulsion technique. Followed by solvent removal from the droplets, the preparation of nanoparticle dispersions via emulsification of a polymer solution necessitates the formation of sufficiently small droplets that are also colloidally stable enough over time to allow for solvent removal without detrimental droplet coalescence. Landfester et al. produced nanoparticles from diverse polymers, including conductive polymers, using the approach depicted in Fig. 3.1a [10].

To make CPNs, it is required to dissolve the polymer in a water immiscible organic solvent and then inject the solution into an aqueous solution containing a suitable surfactant. The mixture is ultrasonically agitated fast to generate stable miniemulsions containing tiny droplets of the polymer solution. To produce a stable dispersion of polymer nanoparticles in water, the organic solvent is evaporated. Depending on the concentration of the polymer solution, the size of the nanoparticles might range from 30 to 500 nm. However, the droplets may be destabilized by Ostwald ripening as well as flocculation generated by droplet coalescence. Proper surfactants are employed to avoid flocculation, and Ostwald ripening can be inhibited by adding a hydrophobic substance (a hydrophobe) to the dispersed phase. The hydrophobe supports the production of an osmotic pressure inside the droplets, which counteracts the Laplace pressure (the difference in pressure between the interior and outside of a droplet) and prevents diffusion from one droplet to the surrounding aqueous medium.

This method has also been used to prepare multiphase particles. Nanoscale phase-separated particles of 40–150 nm size were generated from a miniemulsion comprising poly(9,9-dioctylfluorene-2,7-diyl-co-benzothiadiazole) and poly(9,9-dioctylfluorene-2,7-diyl-*co*–N,N′-bis(4-butylphenyl)-N,N′-diphenyl-1,4-phenylenediamine). The majority of the examples in the literature involve the synthesis of nanoparticles from premade polymers utilizing an oil-in-water solution.

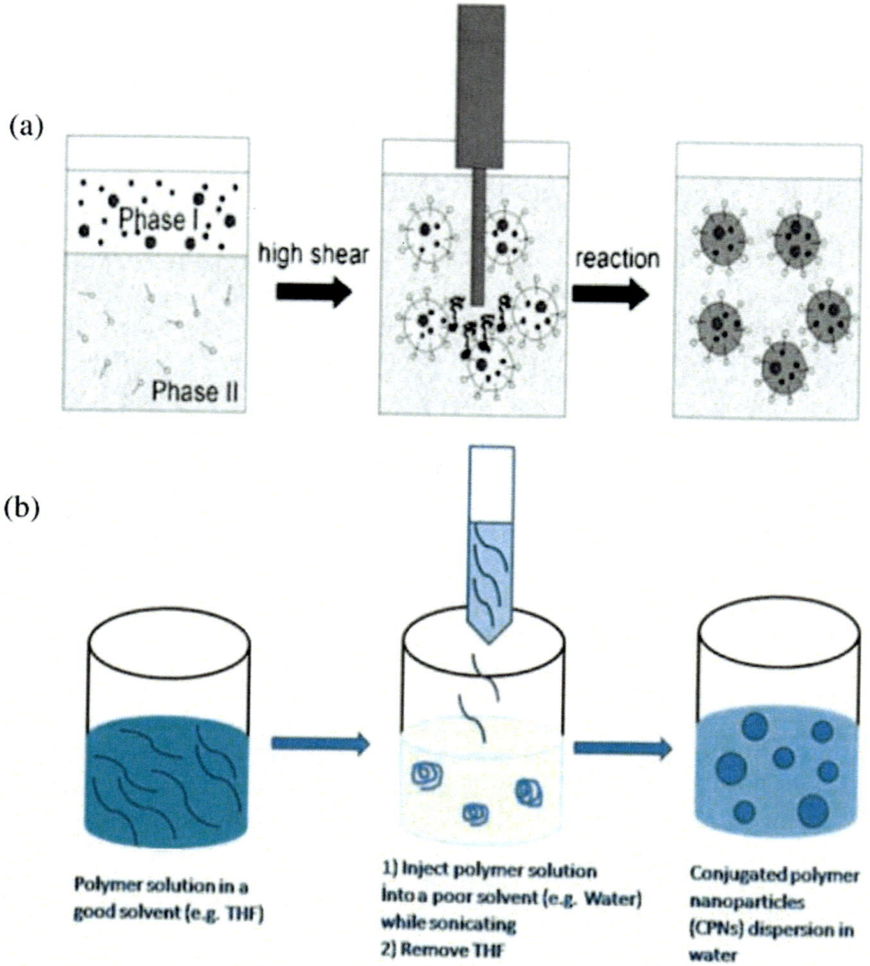

Fig. 3.1 **a** Preparation of nanoparticles using the miniemulsion method. **b** Preparation of polymer nanoparticles by the reprecipitation method. Reprinted from ref. [14], copyright 2010, with permission from the Royal Society of Chemistry.

Reprecipitation

In the process depicted in Fig. 3.1b, a hydrophobic conductive polymer is dissolved in a suitable solvent (e.g., THF) and poured into a poor solvent (e.g., water), which is miscible with the good solvent. To aid the creation of nanoparticles, the resultant mixture is aggressively agitated, often using a sonicator. Following the production of the nanoparticles, the organic solvent is removed, leaving behind water-dispersible

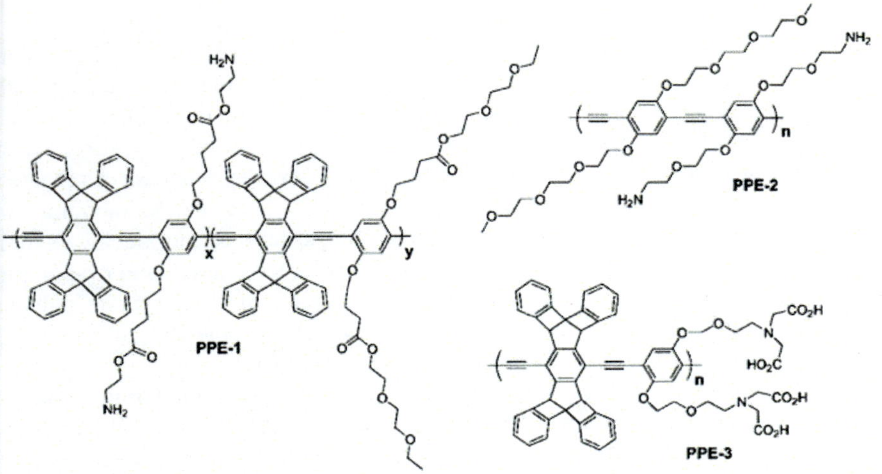

Fig. 3.2 The structures of poly(phenylene ethynylene) (PPE) derivatives utilized in the preparation of CPNs. Reprinted from ref. [14], copyright 2010, with permission from the Royal Society of Chemistry

nanoparticles. The hydrophobic effect is the primary driving force behind the production of nanoparticles. When a polymer solution in an organic solvent is given to water, the polymer chains seek to avoid contact with the water and, as a result, fold into spherical forms to achieve the least amount of exposure. The method uses no additives such as surfactants or hydrophobes and may be used to a wide range of conductive polymers that are soluble in organic solvents. Furthermore, by altering the polymer concentration and utilizing polymers with appropriate molecular weights, it is feasible to tailor the size of nanoparticles using this approach. Nanoparticles having diameters of 5–10 nm comprising single polymer chains, for example, are manufactured.

Moon et al. demonstrated the phase inversion precipitation synthesis of nanoparticles from poly(phenylene ethynylene) (PPE) derivatives, as illustrated in Fig. 3.2. The polymer is dissolved in DMSO and then added to an aqueous SSPE buffer throughout the synthesis (saline, sodium phosphate, EDTA). Electron micrographs showed particles with sizes ranging from 500 to 800 nm. The average particle size determined by dynamic light scattering (DLS) is 400–500 nm. The polymer chemical structure, particularly the kind and density of the hydrophilic side groups, played a role in the particles formation. The side chain with protonated amine and the short chain nonionic diethylene oxide moiety appear to stabilize the developing particle surface and prevent aggregation and precipitation. Smaller nanoparticles (as small as 8 nm) were produced from identical hydrophilic conductive polymers by adjusting the variables in the particle production process, such as pH, acid nature, salt concentration, and mixing parameters.

Methods of reprecipitation are limited to polymers that can be processed in a solution. However, the inflexible backbones of conductive polymers cause low solubilities in organic solvents, and the unsubstituted parent polymers are practically insoluble in any solvent. Solubility necessitates the inclusion of side chains, which affects the electrical characteristics and frequently requires further synthetic effort. The precipitation approach using a precursor pathway was reported to prepare a dispersion of nanoparticles of an unsubstituted conductive polymer. Shimomura et al. created water-soluble PPV precursor polymer nanoparticles by introducing an ionic liquid, a poor solvent on its own, to an aqueous polymer solution and evaporating the water. Due to the high boiling point of the ionic liquid, the precursor polymer was thermally converted to poly(p-phenylenevinylene) nanoparticles. The polymer molecular weights and colloidal stability remain unresolved issues.

Compared to the emulsion/miniemulsion approach, the precipitation technique produced smaller particles, and the polymer solids contents of the resultant dispersions or suspensions were lower. Particle sizes are so small in some circumstances that they correspond to single-polymer-chain particles. It should be noted that this is possibly not due to a restriction of emulsion methods in terms of the accessibility of very tiny particles, but rather emulsion techniques were used when large polymer concentrations of the resultant dispersions were desired [12].

3.1.1.2 Emulsion/Miniemulsion Polymerization

Müllen et al., on the other hand, showed the production of nanoparticles from monomers in non-aqueous emulsions (oil-in-oil). In this example, the continuous phase was cyclohexane, and the dispersed phase was acetonitrile, with polyisoprene-block-poly(methyl methacrylate) (PI-b-PMMA) as the emulsifying agent. Using this approach, nanoparticles of poly(3,4-ethylenedioxythiophene) (PEDOT), polyacetylene, and poly(thiphene3-yl-acetic acid) were produced by catalytic and oxidative polymerization. Furthermore, when the nanoparticles were formed, the emulsifying agent (PI-b-PMMA) was washed away with THF. The particles have a number-average diameter of 43 nm (10 nm).

Mecking et al. reported the production of processable polyacetylene nanoparticles using acetylene polymerization in an aqueous miniemulsion technique in 2006. Pd catalyst was dissolved in a small volume of hexane-ethanol combination and then added to an aqueous solution of surfactant and organic acid for this purpose (sodium dodecyl sulfate and methane sulfonic acid). This combination was then sonicated to generate a miniemulsion. By swirling the miniemulsion in an acetylene environment, a strongly colored, black polyacetylene dispersion was formed. Transmission electron microscopy (TEM) was used to determine the size of polyacetylene nanoparticles, which was found to be around 20 nm.

It was demonstrated by Mecking et al. that, by miniemulsion polymerization of suitable monomers under Glaser coupling conditions, nanoparticles of poly(arylene diethynylene) (arylene=2,5-dialkyoxyphenylenes and 9,9′-dihexylfluorene) derivatives may be directly produced. Gel permeation chromatography (GPC) indicated the molecular weights of these nanoparticles to be in the range of *Mn* 10^4 to 10^5 g mol^{-1}, and TEM determined their diameters to be about 30 nm. In order to produce conductive polymer nanoparticles with the chosen emission wavelength through energy transfer, they additionally covalently added 0.1–2 mol% perylene dye and 2–9 mol% fluorenone dye to the poly(arylene diethynylene), respectively.

3.1.2 *Template-Based Approaches*

The template-based nanostructuring technique enables the creation of simple, efficient, and highly regulated NCPs. A template is used in template-based techniques to direct NCPs to grow in predetermined forms and sizes. There are two types of template approaches: hard templates that rely on physically shaping the CPs into forms and soft templates that rely on CP self-assembly.

3.1.2.1 Hard Template Approach

The hard template approach is a practical, controlled, and frequently used method for producing nanostructured inorganic semiconductors, metals, and polymers. This method uses a physical template as a mold or scaffold to grow nanostructures. Colloidal nanoparticles or nanosized channels can be used as the templates to offer complete control over the size and shape of NCP. Commercially available templates have widely used polycarbonate films prepared by the "track etching" method and porous alumina films produced by an electrochemical approach.

As an advantage of the hard template method, chemical and electrochemical polymerization can be used in the next step. For chemical template synthesis, the monomers were polymerized either on the surface or in pores or channels of the hard template, as is immersed in a solution containing monomers, oxidants, and dopants. The reaction conditions such as template properties, monomer concentration, oxidant type, reaction time, and temperature determined the properties of chemically synthesized NCPs. TiO_2 nanotubes-PSS was employed as a template and dopant by Zhang et al. to synthesize PPy. TiO_2-PSS has the benefits similar to ASPB because they both can catch and control ions in the layers of the PSS chain by means of electrostatic collaboration. PPy was produced by oxidizing the pyrrole monomer by ferric ions in the PSS shell. For electrochemical template synthesis, NCP can be achieved with a template that has a metal film as an electrode. In addition to chemical factors, monomer/dopant concentrations, potential ranges and durations, and charge transport during the polymerization process also have influence on NCP properties.

Preparation of 0D NCP typically applies colloidal nanoparticles as a template. The monomer polymerizes on the surface of the nanoparticles to form a core–shell structure. The most widely used colloidal nanoparticles include metal oxide nanoparticles or polymer microspheres. The final shape and size of the NCP are strongly dependent on the dimensions of the colloidal nanoparticles. This method has several advantages, including uniform size distribution, easy availability in large quantities, and easy synthesis. For example, a hollow octahedral PANI was synthesized using a uniform octahedral Cu_2O crystal (diamond length 0.6–1.3 μm) as a template in a solution containing a dopant and an oxidant. Cu_2O with a new structure can be easily removed by an initiator. However, the post-treatment of template removal is tedious and can affect the final shape of the hollow nanostructures.

For 1D NCP, it was proposed by Martin that the template with nanosized channels or porous membranes can limit deposition/growth for the production of aligned CP nanowires/tubes. Manufacturing CP nanowires/tubes have widely applied anodized aluminum (AAO) and particle track etching films (PTM) with excellent controllability. The diameter of the pores of PTM is as small as 10 nm, and the density of pores ranges from 10^5 to 10^9 pores/cm^2. Numerous NCPs like PANI, PPy, PEDOT have been chemically or electrochemically synthesized as nanowires or nanotubes by various films. Figure 3.3a shows the basic research on PEDOT nanotubes and nanowires prepared in AAO templates. High monomer concentrations and low electrochemical polymerization potential benefit nanowire formation, while high oxidation potential and low monomer concentration favor the creation of nanotubes. This mainly results from the speed competition between monomer diffusion and electrochemical polymerization [1]. In Fig. 3.3b, via an electrochemical polymerization method using AAO as a template, PPy nanowire-based triboelectric nanogenerators were synthesized. PPy nanowires were generated along the AAO pores fixed on Ti electrode by electrochemical polymerization [2]. Ultimately, the nanostructures were achieved by dissolving the AAO template in NaOH solution. Several factors, including diffusion, kinetics, and the interaction of the pore walls with the polymer at higher oxidation potentials have impact on the morphology of the grown NCP. It is feasible to investigate the performance of electrical equipment by controlled nanowire/tube synthesis technology. However, in regard to delivering the NCPs in enormous scope utilizing this technique, the size and pore density of the solid template are still challenging.

As to the main drawback of the hard template method, the necessary removal of the template by post-processing usually disrupts or disorders the nanostructures formed and reduces key properties of CP.

3.1.2.2 Soft Template

As a frequently used technique for producing NCPs, soft template synthesis uses the surfactant micelles to limit the polymerization of CP to low-dimensional nanomaterials of particular shapes and size. In comparison with hard templates, this relatively simple and inexpensive approach allows for synthesizing NCP in large amount. The resultant size and morphology of the NCP depend on the factors such as microstructure, morphology, and concentration of molecular templates. Microemulsion and reverse microemulsion polymerization are commonly used to produce such nanostructured materials. Micelles that function as nanoreactors were formed using cationic surfactants such as octyltrimethylammonium bromide

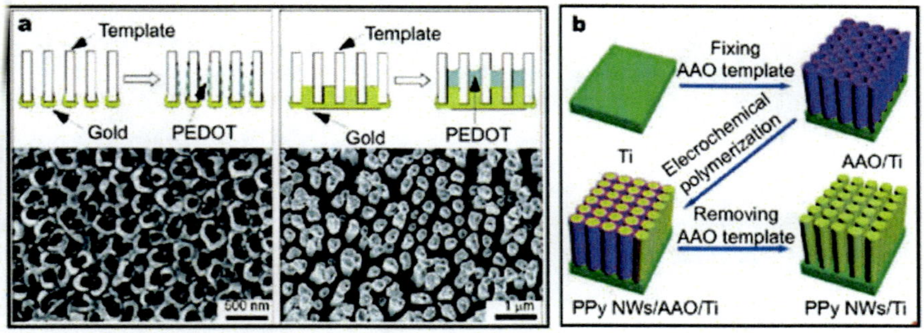

Fig. 3.3 **a** Growth mechanism of PEDOT nanostructures at lower oxidation potentials (<1.4 V): the predominance of electrochemically active sites on sharp electrode edges. Two base electrodes are considered: (left) annular and (right) flat-top electrodes. SEM images present the base electrode morphologies. **b** Schematic depiction of PPy nanowires preparation by electrochemical polymerization with AAO as the template. Reprinted from ref. [17], copyright 2018, with permission from Springer Nature

(OTAB), decyltrimethylammonium bromide (DTAB), and cetyltrimethylammonium bromide (CTAB).

Figure 3.4 illustrated a recent notable example of such a formation mechanism. PPy hollow nanoparticles of different diameters were prepared by surfactant templates and their electrical properties were size dependent. Dat et al. utilized a quaternary mixture containing brine, SDS as a surfactant, 1-pentanol as a cosurfactant, and cyclohexane oil to create a swollen liquid crystal template. Additionally, insoluble complex templates with lamellar mesostructure were created with a noble metal compound and a cationic surfactant such as CTAB. Other surfactants investigated consist of indigo carmine and Pluronic® F127.

CTABs or DTABs with long alkyl chains can yield 1D nanostructures, e.g., PPy nanowires/ribbons 25–85 nm wide, while OTABs or nonionic surfactants with shorter chains can generate non-1D nanostructures. The morphology of the resulting PPy nanostructures is dependent on both surfactant and monomer concentrations. As shown in Fig. 3.5a, it was first reported by Jang et al. that PPy core–shell nanomaterials were synthesized by microemulsion polymerization using different oxidants such as copper (II) chloride and iron (III) chloride. Finally, PPy hollow nanospheres were produced by one-step solvent etching of the linear core and surfactant in which the core and shell were created because the different oxidation potentials of the oxidants ensured the double layer structure. Figure 3.5b shows images of transmission electron microscope (TEM) and scanning electron microscope (SEM) of PPy nanoparticles and hollow spheres. As the surfactant concentration increased, the average size of nanoparticles decreased, and an increase in the amount of monomer (PPy in this case) incremented the micelle size. It should be noted that much effort should be devoted for the above surfactant templates to synthesize hollow nanostructures [8].

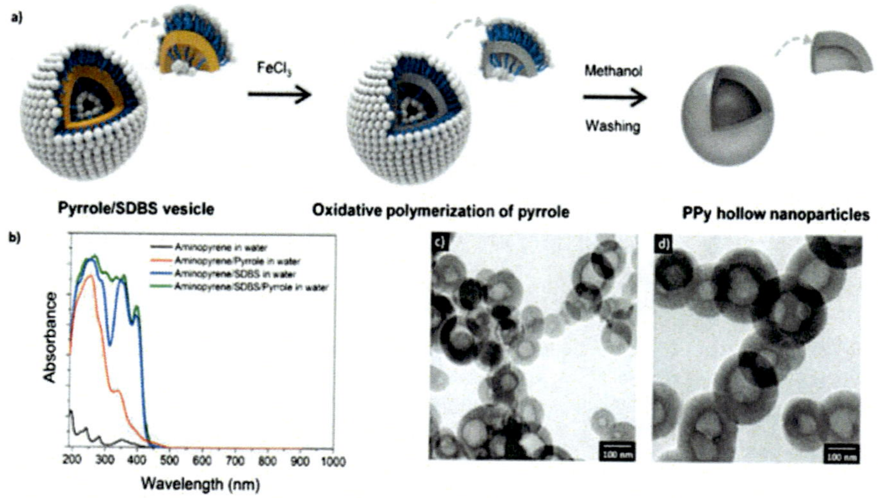

Fig. 3.4 Preparation of PPy hollow nanoparticles. **a** Schematic of the formation mechanism, **b** UV–vis absorption spectra of aminopyrene in different solvents; **c, d** Transmission electron microscopy (TEM) images of the hollow PPy nanoparticles with diameters of 100 and 200 nm, respectively. Reprinted from ref. [4], copyright 2016, with permission from MDPI

To ease that, reverse microemulsion polymerization can be employed as an effective method for producing nanotubes. Unlike microemulsion systems, the aggregation of detergent molecules in reverse microemulsions originates from the solvation of polar groups and hydrogen bonds improved by water. This method allows for incorporating different types of reactants into nanosized aqueous domains. Confined within the reverse micelles, these water domains act as nanoreactors for the growth of NCPs. Ions added to the aqueous solution can help form a rod-like structure. By Jang et al., the PPy nanotubes were fabricated by reverse microemulsion polymerization in polar solvents and the mechanism behind nanotube formation was investigated. Figure 3.5c, d showed that PPy nanotubes possessed a hollow structure with a diameter of 95 nm, a length of 5 μm, and a wall thickness of 22 nm. The conductivity and aspect ratio of the resulting PPy nanotubes are higher than those prepared by the self-assembly method [9]. Similarly, uniform and well-controlled PEDOT nanotubes and nanorods are produced, and these soft templates based on bis(2-ethylhexyl) sodium sulfosuccinate (AOT) reverse micelles are viable for preparing 1D hollow nanostructures.

However, it is challenge to overcome the instability of soft templates and the difficulty of sophisticatedly controlling the final morphology of NCPs. The aforementioned discussion about soft template methods concentrated on the synthesis of 0D or 1D nanostructures with few emphasized strategies for manufacturing 2D or 3D nanostructures. Using cetrimonium cations and peroxydisulfate anion $(CTA)_2S_2O_8$ as templates, Zhang et al. synthesized 2D clipped nanostructures of PPy, PANI, and PEDOT by a new one-step general oxidation template assembly (OTA). Nanoclip

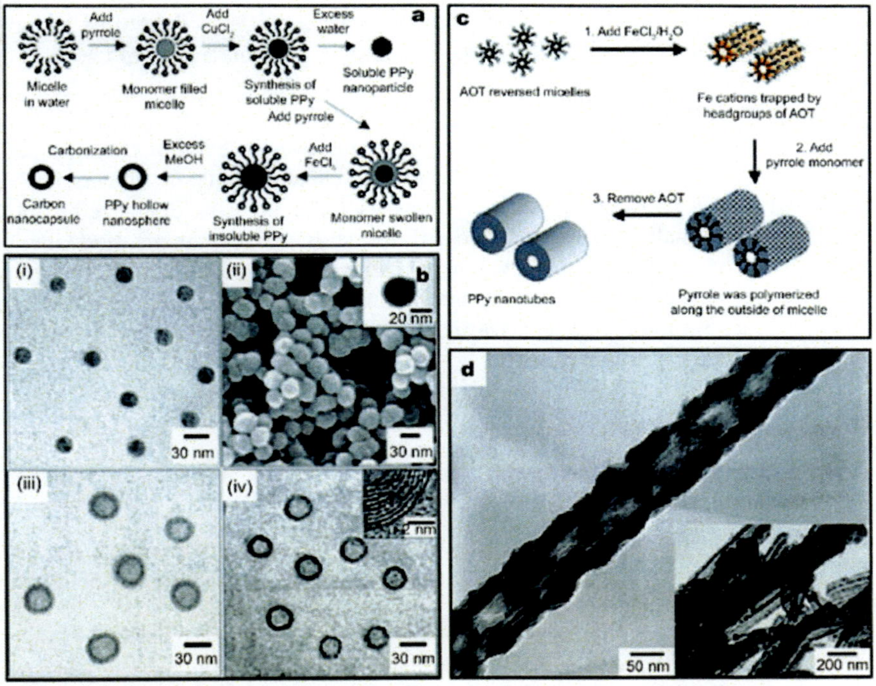

Fig. 3.5 **a** Schematic of the microemulsion fabrication of polypyrrole hollow nanospheres and their carbon derivative. **b** TEM and SEM images of PPy nanoparticles and hollow spheres: (i) soluble polypyrrole nanoparticles; (ii) linear PPy core/shell nanoparticles; (iii) polypyrrole nanocapsules and (iv) carbon nanocapsules. **c** Schematic of PPy nanotubes fabrication using reverse microemulsion polymerization, **d** TEM image of PPy nanotubes. Reprinted from ref. [17], copyright 2018, with permission from Springer Nature

morphology of PANI has a diameter of about 100 nm, while the diameter of nanoclip PEDOT is about 200 nm. This difference in diameter is attributed to the size of the individual polymer chains and the filling patterns of the different polymers. By this OTA approach, bulk morphology can be controlled and large-scale production of NCPs is permitted.

More investigation on other new developments for combining template techniques has also been conducted. Rivero et al. reported that self-doping copolymer poly (aniline-co-aminobenzoic acid) was synthesized in the presence of CTAB by a new approach using mesostructured SiO_2 as a solid template. Upon the incorporation of CTAB to the synthetic medium, the silica template demonstrated well-arranged growth with parallel channels about 2–3 nm wide. On the other hand, microporous silica deposits were obtained in the precursor solution without CTAB. It should be noticed that the as-prepared film presented the redox current response only after the removal of the surfactant molecules. Furthermore, the effect of the time to introduce the anionic surfactant SDS during the template polymerization of CP was also investigated.

Template technology has been developed for the synthesis of CP nanomaterials and various important processing factors are being investigated. For chemical pathways, these consist of oxidant/monomer ratios and post-cleaning treatments, solvent chemical effects (solvent-free vs. solvent-restricted synthesis), dopant effects (D-tartrate acid, aromatic and amphipathic sulfonic acids), and polymerization kinetics. The electrochemical approach involves investigating the deposition potential and temperature, and chloroform as an environment. Finally, sulfanilic acid azochromotrop, as a new anionic dopant, was investigated for the combination of chemical and electrochemical polymerization.

3.1.3 Template-Free Approaches

Compared to the template-based approach, the template-free approach without need of external templates or post-treatment is considered as a simple self-assembly and economical method. In spite of various approaches to synthesize NCP, it is necessary and desirable to only use the mixture of ingredients to achieve an orderly structure for the pure, homogeneous, template-free CP nanostructures. Controlling synthesis conditions such as temperature and molar ratio of monomer to dopant can create NCP. The template-free approaches most widely include interfacial, seeding, electrospinning, and other nanomanufacturing methods.

3.1.3.1 Interfacial Approach

By one-step easy aqueous/organic interfacial polymerization, CPs in 1D and 2D nanostructures with uniform diameters, i.e., PANI and PPy nanofibers/nanoneedles/nanoclips, may be produced in bulk. The non-template technique of interfacial polymerization exploits high local concentrations of both monomer and dopant anions at the liquid–liquid interface to encourage the production of monomer-anion (or oligomer-anion) aggregates. The monomer is dissolved in an organic solvent, while the oxidant (ammonium peroxydisulfate) is dissolved in an acidic aqueous solution. The interfacial technique, which differs from standard aqueous-phase synthesis, is comparable to Nylon synthesis and is accomplished in an immiscible organic/aqueous biphasic system to permit the start and propagation of the polymer reaction with fewer nucleation sites. The interaction between initiator molecules and monomers occurs mostly in a 2D region (the aqueous/organic interface). NCPs develop at the contact and rapidly disperse into the aqueous phase, freeing up the interface for further reaction. Several organizations have used this method to create PANI nanofibers.

Aniline, for example, is dissolved in an organic solvent (chloroform or toluene, both of which are insoluble in water), whereas ammonium persulfate (APS) as an oxidant is dissolved in an aqueous solution containing strong acid as a CPs dopant (HCl or $HClO_4$). The oxidation of aniline takes place at the interface of two immiscible solvents. Within a few minutes, PANI nanofibers develop at the interface. As the reaction progresses, the color of the organic phase darkens and eventually ceases to change, suggesting the completion of the reaction. Matsui et al. were the first to disclose the use of an interfacial polymerization crystallization technique to synthesize PEDOT in nanoneedles. The aqueous/organic interface is composed of organic solvent including 3,4-ethylenedioxythiophene and ferric chloride as an oxidant in the aqueous phase. Using the same procedure, they also produced PANI (63 × 12 nm) and PPy (70 × 20 nm) in a rice-like structure at the liquid/liquid interface. The crystalline polymers have a quick conductance switching time between the insulating and conducting states in the millisecond range. The fundamental benefit of this liquid–liquid interfacial approach over other synthetic methods is that it produces homogeneous and needle-shaped CPs via oxidative coupling processes in a single crystal state.

3.1.3.2 Seeding Approach

The seeding methodology, initially presented by Zhang and Manohar, is a relatively new method for rapidly synthesizing large quantities of diverse CP nanowires/fibers in a single step while controlling bulk nanoscale polymer morphologies. The type and form of nanofibrillar seeds are critical in this strategy for nanofiber production, which may be chosen from a variety of organic, inorganic, and biological systems: (a) PANI nanofibers, (b) SWCNT bundles, (c) nanofibrous hexapeptide, and (d) V_2O_5 nanofibers. To generate CP nanofibers and nanotubes, this strategy employs two techniques: "nanofiber seeding" and "sacrificial template." The nanofiber seeding approach allowed for perfect control of morphology in bulk amounts of PANI nanofibers (20–60 nm in diameter and 2–10 S cm^{-1} in conductivity), PPy nanofibers (60–90 nm in diameter and 50 S cm^{-1} in conductivity), and PEDOT nanofibers (100–180 nm in diameter and ~16 S cm^{-1} in conductivity). These NCPs were created in a single step without the need of huge amounts of organic dopants, surfactants, or insoluble templates. Seeding a standard chemical oxidative polymerization of the monomers with extremely modest amounts of nanofiber seeds (about 1%) drastically transforms the morphology of NCPs from particles to almost exclusively nanofibers. This method solves a significant synthetic obstacle in the management of the nanostructure of CPs by revealing an essential chemical feature for seed templates to efficiently orchestrate the growth of polymer fibers.

Furthermore, Zhang et al. proposed a one-step and "green-nano" strategy to manufacture bulk amounts of PPy nanofibers (80–100 nm in diameter) using V_2O_5 nanofiber seeds and ferrous chloride ($FeCl_2$), as well as the green oxidant H_2O_2. Because the byproduct of the reaction is only water, H_2O_2 is a green oxidant, and its combination with other catalysts will provide varying oxidative strengths. To summarize, the seeding methodology allows for the fast synthesis of large amounts of NCPs in a single step, which is more simple and efficient than template-based approaches. The specific synthesis method for these nanostructures by this approach should be researched in the future for a better understanding.

3.1.3.3 Electrospinning Approach

Electrospinning is regarded as one of the simplest and most successful physical processes for creating nanoscale 1D structures (nanofibers and nanorods) without the need of a template. As illustrated in Fig. 3.6a, the fundamental setup of an electrospinning system consists of three key components: a high-voltage power source, a spinneret, and an electrical conductive collector [11]. When the repulsion force of a charged solution overcomes the surface tension of the solution in a strong electrostatic field between the spinneret and collector, electrospinning occurs. The electrical force may concurrently pull, stretch, elongate, and bend the jet, resulting in the formation of nano/microfibers by fast solvent evaporation. The fundamental benefit of the electrospinning technology is that it can mass-produce continuous long nanofibers that are difficult to create using conventional synthetic methods. Electrospinning fibers have unique qualities as compared to traditional fibers, such as a high aspect ratio, a high surface to volume ratio, regulated length, pore size, and improved mechanical capabilities.

Many research groups have manufactured PANI nanofibers using various ways; nevertheless, producing clean PANI nanofiber by electrospinning proved a significant problem due to PANI low solubility in common solvents. For gas sensing applications, electrospun PANI fibers doped with varying quantities of acid were created. The conductivity of the fibers varies (2×10^{-6} to 50 S cm^{-1}), yet they have comparable sizes (500–850 nm). Figure 3.6b depicts a typical morphology. PANI is usually doped with organic acids to improve its solubility in organic solvents [18]. Previous research described the synthesis of PANI with another insulating copolymer,

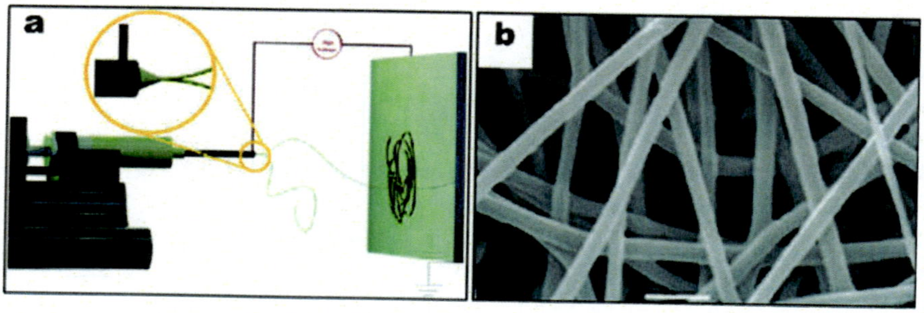

Fig. 3.6 **a** Schematic of electrospinning. **b** SEM images of electrospun PANI fibers (scale bar: 2 μm). Reprinted from ref. [17], copyright 2018, with permission from Springer Nature

polyethylene oxide (PEO), to create nanofiber with enhanced flexibility. The conductivity of PANI/PEO, on the other hand, was dramatically lowered, and the physical and chemical characteristics were changed.

Li et al. created an electrospun conducting PPy fiber with a diameter of 3 μm and a conductivity of 0.5 S cm^{-1}. APS was used as an oxidant and dodecylbenzene sulfonic acid (DBSA) as a dopant source to prepare highly conductive PPy. The electrical conductivity and solubility of a polymer were determined by polymerization circumstances such as oxidant concentration, polymerization temperature, and time. Different dopants and oxidants were used to electrospun PPy nanofibers (70–300 nm in diameter).

3.1.3.4 Other Template-Free Approaches

Other template-free approaches for producing NCPs have also been described. Radiolysis has been utilized to manufacture NCPs under ambient temperature and pressure using γ-irradiation. This method has various benefits over other methods, including the lack of foreign matter, polymerization at ambient temperature, homogeneous reaction by γ-irradiation, and simple control of reaction rate. Pillalamarri et al. discovered the radiolytic synthesis of PANI nanofibers (50–100 nm in diameter and 1–3 μm in length) and nanorods (250–500 nm in diameter and 5–10 μm in length) in the presence of initiator using γ-ray irradiated aqueous solutions of aniline. After washing cycles, the nanofibers may be extracted, and the polymer can be doped with acids such as HCl, $HClO_4$, and so on. Remita et al. conducted a thorough examination of the radiolytic production of PEDOT nanostructures and the mechanisms involved.

The electrochemical nanowire assembly method, which has previously been used to generate metal nanowires, may also be used to manufacture and regulate the shape of CPs nanostructures. Li et al. described a unique template-free method for producing PANI nanofibrils by controlled multipotential electropolymerization. PANI nucleation sites were initially created on an electrode at a greater potential, and then nanofibrils were created at a lower potential. Kwon et al. effectively created PPy nanowires from an aqueous solution using cathodic electropolymerization without the need of templates or chemical additives. The nanowires were immediately deposited on the substrate in the form of a nanoporous, interconnected network structure. By extending directed electrochemical nanowire assembly to grow PPy (560 nm in diameter) and PEDOT nanowires (340 nm in diameter) following the expected pathway, a new electrochemical technique was established.

Liao et al. effectively produced PPy nanostructures of various morphologies on a pre-nucleated PPy thin film using a template-free electrochemical polymerization process and investigated their development mechanism using 3D AFM imaging and field-emission scanning electron microscopy (FE-SEM). Surprisingly, the use of β-naphthalene sulfonic acid as a dopant resulted in changes in the self-assembly behavior of further expanding PPy chains at distinct redox states. The final morphology of the PPy nanostructure was dictated by these differences.

In electrophoretic assembly, charged "particles" are deposited onto electrodes in a solution by applying a homogenous electric field. In contrast, the "particles" in dielectrophoretic assembly are exposed to a non-uniform electric field and hence do not need to be charged. The dielectrophoresis method offers advantages over electrophoresis in that it avoids oxidation of the template surface, allowing the template to be reused and the components to be assembled without charging. As a result, PANI was consistently deposited onto 100-nm-wide electrodes with different length scales.

Soft lithography is another low-cost, high-resolution, high-throughput method for creating nanoscale patterns with a micromold. PEDOT:PSS nanowires (278–833 nm in diameter) were patterned by micromolding in capillaries on a glass or Si wafer, according to Zhang et al. By applying adequate force to the stamp during micropatterning, the height of these generated nanowires may be regulated. Hu et al. created a large CP nanowires array with internal preferential alignment using a simple residue-free embossing procedure. Another set of researchers investigated the direct patterning of functional CPs using a nanoimprint lithography approach. For patterning CPs, an approach based on nanoimprint lithography and a liftoff procedure was developed. At the nanoscale, patterned CPs from aqueous solutions or organic solvents allow these materials to be used in flexible optoelectronic devices.

Solid phase and vapor phase polymerization are the two basic approaches for non-liquid phase polymerization. Although it is fairly simple, solid-phase polymerization is rarely employed for CP synthesis. The first investigation of PANI nanowires generated using a mechanochemical technique was given by Bhandari et al., but the benefits of such a process are unclear. Vapor-phase synthesis has opened up new opportunities for fine-tuning the synthesis of CPs, allowing for the fabrication of multidimensional CP nanostructures. Co-vapors can be injected into the VPP process to vary the morphology and characteristics of the final NCP product, according to new research (Fig. 3.7).

As seen in Table 3.2, each strategy has advantages and downsides. The size, shape, and qualities of finished CPs are determined by the preparation procedures and the technique capability. For instance, the solid template is required for the hard template, which mostly controls the morphology of NCPs, while it is not required for a soft template in which the morphology of the NCPs can be changed by varying synthetic circumstances. The best approach for obtaining long nanofibers is thought to be electrospinning. However, some non-conducting polymers (e.g., PEO) are frequently utilized as a helper to adjust the mixture viscosity. Many elements influence how to choose an approach, including the size and morphology of final NCPs, physical and chemical qualities, initial raw materials, NCP quantity, and current experimental conditions.

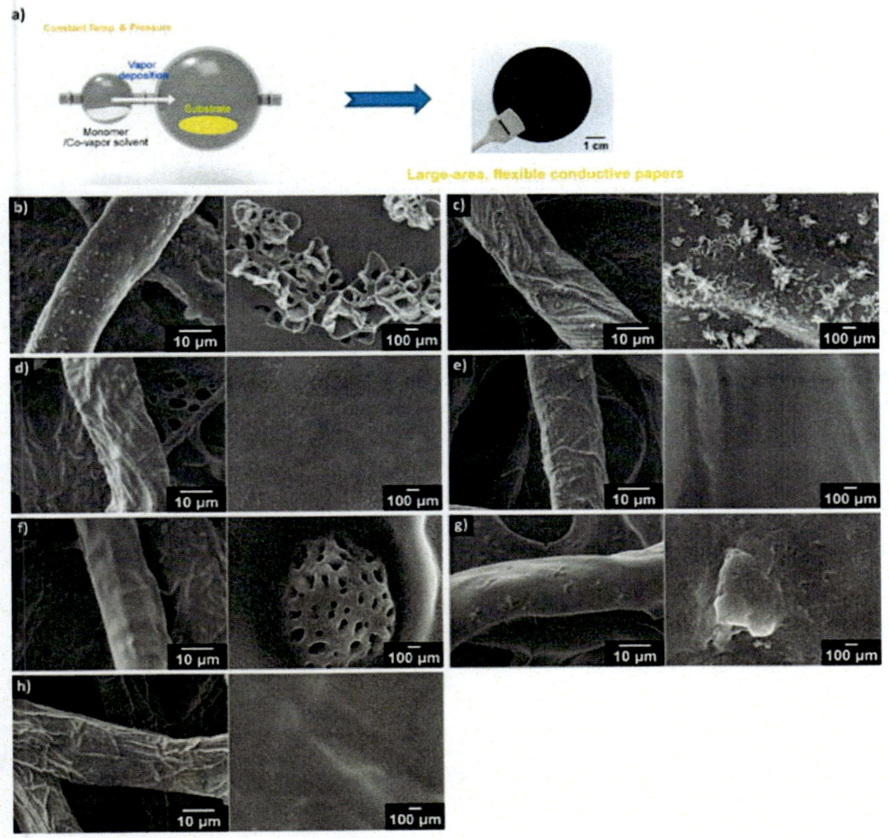

Fig. 3.7 **a** Vapor-phase polymerization (VPP) of pyrrole with co-vapors on a cellulose substrate and a typical photograph of as-prepared paper; **b–h** Scanning electron microscopy (SEM) images of papers prepared with covapors: **b** methanol, **c** ethanol; **d** water; **e** hexane; **f** toluene; **g** benzene; and **h** no vapors (left: low magnification, right: high magnification). Reprinted from ref. [4], copyright 2016, with permission from MDPI

3.1.4 Electrosynthesis

Chemical and electrochemical methods can be used to make CP nanoparticles. Chemical reactions frequently result in powdery nanoparticles that are easy to scale up. Electropolymerization, on the other hand, limits reactions on electrode surfaces, and CP nanostructures are often deposited as films on the electrode surface. Furthermore, nanostructures often produce directed structures by growing in the direction of the electric field. The applied voltage or current density may readily control the rate of electrochemical polymerization, and the amount of product can be controlled by the integrated charges utilized for electrosynthesis. Electropolymerization conditions

Table 3.2 Advantages and disadvantages of synthetic approaches for nanostructured conductive polymers

Synthetic approaches	Advantages	Disadvantages
Hard template	Feasible for almost all CPs. The size and morphology of CPs can be well-controlled	Additional process to remove the template is required
Soft template	A simple and inexpensive method. Synthesize NCPs in large quantities	Weak control the size and morphology of NCPs
Interfacial	Simple and facile synthesis without template	Restrict to certain precursors
Seeding	Synthesize NCPs in one step rapidly with bulk quantities. The seeds play a dual role as both the templates and the reactive oxidants	Difficulty in fabricating hollow nanotubular structures
Electrospinning	Simple method to synthesize CP nanofibers	Only available for soluble and thermoplastic CPs
Radiolysis	Synthesize NCPs at room temperature and in ambient pressure. The process is easily controlled and adaptable without inducing impurities	Weak control on the size and morphology of NCPs. Precise control of dose rate and time of radiolysis is required
Electrochemical assembly	The size and morphology of NCPs can be well-controlled	Lacks a precise control in the morphologies and dimensions of NCPs
Soft lithography	Low cost, high resolution, and high throughput	Expensive facilities are required

Reprinted from ref. [17], copyright 2018, with permission from Springer Nature

can also affect the form and characteristics of nanomaterials. As a result, electrochemical polymerization is a simple and efficient method for making CP nanomaterials with controlled morphologies and properties.

CPs are often deposited as compact films with a cauliflower-like shape on the working electrode surfaces. However, CP nanostructures can be produced if the polymerization is directed by a solid template (e.g., nanoporous membranes), a soft template (e.g., micelles), or selective nucleation. The form of the nanostructures is determined by the nature of the CPs and the electrochemical polymerization conditions, particularly the volume limitation of the templates. Various CP nanostructures have been described to date, including nanowires, nanotubes, nanonetworks, nanosheets, and nanospheres or microcontainers.

3.1.4.1 Nanowires and Nanotubes

Nanotubes and nanowires are essential nanostructures of materials that have received a lot of interest, particularly since the discovery of carbon nanotubes. Depending on whether a solid template is used, there are two basic electrochemical techniques for producing CP nanowires and nanotubes. However, the one-dimensional (1D) nanostructures are roughly aligned on the working electrode surfaces in both circumstances. The template is used to specify the morphology and size of materials in a template-guided growth process. Anodized aluminum oxide (AAO) and track-etched polycarbonate (PC) membranes are commonly employed as templates in the electrosynthesis of nanotubes or nanowires. Because these membranes are insulators, they cannot be employed directly as working electrodes. As a result, as the working electrode, a thin conductive metallic layer such as gold must be sputtered or vapor formed on one side of the membrane. Penner and Martin were the first to employ porous membranes as templates for the electrosynthesis of CP nanowires and nanotubes, and they first used pyrrole as the monomer.

This approach has been widely used to a wide range of CPs, including PANI, PTh, poly(p-phenylene) (PPP), and its derivatives. During the development process, polymers preferentially nucleate on the pore walls of the membrane, forming polymer tubes that can be converted into wires if a significant total charge is utilized for the electrosynthesis.

The template-free electrochemical development of CP nanowire arrays typically consists of two steps: nucleation and epitaxial polymerization. This technology was used to create large arrays of PANI nanowires. Aniline was initially polymerized at a high current density in this technique to produce nanosized PANI nuclei. The nuclei were then expanded into PANI nanowires by performing a second or third deposition process at a lower polymerization current intensity. PTh single molecule wires were also created via electrochemical epitaxial polymerization. The electrolyte containing thiophene monomer and iodine ions was electrolyzed using voltage pulses. Iodine nanoparticles were formed by reducing iodine ions and were first deposited on the Au(111) surface of the working electrode, where they served as nucleation sites for the growing PTh. It was possible to create high-density arrays of single CP wires as long as 75 nm. This method may pave the way for the mass manufacture of molecular-scale electronics based on CPs.

3.1.4.2 Nano-networks

Electrosynthesis may also be used to create networks by the connectivity of nanowires or nanopores of CPs. These networks are very conductive and have a large specific surface area. Micelle or soft template-guided growth is a popular method for creating CP nanoworks. In aqueous solution, 1D rod-like J-aggregates of 5,10,15,20-tetrakis(4-sulfonatophenyl)porphyrin (TPPS) or rod-like micelles of 1-pyrenesulfonic acid (PSA) were employed as soft templates for the electrosynthesis of PEDOT and PPy nanonetworks, respectively. CPs were developed along the 1D nanostructures of TPPS and PSA during electropolymerization, and these template molecules were entrapped in the resultant PEDOT and PPy films, resulting in the creation of CP nanofiber networks on the electrodes.

Using phosphate buffer solution, a simple template-free electrochemical technique was established to create a superhydrophilic PPy nanofiber networked film (PBS). It was hypothesized that the presence of hydrogen bonding between phosphate and pyrrole oligomers is required for the formation of the 1D nanofiber structure. On the other hand, electrostatic interactions between ion dopant and pyrrole oligomers result in an uneven nanostructure. A relatively high concentration of PBS, a high pH, a low concentration of pyrrole, a low concentration of dopant ClO_4^-, and a low oxidation potential all enhance the synthesis of 1D PPy nanofibers, according to systematic experiments.

An interfacial electropolymerization approach employing ionic liquid as the electrolyte was also developed to create PPy nanonetworks. The working electrode, in this case a Pt wire, was inserted vertically throughout the electrolyte, with one end in contact with the bottom of the electrochemical cell. The electrolyte was a deaerated ionic liquid containing 0.1 M pyrrole and 1-ethyl-3-methylimidazolium bis(trifluoromethanesulfonyl)amide. As the counter electrode, a Pt-coated poly(vinyl difluoride) membrane was wrapped around the inner wall of the electrochemical cell, fully circling the working electrode. Potential pulses (0.8 V, vs. Ag) were used to build the PPy film, with the same time resting at 0 V between nearby pulses. It was discovered that PPy was first formed as fibrils, which later expanded to create a nanofiber networked film along the solution surface.

Electrochemical polymerization within the interstitial voids in assemblies of polystyrene (PS) nanospheres precoated with multilayers of poly(diallyldimethylammonium chloride) and poly(sodium 4-styrene sulfonate), followed by toluene extraction of the PS cores, resulted in two-dimensional (2D) PANI networks with honeycomb structures. It was discovered that depositing polyelectrolyte multilayers onto PS particles considerably improved the consistency of the PANI attachment onto the PS particle surfaces. PANI honeycomb films with suitable pore sizes, pore walls, and film thicknesses may be produced by systematically varying the polyelectrolyte layer number and electropolymerization parameters.

CP inverse opals are three-dimensional (3D) nanonetworks. Polymerization of the matching monomers in the interstitial voids of a PS colloidal crystal template was used to create these 3D networks. The film thickness may be controlled using electropolymerization time and applied voltage in this method. Given that CPs can display metallic or insulating behavior depending on doping level, the stop band of such inverse opals can be adjusted by adjusting the applied voltage or the temperature, which provides varying the spectral position of the stop band.

3.1.4.3 Nanospheres

Nanospheres or microcontainers are spheres or containers with diameters in the micro- or submicrometer range and skin thicknesses in the nanometer range. They can be used to encapsulate and release organic species such as dyes and medicines in a reversible manner. Polyelectrolyte capsules can be used as a template to create PPy nanospheres or capsules. Shchukin et al. created poly(diallyldimethylammonium)/poly(styrene sulfonate) capsules by alternating the absorption of the two components onto the surface of silica particles. The PPy capsules were then created by potentiostatically oxidizing pyrrole on an ITO electrode in an electrolyte comprising polyelectrolyte capsules and monomer. When wet, the PPy spheres are circular, but after drying, they collapse into bowl-like containers.

3.2 Conductive Polymer Nanostructure with Different Dimensions

3.2.1 Conductive Polymer Nanoparticles

In recent years, conductive polymer nanoparticles (CPNs) have been proposed and developed as a promising and simple technique. Because of its significant qualities such as outstanding photostability, high brightness, low cytotoxicity, strong biocompatibility, tiny particle size, and excellent capacity to sensitize the generation of reactive oxygen species, CPNs are emerging as a viable contender in the biological area. These CPN features are easily tunable by using alternative backbone architectures, mixing different CPs, and altering the surface. These qualities are greatly influenced by the nature, chain conformation, aggregation extent, and size of the chromophore for a specific CPN. Furthermore, surface charges and other functional groups have a major influence on the characteristics of CPNs. In general, when intra- and interchain contacts develop, the absorption and emission spectra of a specific conductive polymer redshift due to high order aggregation and efficient energy transfer to low-energy emitting sites. Compared to the polymer in a good solvent, CPNs generated by miniemulsion and nanoprecipitation techniques invariably exhibit a blueshift in the absorption spectra, which is caused by the shorter conjugation length and contractively compressed and disordered conformation. In contrast, the absorption spectra of CPNs formed by the self-assembly process were clearly redshifted due to a relaxed and organized structure. This situation is most likely the outcome of CP aggregation via enhanced interchain interactions.

Typically, the brightness and photostability of CPNs depend on the particle size and improve as particle size increases due to greater absorption cross section and particle volume. However, the fluorescence quantum yield of CPNs drops as particle size increases because bigger particles have a higher efficiency of energy transfer to diverse fluorescence quenching sites than small particles. CPNs have been utilized to mark cells and achieve fluorescence imaging in vitro and ex/in vivo due to these benefits. Researchers created multifunctional CPNs that can transport genes and medications as well as target and destroy microorganisms or tumor cells. CPNs demonstrated high photostability and minimal cytotoxicity in these experiments [5].

3.2.2 One-Dimensional Conductive Polymers

Polymer characteristics have been shown to be substantially connected to morphologies and structures. Because of their highly ordered structure, large surface area, and superior carrier transport, nanoscaled conducting polymers, particularly those with one-dimensional (1D) nanostructures (such as polymer nanowires, nanotubes, nanofibers, nanorods, and so on), have distinct advantages over their corresponding bulk materials. Several techniques for synthesizing 1D conducting polymers have been developed, including hard or soft template-based approaches as well as template-free methods such as inverse microemulsion polymerization and electrospinning. Significant progress has been made in the 1D engineering of conducting polymers for a broad range of applications, including batteries, supercapacitors, sensors, transistors, electrochromic displays, photovoltaics, and so on.

3.2.2.1 PEDOT

The production and analysis of nanostructured PEDOT for various applications are gaining significant attention. Taggart et al. investigated the thermoelectric properties of PEDOT nanowires in 2011, which were synthesized by oxidative electropolymerization of 3,4-ethylenedioxythiophene (EDOT) monomer in aqueous $LiClO_4$ solution via a hard template using a lithographically patterned nanowire electrodeposition (LPNE) approach with non-dependent control over the nanowire thickness and width. Zhang et al. recently described a three-dimensional (3D) assembly of PEDOT nanowires generated by electrochemical polymerization in 3D-alumina templates using the hard template approach. After removing the template, the network became self-contained. Different oxidants, $FeCl_3$ and $Fe(Tos)_3$, were used to create two kinds of PEDOT nanowires (C-PEDOT and T-PEDOT). However, the lengths are varied (0.5–5 μm for C-PEDOT and 0.2–0.7 nm for T-PEDOT).

In 2015, Hu et al. reported the thermoelectric performance of nanostructured PEDOT with diverse morphologies (Fig. 3.8a–e) produced by chemical oxidation polymerization using sodium bis(2-ethylhexyl) sulfosuccinate (AOT) as the surfactant in reverse microemulsions. Tuning the amount of oxidizing agent applied and the rate of polymerization produced PEDOT with various nanostructures, which were shown to have significant influence on thermoelectric capabilities (Fig. 3.8f–h). In the order of bulk PEDOT < globular nanoparticle < nanorod < nanotube < nanofiber, thermoelectric properties in terms of electrical conductivity, Seebeck coefficient, and power factor increased [6].

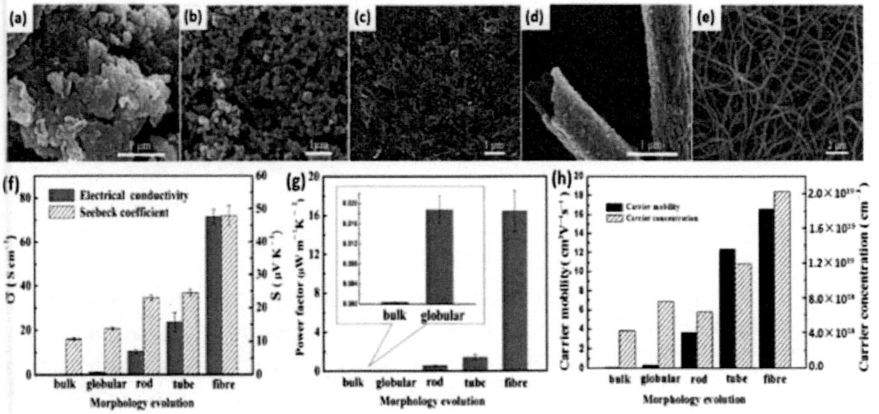

Fig. 3.8 SEM images of **a** bulk PEDOT power; **b** globular nanoparticles; **c** nanorods or ellipsoidal nanoparticles; **d** nanotubes; **e** nanofibers; **f** Seebeck coefficient and **g** power factor of bulk PEDOT and PEDOT nanostructures; **h** Carrier mobilities and carrier concentrations of bulk PEDOT and PEDOT nanostructures. Reprinted from ref. [13], copyright 2019, with permission from MDPI

3.2.2.2 Poly(3-alkylthiophene)

Among many polythiophenes, poly(3-alkylthiophene) (P3AT) is a notable family that has received a lot of attention for its electrical transport mechanisms, processing methods, and device applications in solar cells, sensors, optoelectronics, and so on. Various ways have been used to create a range of 1D nanostructured materials based on P3AT. Continuous long P3AT nanofibers have been reported to be electrospun, but shorter P3AT nanofibers are generally created by self-assembly in solution with high crystallinity due to strong contacts along their longitudinal axis. Other methods, such as vapor-assisted imprinting, electrochemical synthesis, and solution casting using sacrificial templates, have also been used to generate P3AT nanowires or nanotubes for a variety of applications.

Samitsu et al. employed anisole as a solvent to create high-aspect-ratio P3AT nanofibers using the Whisker technique. The length of the alkyl chain and the molecular weight of the polymer have a substantial impact on fiber production and characteristics. In a typical experiment, a P3AT polymer is thoroughly dissolved in anisole while being heated and stirred. The solution can then generate nanofibers during a fast cooling procedure to room temperature. Endrodi et al. created poly(3-hexylthiophene) (P3HT) nanofibers (50–60 nm wide and several microns long) using the Whisker process, which self-assembled to form networks by drop casting and subsequent doping with silver perchlorate to regulate the amount of oxidation. Because of the very regularly spaced, crystalline supramolecular structure, the 1D nanofiber shape, and the intimate electrical contact between each nanofiber inside the nanonet structure, the nanonet structure has a ZT value (0.0026) that is six times larger than its bulk polymer equivalent. Hiura et al. used a similar process to create

P3HT nanofiber mats with great crystallinity. Despite having comparable electrical conductivity, they discovered that the thermal conductivity of the nanofiber mat was twice lower than that of the non-fibrous P3HT polymer, which is attributable to the existence of a more efficient electron transport channel and a high vacancy percentage in the fibrous mats.

3.2.2.3 PPy

Wu et al. were the first to disclose the production of free-standing PPy nanotube films for thermoelectric applications in 2014. Two types of PPy nanotubes (PPy-1 and PPy-2) were synthesized using self-degraded template procedures that entail the oxidation of pyrrole monomers in the presence of methyl orange as soft templates. As indicated in the SEM and TEM pictures (Fig. 3.9c–f), the as-prepared PPy nanotube suspensions were dried in Petri dishes to yield free-standing flexible films by self-assembly (Fig. 3.9a, b) [16].

Through a chemical oxidative polymerization approach, Du et al. successfully prepared PPy of various textures and topography (nanoparticles, nanotubes, and nanowires) with or without various surfactants (methyl orange (MO), cetyltrimethyl ammonium bromide (CTAB)) and oxidants (ammonium peroxydisulfate (APS), $FeCl_3$) in 2018 (Fig. 3.10). The impact of various morphologies on thermoelectric characteristics have been investigated, and the highest electrical conductivity

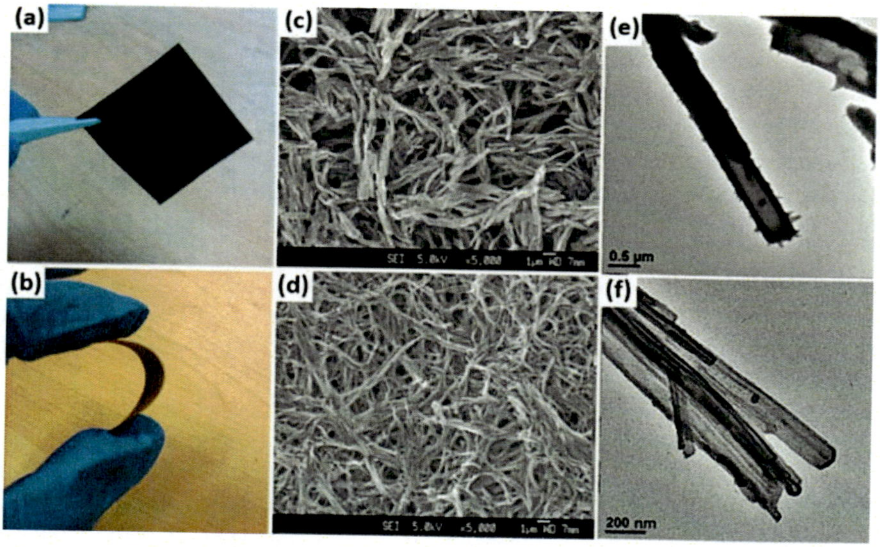

Fig. 3.9 **a**, **b** Photographs of the flexible free-standing PPy nanotube film. SEM images of the free-standing **c** PPy-1 films and **d** Ppy-2 films. TEM images of **e** PPy-1 nanotubes and **f** PPy-2 nanotubes. Reprinted from ref. [13], copyright 2019, with permission from MDPI

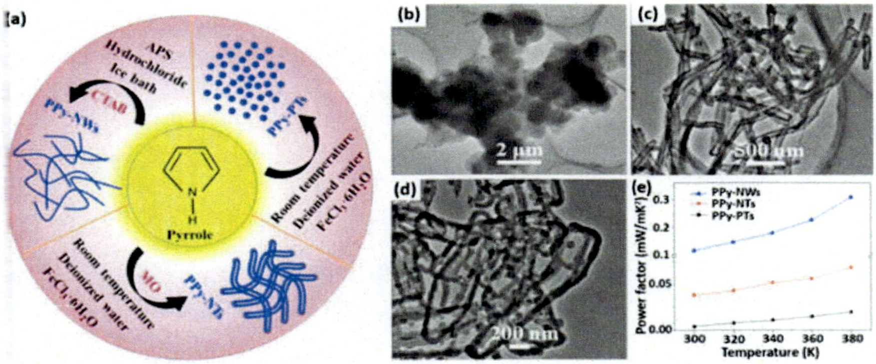

Fig. 3.10 **a** Schematic illustration of the formation procedure of PPy nanoparticles (PTs), nanotubes (NTs), and nanowires (NWs). TEM images of **b** PPy-PTs, **c** PPy-NTs, and **d** PPy-NWs. **e** Thermoelectric power factor of PPy-PTs, PPy-NTs, and PPy-NWs. Reprinted from ref. [3], copyright 2019, with permission from MDPI

and Seebeck coefficient of PPy nanowires with a diameter of 30–60 nm resulted in an optimal power factor of 0.3 $\mu Wm^{-1} K^{-2}$ at 380 K [3].

3.2.2.4 PANI

PANI is another essential conductive polymer because of its low cost, high stability, simple synthesis procedure, and changeable electrical characteristics for a variety of applications. Various protonic acids (such as H_2SO_4, HCl, $HClO_4$, polyacrylic acid, sulfosalicylic acid, p-toluenesulfonic, etc.) have been utilized as dopants for PANI to improve transport characteristics, and the conductivities are strongly impacted by the doping amount and dopants. Sun et al. created template-free β-naphthalene sulfonic acid (β-NSA) doped PANI nanotubes by oxidative polymerization of aniline monomer using APS as the oxidizing agent and β-NSA as the dopant and soft template. The tubes range in length from 5 to 8 μm and have an average diameter of 200 nm. Wu et al. also published the production and thermoelectric performance analysis of PANI nanowires or nanorods doped with various acids. The soft template approach was used to create the 1D nanostructured PANI, with APS as an oxidizing agent, CTAB as a soft template, and several acids (HCl, acetic acid, and p-toluenesulfonic (p-TSA)) as dopants.

3.2.3 Conductive Polymer Nanowire Arrays

Using in situ chemical polymerization or electrochemical polymerization without a template, aligned conductive polymer nanowire arrays and their composites with nanocarbon materials may be created by managing the nucleation and growth phase of polymerization. This type of nanostructure (such as polypyrrole and polyaniline nanowire arrays) has a high capacitance, higher rate capability due to the huge electrochemical surface area, and an excellent ion diffusion channel in the ordered nanowire structure, as has been shown to be an appropriate electrode material for high-performance supercapacitors.

3.2.3.1 Nanowire Arrays as a New Strategy to Improve Electrochemical Performance

When material sizes are lowered to the nanoscale, their physical and chemical characteristics alter substantially. Because the size reduction of materials enhances the contact surface area between the electrode and the electrolyte and minimizes the transport channel length for both electrons and ions, nanomaterials are projected to provide significant advances to energy storage devices. Pseudocapacitors store charge on the surface layer, where electrolytes ions can enter the active materials matrix, and size reduction enhances the utilization of the active material, which improves charge storage capabilities. Furthermore, because the transport channel length for both electrons and ions is reduced, this results in a high charge–discharge rate capability. One-dimensional (1D) nanostructures (such as nanowires, nanotubes, and nanoribbons) are developed as a representative family of novel energy materials among the diverse nanomaterials because they represent the low-dimensional structures for effective charge transfer in energy storage devices.

Vertical nanowire arrays have been proven as one of the best nanostructured materials for supercapacitors, outperforming disordered nanowire networks as electrodes for energy storage devices. To begin, each nanowire is electrically coupled to the conductive substrate, ensuring that all nanowires contribute to the capacity. Second, the nanowire arrays contain direct 1D electronic channels that enable efficient charge transport and shorter ions-transport lengths, resulting in excellent rate performance. Third, the gap between orientated nanowires may allow substantial volume changes without causing fracture in bulk or micron-sized materials. Iijima et al. were the first to demonstrate that aligned single-walled carbon nanotube (SWCNT) arrays had reduced capacitance drop during high-power operation due to improved ion diffusion from the aligned pore structures as compared to active carbon. Aligned nanowire (-tubes) arrays are thought to be a good material for compact, high-power delivery, and high-energy–density storage, and would provide a promising active material for next-generation supercapacitors. The capacitance, charge–discharge rate, and cycle life of supercapacitors rise considerably when conducting polymers have ordered, one-dimensional nanostructures.

3.2.3.2 Preparation of Ordered Conductive Polymer Nanowire Arrays in a Large Area

Methods for creating 1D conductive polymer nanostructures (nanowires, nanotubes, nanorods, nanoribbons, nanocones, and so on) can be classified into two categories: the template method and the template-free method. The template and template-free approaches are also used to create an essential 1D nanostructure, organized conductive polymer nanowire arrays.

Because of its simplicity, adaptability, and controllability, Martin and Parthasarathy's template approach has been widely employed in the fabrication of conductive polymer nanowire (-tube) arrays. As a template, a thin porous film of aluminum oxide (AAO), particle track-etched membranes (PTM), or porous diblock copolymer films are often used. To generate nanotubes or nanowire arrays, materials are deposited into the pores of these films using techniques such as pressure injection, vapor deposition, chemical deposition, and electrodeposition. Furthermore, some inorganic nanowire (-tubes) arrays (such as TiO_2 and MnO_2 nanotubes) can serve as sacrificial templates.

Although template techniques may produce well-defined conducting polymers (such as nanotubes or nanowires) with narrow distributions in sizes ranging from 20 nm to a few micrometers, the cost of templates is substantial and the handling operations are very complicated. Furthermore, removing the template frequently breaks the alignment and causes the polymer to collapse into structures that lack the ideal orientation, particularly for polymer nanorods or tubes with diameters less than 100 nm.

In addition to template approaches, template-free methods such as self-assembly, electrochemical polymerization, interface polymerization, fast mixing reaction, dilute polymerization, electrospinning have been used to create 1D conducting polymers. Electrochemical polymerization and dilute polymerization have been reported as ways for producing ordered nanowire arrays. Debiemme-Chouvy described a one-step electrochemical approach for direct synthesis of orientated PPy nanowire arrays without the need of a template. The PPy nanowire array was created in the presence of weak-acidic and/or non-acidic anions. The suggested response mechanism consists of three steps: (1) electrodeposition of a thin PPy insulated film due to weak-acidic conditions; (2) oxidation of water leading to overoxidation of the polymer due to hydroxyl radicals and the formation of O_2 nanobubbles, which can prevent further overoxidation of the polymer; (3) formation of PPy nanowires in places where O_2 is evolved.

Debiemme-Chouvy developed a one-step electrochemical procedure for direct creation of orientated PPy nanowire arrays in order to create large area PPy nanowire arrays using a simple synthesis approach. Using a biphasic electrochemical technique, PPy nanofiber arrays were first synthesized in a one-step electrochemical process. The electropolymerization was carried out in a biphasic aqueous–organic system. The pyrrole monomer was dissolved in the organic phase (a buffer phase) in this procedure, and it can slowly diffuse from the buffer phase into the water phase across the interface. In the aqueous electrolytic phase, the dopant was dissolved. Due to the presence of electrolytes in this phase, polymerization took place mostly in the aqueous phase. Only a very low concentration of monomer existed near the electrode in the aqueous phase due to interfacial diffusion, indicating improved controllability in the production of conductive polymer nanofibers. The nanowires have sizes ranging from 80 to 100 nm, and their length may be varied between 1 and 4 μm by varying the polymerization time. Aligned PPy nanowire arrays demonstrated greater capacitance and stability than conventional films and disordered nanowire networks.

A simplified, one-step electropolymerization method in homogeneous solution was also reported to synthesize a large area of well-defined PPy nanowire arrays, and the enhanced capacitance performance of PPy nanowire arrays was demonstrated when compared to disordered nanowire networks or compact films. First, an ion-sputtered compact Au nanoparticle layer was placed on the working electrode surface as pseudonuclei, which may stimulate PPy nucleation and reduce the interfacial energy barrier for subsequent PPy development [7]. The compact PPy film, PPy nanowire networks, and nanowire arrays were controllably produced by modifying the solution composition and electrochemical techniques (Fig. 3.11).

Polyaniline (PANI) has the highest theoretical pseudocapacitance of any conducting polymer due to its numerous redox states. Liang et al. used a stepwise electrodeposition process to create vast arrays of PANI nanowires with diameters

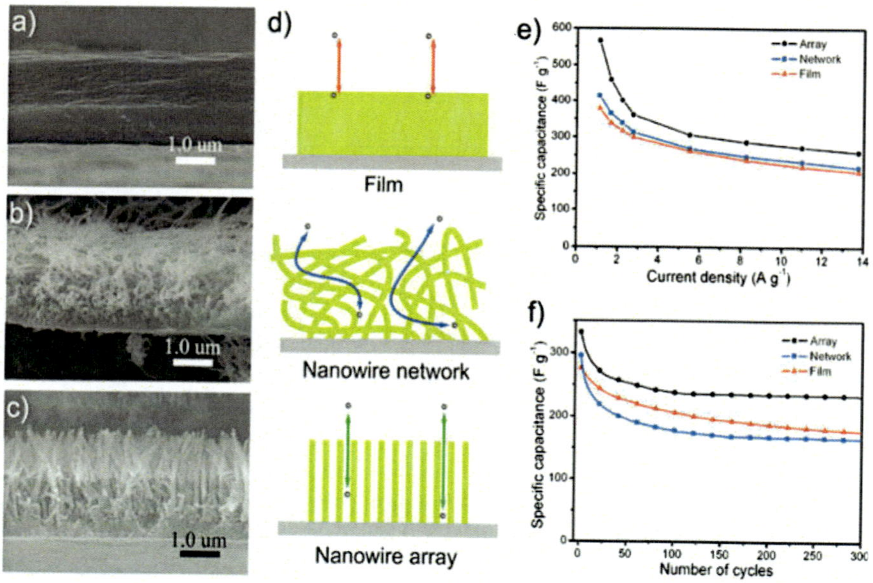

Fig. 3.11 Conducting polypyrrole nanowire arrays. SEM images of **a** PPy film, **b** PPy nanowire network, and **c** PPy nanowire arrays. **d** Ion transport pathways, **e** Specific capacitances, and **f** Capacitance as a function of the cycle number of PPy with different morphologies. Reprinted from ref. [15], copyright 2014, with permission from John Wiley and Sons

considerably less than 100 nm on a variety of substrates (Pt, Si, Au, C, SiO_2). PANI was developed in three current density increments using a programmed constant-current technique. A strong current (0.08 mA cm^{-2}) was utilized to produce the nucleation sites on the substrate in the initial stage, followed by a reduced current (0.04 mA cm^{-2}) for 3 h and then another 3 h at 0.02 mA cm^{-2}. The nanowires are thought to be generated by the following mechanisms: PANI nucleation on the substrate occurred at a high current density, followed by continuing nucleation and growth at a low current density. If an appropriate current density was used, Wang et al. discovered that PANI nanowire arrays could also be manufactured in a one-step procedure. Even at low current density (0.04 mA cm^{-2}), uniform nucleation can occur on the working electrode. Due to the strong electric field generated by edge effects, further deposition occurs on the ends of the nuclei, resulting in an extended length along the original nuclei and then the formation of PANI nanowires.

Chemical polymerization can also be used to create PANI nanowire arrays. Chiou et al. created a dilute polymerization process that used much lower quantities of aniline and oxidant than normal. Aligned PANI nanofibers were concurrently produced on diverse substrates, including non-conducting and conducting substrates, using the dilute polymerization process. They hypothesized that the alignment process of PANI nanofibers is comparable to the model given by Liu et al. to account for stepwise electrochemical polymerization. When extremely dilute aniline (or oxidant) is used, heterogeneous nucleation first occurs on solid substrates. The majority of active nucleation centers develop on solid substrates, which reduces the interfacial energy barrier for the subsequent growth of PANI on solid substrates. However, PANI nanofibers are generated in the bulk solution at the same time and consume some reactive aniline cation-radicals and oligomeric intermediates, reducing the rate of PANI development on solid surfaces. As a result, PANI can only develop along the active nucleation sites formed during the first polymerization step. This approach appears to be appropriate for a variety of substrates (both conductive and non-conductive substrates), implying that the method given here is quite resilient and might lead to a wide range of applications. The sole limitation of this approach is that the substrate must be stable and not dissolve in acidic medium since the polymerization step is generally carried out in acid solution (pH 4) to produce PANI with high electrical conductivity.

3.2.3.3 Superior Electrochemical Performance of Conducting Polymers Nanowire Arrays

According to Huang et al., the electrochemical characteristics of PPy nanowire arrays, nanowire networks, and films were systematically studied. PPy nanowire arrays had a higher capacitance of 566 Fg^{-1} t a discharge current density of 1.10 Ag^{-1}, at the same current density, the capacitance of nanowire networks and PPy film was approximately 414 Fg^{-1} and 378 Fg^{-1}, respectively. Furthermore, PPy nanowire arrays kept 70% of their original capacitance after hundreds of successive charge–discharge cycles. This performance was superior to disordered NW networks due to the easy strain relaxation of the nanowire arrays. The electrochemical characteristics were significantly improved because the aligned nanostructures are more accessible to counterions when diffusing into the nanowire interstices. Based on the mass of active electrode material, the energy densities of the electrodes of the PPy film, nanowire network, and nanowire array were predicted to be 68 Wh kg^{-1}, 74 Wh kg^{-1}, and 102 Wh kg^{-1} at a power density of 900 W kg^{-1}. At the same power density, PPy nanowire arrays had greater energy densities than film and nanowire networks. Although PPy nanowire arrays demonstrated significantly higher capacitance than PPy nanowire networks and compact supercapacitor sheets, the capacitance of PPy proved challenging to increase due to its low theoretical capacitance.

When compared to PPy, PANI has a substantially larger theoretical capacitance. Furthermore, because of their unique structure, orientated PANI nanowire arrays are projected to have improved capacitance performance. The capacitance of PANI nanowire arrays generated by one-step electropolymerization is 950 Fg^{-1} and may be increased to 780 Fg^{-1} at 40 Ag^{-1}. The large capacitance and impressive rate capabilities suggested that well-defined nanostructures may significantly increase storage performance of functional nanomaterials. Due to the bulk of active electrode materials, the energy density of PANI nanowire arrays is 130 Wh Kg^{-1} at a power density of 700 W Kg^{-1} in $HClO_4$ solution. Even at a power density of 28,000 W Kg^{-1}, the energy density is still at 100 Wh Kg^{-1}, indicating a strong power production capacity while retaining a high energy density.

References

1. Cho SI, Lee SB (2010) Fast electrochemistry of conductive polymer nanotubes: synthesis, mechanism, and application. Acc Chem Res 41(6):699–707
2. Cui S, Zheng Y, Liang J et al (2016) Conducting polymer PPy nanowirebased triboelectric anogenerator and its application for self-powered electrochemical cathodic protection. Chem Sci 7:6477–6483
3. Du Y, Niu H, Li J, Dou Y, Shen S, Jia R, Xu J (2018) Morphologies tuning of polypyrrole and thermoelectric properties of polypyrrole nanowire/graphene composites. Polymers 10:1143
4. Duong N, Hyeonseok Y (2016) Recent advances in nanostructured conducting polymers: from synthesis to practical applications. Polymers 8(4):118
5. Feng L, Zhu C, Yuan H, Liu L, Lv F, Wang S (2014) Conjugated polymer nanoparticles: preparation, properties, functionalization and biological applications. Chem Inform 44(41):6620–6633
6. Hu X, Chen G, Wang X, Wang H (2015) Tuning thermoelectric performance by nanostructure evolution of a conducting polymer. J Mater Chem A 3:20896–20902
7. Huang J, Wang K, Wei Z (2010) Conducting polymer nanowire arrays with enhanced electrochemical performance. J Mater Chem 20:1117–1121
8. Jang J, Li XL, Oh JH (2004) Facile fabrication of polymer and carbon nanocapsules using polypyrrole core/shell nanomaterials. Chem Commun 7:794–795
9. Jang J, Yoon H (2003) Facile fabrication of polypyrrole nanotubes using reverse microemulsion polymerization. Chem Commun 6:720–721
10. Landfester K (2009) Miniemulsion polymerization and the structure of polymer and hybrid nanoparticles. Angew Chem Int Ed 48(25):4488–4507

11. Mirabedini A, Foroughi J, Wallace GG (2016) Developments in conducting polymer fibres: from established spinning methods toward advanced applications. RSC Adv 6:44687–44716
12. Pecher J, Mecking S (2010) Nanoparticles of conjugated polymers. Chem Rev 110(10):6260–6279
13. Shah KW, Wang SX, Soo DXY, Xu J (2019) One-dimensional nanostructure engineering of conducting polymers for thermoelectric applications. Appl Sci 9(7):1422
14. Tuncel D, Demir HV (2010) Conjugated polymer nanoparticles. Nanoscale 2:484–494
15. Wang K, Wu H, Meng Y, Wei Z (2014) Conducting polymer nanowire arrays for high performance supercapacitors. Small 10:14–31
16. Wu J, Sun Y, Pei W-B, Huang L, Xu W, Zhang Q (2014) Polypyrrole nanotube film for flexible thermoelectric application. Synth Met 196:173–177
17. Zhang L, Du W, Nautiyal A, Liu Z, Zhang X (2018) Recent progress on nanostructured conducting polymers and composites: synthesis, application and future aspects. Sci China Mater 61:303–352
18. Zhang Y, Kim JJ, Chen D et al (2014) Electrospun polyaniline fibers as highly sensitive room temperature chemiresistive sensors for ammonia and nitrogen dioxide gases. Adv Funct Mater 24:4005–4014

Chapter 4
Conductive Polymer Composites

CP hybrids are a growing topic in the field of CPs. The integration of a co-component into CPs may equip them with improved qualities over individual ones. Other functional materials, including carbon-based materials (carbon nanotubes and graphene), metals and metal oxides, and biological materials, have been added into CPs.

4.1 Conductive Polymer–Noble Metal Nanoparticle Hybrids/Composites

Noble metal nanoparticles have high conductivity, catalytic activity, a large surface-to-volume ratio, unique optical characteristics, and can be used in disciplines such as electronics, photonics, catalysts, nano- and biotechnology. NMNPs such as Au, Ag, Pt, Pd, and others, have been a research focus for material scientists for decades. Their high surface-to-volume ratio provides a diversity of active sites for antibody immobilization and exceptional catalytic ability. Noble metal nanomaterial catalysis is an excellent approach for signal amplification and high sensitivity, as is ensured by excellent conductivity. These advantages provide up a plethora of opportunities for increasing tumor marker immunoassay performance. NMNPs are often capped with a stabilizer composed of alkanethiols, amines, nucleotides, and polymers. The primary idea is to guide the nucleation and development of NMNPs while avoiding NMNP aggregation caused by van der Waals attraction. As previously stated, CPs have specific multifunctional groups that align well with the essential criteria for NMNP stabilization.

The combination of CPs and noble metal nanoparticles (NMNPs) has piqued the curiosity of researchers. Because of this synergistic effect, a fresh light has been thrown on conductive polymer-based metal nanomaterial hybrids. They are easily polymerized when a noble metal salt is used as a chemical oxidant and dopant. Because of the superconductivity, large specific surface area, and outstanding catalytic activity of noble metal nanoparticles, this approach improves conductivity

Y. Wang and W. Feng, *Conductive Polymers and Their Composites*,
https://doi.org/10.1007/978-981-19-5363-7_4

and catalytic ability even more. Antibodies may also be readily fixed onto the surface of composites. Conductive polymers, in particular, are compatible with a wide range of biomolecules, including DNA, proteins, antibodies, and antigens, due to their intrinsic nature and ease of modification. In terms of strong electrochemical signal and sensitivity, high conductivity plays an essential role in lowering the interface resistance between the electrolyte and the electrode. Furthermore, conductive polymers are easily changed with certain groups containing amino and carboxyl groups, which are required for antibody immobilization, as well as mercapto groups. This is utilized to form an Au–S connection between noble metal nanoparticles such as gold nanoparticles. These characteristics make them ideal candidates for biomolecule immunoassays.

Conductive polymer-based noble metal nanomaterial hybrids combine the benefits of conductive polymers with noble metal nanostructures. Amperometric immunoassay of tumor marker based on conductive polymer has achieved significant progress. A conducting long chain polythiols-based immunosensor was created using the feature of simple modification of polyaniline as well as the stubborn attachment of gold nanoparticles with mercapto group modified glass carbon electrode, as shown in Fig. 4.1. Poly(2-aminothiophenol) was connected to the electrode in the immunosensor by producing a thin layer [11]. Gold nanoparticles were connected to the poly(2-aminothiophenol) modified electrode through an Au–S bond due to the lengthy chains and numerous thiol groups. As a result, several sites for antibody mobilization were created [14].

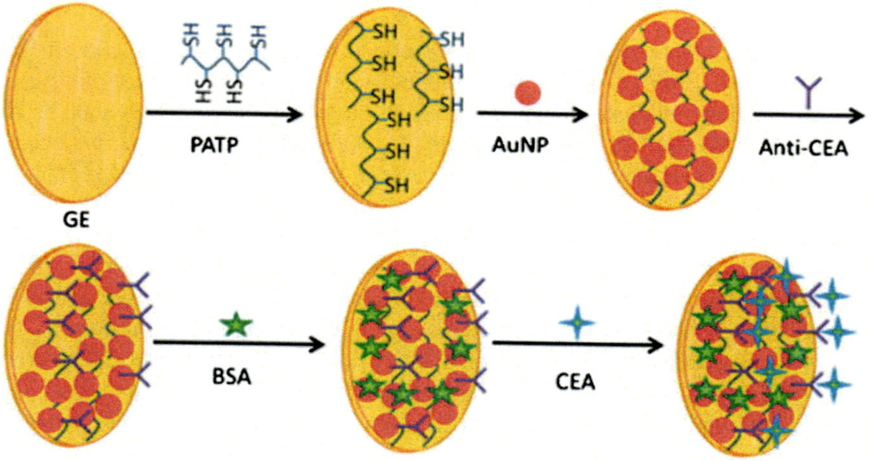

Fig. 4.1 Schematic illustration of fabrication of label-free immunosensor based on poly(2-aminothiophenol). Reprinted from ref. [11], copyright 2013, with permission from Elsevier

4.1.1 CP and NMNP Composites

4.1.1.1 Mixing CP and NMNP Solutions

NMNPs are often produced and suspended in a solvent, whereas CPs are typically dissolved in another solvent. Solvents are evaporated after the two solutions are combined, resulting in the production of a CP/NMNP hybrid film. The CPs may be made using APS or H_2O_2 as an oxidant, while the NMNPs can be made by reducing noble metal salts using trisodium citrate, sodium borohydride, hydrazine hydrate, and other reducing agents. The compositions of hybrids may easily be changed by varying the concentration of solutions. Izumi et al. reported that the concentration of emeraldine base PANI was crucial in determining the degree of Au nanoparticle aggregation, and the interaction of emeraldine base PANI with Au or Ag colloids was investigated using surface-enhanced resonance Raman scattering (SERRS).

4.1.1.2 CP Prepared by Electrochemical Route

To produce one-dimensional CP/NMNP hybrids, CPs with one-dimensional nanostructures of typical nanofiber/nanowire and nanotube have been employed to capture NMNPs. In this method, CPs are synthesized through oxidative electropolymerization of monomers, while NMNPs are synthesized via electrodeposition of noble metal salts. The number of cycles is changed to produce films of varying thickness and NMNPs of varying amounts. To capture NMNPs on the CP surface, CPs can be electropolymerized on an electrode surface and then submerged in an NMNPs colloidal solution for a given period. Chemical bonding was used to connect the Au and Ag nanoparticles with the sulfur atoms. Similar findings have been found elsewhere. Electrostatic interactions, in addition to chemical connections, have been employed to catch NMNPs. Guo et al. electrochemically produced doped PANI nanowires on glassy carbon electrodes; trisodium citrate-protected negatively charged Au nanoparticles (20 nm) were electrostatically adsorbed on the surfaces of the PANI nanowires.

The monomers can be cast on electrode or NMNP surfaces in addition to dissolving in the reaction media. Electrochemical processes at the electrode surface with the use of a template can form as-synthesized CP/NMNP hybrids. The electrochemical reaction occurs at the electrode/solution interface, as opposed to the solution-based chemical reaction pathway that generally occurs in a bulk solution. A porous membrane is always applied to electrode surfaces in order to shape the ensuing deposited CPs and NMNPs. Porous anodic aluminum oxide (AAO) and polycarbonate track-etched membranes have long been used as electrochemical templates for CPs nanorods and nanotubes.

4.1.2 CPs and Noble Metal Ions

In situ reduction of noble metal ions on pre-synthesized CPs with predetermined nanostructures can result in CP/NMNP hybrids. Nanostructured CPs not only serve as a support for NMNPs, but they also function as an effective reductant in the reduction of noble metal ions to create NMNPs. Kang et al. first proposed the notion of creating CP/NMNP hybrids via oxidation–reduction interactions between CPs and noble metal ions, where the CP of PANI was used to recover Au from chloroauric acid solution.

PANI has a somewhat lower reduction potential than Au, Ag, Pt, and Pd ions. As a result, the presence of Au ions, for example, causes spontaneous deprotonation by oxidizing PANI from emeraldine salt to pernigraniline and reducing Au ions to metallic state (Fig. 4.2a). In turn, the pernigraniline is easily deprotonated and reduced to emeraldine salt in an acidic media (reverse reaction of Fig. 4.2b), resulting in a redox process recycle. Similar reactions take place with other CP and metal ion couples, such as PPy and Pd ions. The inherent oxidation state of PANI, pH, reaction temperature, and surface area of PANI polymer all had a major influence on the Au reduction rate. The size of NMNPs is heavily influenced by the nature and shape, doping status, and even surface texture of CP nanostructures, while the morphology of CP/NMNP hybrids is mostly determined by the configuration of CP nanostructures.

4.1.2.1 Without Additive

Wang et al. reported that following alkali dedoping in the emeraldine base state, PANI powders were employed to reduce $AuCl_3$ in either aqueous or NMP environments to yield Au nanoparticles in PANI. The size of the Au nanoparticles was determined by the reaction media and the initial metal ion-to-PANI ratio. When $AuCl_3$ was reduced in PANI NMP solutions and PANI powder formed in aqueous media, the

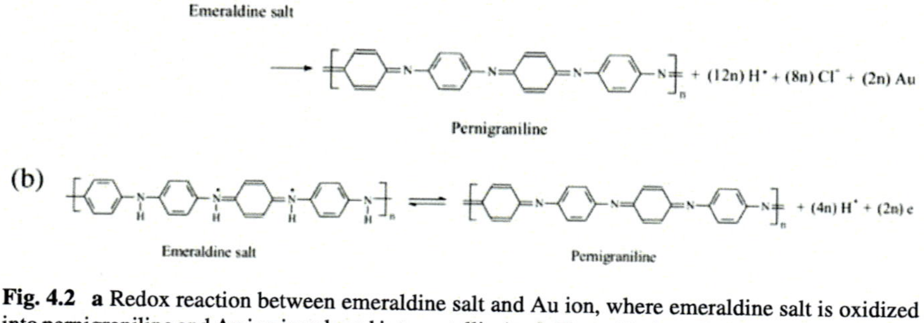

Fig. 4.2 **a** Redox reaction between emeraldine salt and Au ion, where emeraldine salt is oxidized into pernigraniline and Au ion is reduced into metallic Au. **b** The oxidation of PANI from emeraldine salt to pernigraniline. Reprinted from ref. [5], copyright 2017, with permission from Elsevier

Au nanoparticles were around 20 nm and 50–200 nm, respectively. With a higher molar ratio of $AuCl_3$ to PANI, larger Au nanoparticles were produced. The restricted Au accumulation locations accounted for the increase in Au size. Furthermore, the biggest Au nanoparticles were formed when an acidic medium was utilized, which might be attributable to agglomeration caused by the higher reaction rate. Several groups reported that regulating the time and temperature of the reaction between dedoped PANI nanofibers and chloroauric acid resulted in a remarkably homogeneous dispersion of Au nanoparticles. Furthermore, PANI with different morphologies, such as nanotube, nanofiber, and hybrid with other materials, has been used as reactive support for loading NMNPs.

4.1.2.2 With Additive

Additional stabilizers are always required in the synthesis to get NMNPs with lower sizes (10 nm) and improved stability. CPs interact with NMNPs exclusively on the contact side throughout the growing phase, leaving the other regions of the NMNPs unprotected. As a result, the surrounding NMNPs will aggregate with each other and form even, micrometer-sized noble metal particles. As a result, if the other surfaces of the as-formed noble metal nanoclusters are successfully capped by additional stabilizer, the development of noble metal nuclei will be slowed and the aggregation phenomena will be prevented. This is expected to result in uniformly deposited nanosized NMNPs on the surfaces of CPs.

Functional Doping Acid

When a doping acid with functional groups that have a high affinity for NMNPs is doped to CPs, the development and aggregation of noble metal nuclei is slowed, resulting in smaller NMNPs and CP/NMNP hybrids with great uniformity. The standard polyelectrolyte polyacrylic acid (PAA), for example, was chosen to hybridize with PANI because PAA not only rendered the colloidal particles hydrophilic, but it also bound and dragged noble metal ions to the PANI colloids. Another example is the incorporation of thioglycolic acid (TA) into PANI nanofibers to produce Au nanoparticles with a size limit of 10 nm. For comparison, several doping acids, such as citric acid, camphorsulfonic acid, and TA, were used. When compared to dedoped PANI nanofibers, the homogeneity of Au nanoparticles improved when citric acid-doped PANI nanofibers were utilized, and it improved even more when camphorsulfonic acid-doped PANI nanofibers were employed. In the case of TA-doped PANI nanofibers, practically all Au nanoparticles (2–10 nm in a regulated way) were equally deposited on PANI nanofiber surfaces, and the uniformity of such hybrid nanofibers was near 100% (Fig. 4.3). In comparison with the usage of dedoped or citric acid or camphorsulfonic acid-doped PANI nanofibers, Au nanoparticles generated during the early stages of the reaction would be situated on PANI nanofiber surfaces, and

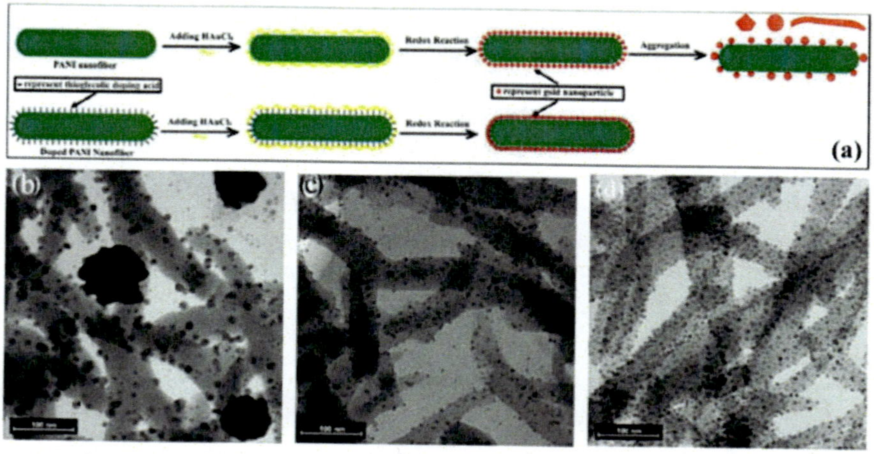

Fig. 4.3 **a** Schematic illustration for the formation of PANI nanofiber/Au nanoparticle hybrids. **b–d** TEM images of Au nanoparticles supported on PANI nanofibers with different doping acids: **b** citric acid, **c** camphorsulfonic acid, and **d** TA. Reprinted from ref. [4], copyright 2010, with permission from the American Chemical Society.

other surfaces, in addition to contact sides, were well-shielded by TA. It was determined that introducing a doping acid with a high affinity for Au nanoparticles into PANI nanofibers will successfully inhibit aggregation of Au nanoparticles, resulting in the creation of uniform hybrids containing Au nanoparticles evenly deposited on surface of PANI nanofibers. TA-doped PANI nanotubes were also loaded with Au nanoparticles with a size restriction of 10 nm using a similar procedure.

PVP

PVP, a routinely used Au nanoparticle protection reagent, has the ability to produce Au nanoparticles on CPs. Han et al. created well-defined Au nanoparticles with adjustable sizes (3–20 nm) supported on the shells of poly(o-phenylenediamine) (POPD) microspheres. In the absence of PVP, apart from the formation of Au nanoparticles supported on POPD hollow microspheres, some micrometer-sized Au aggregates separate from the microspheres, whereas in the presence of PVP, Au nanoparticles of 3 nm were exclusively located on surfaces of POPD hollow microspheres. POPD interacted with Au nanoclusters only on the contact side throughout the development process, and the other regions were not successfully protected, resulting in nearby Au nanoclusters aggregating with each other and even, micrometer-sized Au particulates in the absence of PVP. In the presence of PVP, the remaining regions of the Au nanoclusters were efficiently capped by PVP, which prevented the aggregation phenomena, resulting in equally deposited Au nanoparticles on the surfaces of the POPD hollow microspheres.

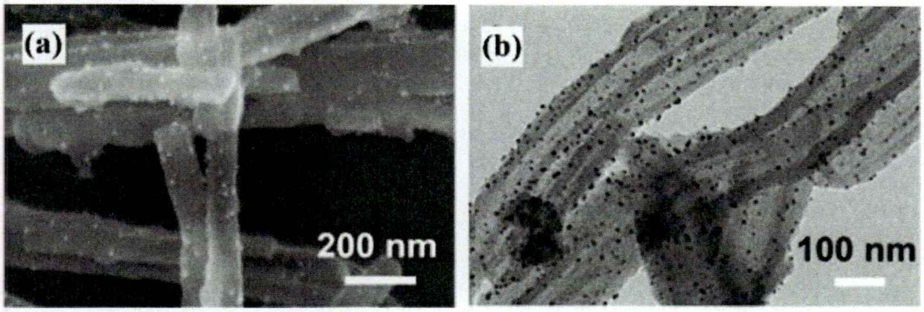

Fig. 4.4 **a** SEM and **b** TEM images of the PPy-Au hybrids using Tween-80 as a stabilizer. Reprinted from ref. [15], copyright 2009, with permission from John Wiley and Sons

Surfactant

Xu et al. studied the influence of various surfactants on the generation of Au nanoparticles by an in situ chloroauric acid reduction process on PPy nanotubes. Surfactant included the cationic surfactant cetyltrimethylammonium bromide (CTAB), the anionic surfactant sodium dodecyl sulfonate (SDS), and the nonionic surfactant polyoxyethylene sorbitan monooleate (Tween-80). Although some tiny Au nanoparticles with an average diameter of 13 nm were scattered uniformly over the nanotubes in the absence of a surfactant, massive Au nanoparticles with an average diameter of 80 nm were also found due to Au nanoparticle overgrowth. Because of the repulsive electrostatic interactions between CTAB and PPy, hardly no Au nanoparticles were deposited on the PPy nanotubes in the presence of CTAB. When SDS was utilized as a stabilizer, massive Au nanoparticles with diameters of about 50 nm developed on the surface of the PPy nanotubes, indicating that SDS protective role was not very good in this case. Tween-80, a nonionic surfactant, was used to equally distribute Au nanoparticles with an average diameter of 13 nm on the PPy nanotubes (Fig. 4.4).

With Physical Barrier

The inclusion of a functional stabilizer to manage NMNP aggregation on CPs has significant drawbacks. For example, the functional doping acid used to create nanosized NMNPs exists exclusively in acidic circumstances and will be dedoped under basic conditions. As a result, removing functional doping acid in a basic environment causes NMNPs to become unstable, limiting their applicability, particularly in catalysis under basic circumstances. As a result, pH-insensitive CP/NMNP hybrids are extremely attractive. The primary concept is to build nanobarriers on the surfaces of CPs. The majority of documented CPs have spherical, fibrous, or tubular nanostructures, and the surfaces on the outside are always smooth. As previously stated, during the in situ creation of NMNPs, additional surfaces other than the contact sites of noble metal nuclei with polymer matrix are not protected, and so surrounding noble metal nuclei aggregate, resulting in NMNP overgrowth. Overgrowth of noble metal nuclei may be controlled if nanobarriers adequately isolate them. One way is to roughen the surface of the CPs. Another approach is to apply mesoporous SiO_2(m-SiO_2) shells containing nanochannels to the surfaces of CPs.

4.1.3 Monomers and NMNPs

In this method, NMNPs of desired size and shape are synthesized ahead of time, and then a frequently used oxidant is introduced to a colloidal solution comprising NMNPs and monomers, followed by monomer polymerization to generate CP/NMNP hybrids. During polymerization, monomers, oligomers, or polymer chains adsorb onto the surfaces of NMNPs via chemical and/or electrostatic interactions, resulting in CP/NMNP hybrids of conventional core–shell nanostructures. Chemical oxidation polymerization with noble metal salt as an oxidant is a simple yet elegant technique. The monomer of aniline derivative containing reducing groups is employed as a reductant in this process, and a noble metal salt is used as an oxidant and dopant. The composites successfully combine the redox activity of polyaniline derivatives with the catalytic capabilities of noble metals.

4.1.3.1 Without Additive

Citrate-stabilized Au nanoparticles produced by a well-established citrate reduction process are the most extensively employed NMNPs due to the easy synthesis conditions and the fine-tuned diameter (10–25 nm) and shape of Au nanoparticles. When APS was introduced to a colloidal solution containing citrate-stabilized Au nanoparticles and monomers, only aggregated Au nanoparticles coated with CPs and bare CPs were formed. This can be explained as follows: (1) Citrate is not an effective stabilizer for Au nanoparticles so that the observed aggregated Au nanoparticles form after several days of aging; (2) Monomers will be present in the whole water phase, but not specifically on the surfaces of Au cores. Following that, polymerization happens not only on the surfaces of Au nanoparticles but also in the solution phase, resulting in aggregated Au nanoparticles coated with CPs and CPs devoid of Au nanoparticles. Surfactant or functional stabilizer is usually required in the synthesis to create well-defined homogeneous CP/NMNP core–shell nanostructures.

4.1.3.2 With Additive

Nonionic Surfactant F127

Surfactant added to a colloidal solution of Au nanoparticles will form a monolayer on the Au/water solid/liquid interfaces to lower interface energy. As a result, the stability of Au nanoparticles will improve. Owing to the hydrophobic properties of the most widely used monomers, they will settle around the hydrophobic areas after being added (with aromatic moiety exhibiting low solubility in water). Monomers on the surfaces of Au nanoparticles were triggered by an oxidant and polymerized to result in the creation of homogenous CP/Au core–shell hybrids. In a similar synthesis, the anionic surfactant SDS was used as a stabilizer. Han et al. investigated the influence of surfactant concentration on the morphology of CP/Au hybrids. Pluronic F127, a nonionic surfactant triblock copolymer of poly(ethylene oxide)-poly(propylene oxide)-poly(ethylene oxide), was chosen as an example. The inclusion of F127 at a concentration less than the critical micelle concentration (CMC) was essential for the formation of well-defined homogeneous POT-Au core–shell hybrids.

Anionic Surfactant SDS

Another typical surfactant employed in the creation of CP/Au hybrids is the anionic surfactant SDS. CP/Au rod-like core–shell hybrids using Au nanorods as cores, such as PANI/Au nanorod core–shell hybrids and PPy/Au nanorod hybrids, have been described, similar to the production of CP/Au spherical core–shell hybrids in Au colloidal solution with the assistance of SDS. If the oxidant is changed from APS to H_2O_2 during the production of PANI/Au core–shell hybrids in Au colloidal solution with SDS, only oligoaniline (OANI) forms on the surfaces of Au nanoparticles (OANI/Au core–shell hybrids). Triple-layer PPy/(Au@Ag) core–shell hybrids were created by initiating the polymerization of pyrrole using $AgNO_3$ rather than APS in Au colloidal solution with the help of SDS.

A unique hybrid vesicle Au@vesicle was generated via the polymerization of thiophene in Au colloidal solution with the help of SDS, which was triggered by H_2O_2 and $FeCl_3$. Through in situ polymerization of monomer in NMNP colloid solution with the assistance of a surfactant, the SDS content was discovered to have a decisive role in the hybrid shape. As CP/NMNP core–shell hybrids, CPs are well-coated on the surfaces of NMNPs. The arrangement of NMNP seeds greatly influences hybrid morphology, and spherical and rod-like morphology is typical in CP/NMNP core–shell hybrids. If NMNPs are initially coated with other materials and then used as seeds for further coating of CPs by a similar way, the shape of the ensuing hybrids will be governed by the middle component, and removal of the middle component will result in the development of CP/NMNP yolk–shell hybrids.

Diblock Polymer

By treating spherical PANI/Au core–shell hybrids with the diblock copolymer polystyrene$_{154}$-block-poly(acrylic acid)$_{60}$ (PSPAA), a multiple-petal morphology with Au nanoparticles encased by hybrid polymers (PANI + PSPAA) was achieved. PANI formed a petal-like shell on the Au surface as a result of the inserted PSPAA assembly. It was hypothesized that the shell morphology underwent a metamorphosis to minimize surface energy to the greatest extent possible by producing a petal-like shell to increase volume and, lastly, to minimize the surface energy-to-volume ratio.

4.1.4 Monomers and Noble Metal Ions

The coprecipitation of CPs and NMNPs to hybrids indicates that metal ion reduction and oxidative polymerization of organic monomers occur simultaneously. The most commonly used approach for realizing the simultaneous formation of CPs and NMNPs nanostructures is that oxidation polymerization of monomers to CPs is induced by the reduction of noble metal ions to NMNPs, where CPs and NMNPs will be generated and deposited at the same time, resulting in the formation of CP/NMNP hybrids. Polymer oligomers attach to the surfaces of freshly generated noble metal nuclei to reduce surface energy. As a result, in the CP/NMNP hybrids developed, NMNPs often appear as the core and CPs as the shell. Well-dispersed NMNPs in bulk CPs may be generated after optimizing the synthetic conditions. The morphology of the hybrids produced by this method is likely to range from sphere-shaped structures to films and perhaps hierarchical nanostructures. The noble metal salts and monomers should be chosen in such a way that the noble metal salt may oxidize the monomer to build a polymer while also using the electrons generated during oxidation to reduce the noble metal salt to form NMNPs.

4.1.4.1 Solution Polymerization

Without Additive

If the oxidant APS is converted into another noble metal salt, such as $AuCl_4$, PANI polymer and NMNPs can be generated at the same time. Adding $HAuCl_4$ aqueous solution to an aqueous acidic solution containing aniline typically results in the reduction of $AuCl_4$, followed by oxidative polymerization of aniline, resulting in PANI nanofibers with a diameter of 35 nm and aggregated Au nanoparticles that precipitate from the liquid phase during the reaction. When the reaction media was changed from aqueous solvent to organic or even ionic liquid solutions, identical results were obtained with NMNPs of restricted size (usually 20 nm) embedded in CP matrix if the noble metal salt solution was introduced drop-wise to monomer solution. During its early oxidative stage, the reduction potential of aniline is pH-dependent, lowering as the pH of the solution increases owing to deprotonation (pKa of aniline is 4.6), potentially causing the reduction of $AuCl_4$ to proceed violently.

With Additive

The inclusion of functional molecules such as functional doping acid and surfactant can change the configuration of the resulting CP/NMNP hybrids involved in the redox reaction between aniline monomer and chloroauric acid. With CSA added in the reaction system of aniline monomer with chloroauric acid in aqueous solution, one-dimensional PANI/Au coaxial nanocables with average diameters of 50–60 nm and lengths of 1–2 μm were formed. Using added TSA, PANI nanospheres (230 $\pm$ 22 nm) with Au nanoparticles (6 $\pm$ 2 nm) inserted were produced. Citric acid is well-known as a frequent stabilizer for Au nanoparticles. Micelles made of citric acid–aniline salt can likewise operate as soft templates, although the shape is very different. Chemical polymerization of aniline with chloroauric acid as the oxidant in the presence of citric acid resulted in PANI hollow nanospheres with controlled incontinuous nanocavities ranging in size from 10 to 50 nm as an unique hollow nanostructure. It was hypothesized that micelles made of citric acid–aniline salt served as a template for the initial oligomer-Au core–shell structure, and that the inserted Au nuclei were drawn to micelle shells with hydrophilic –COOH groups because the contact between Au nuclei and citric acid is more advantageous than the interaction between Au nuclei and PANI with polymerization occurring, in contrast to the usage of CSA and TSA as doping acids instead of citric acid under comparable synthetic circumstances. The creation of multicavities in the PANI matrix was attributed to the aggregation and fusing processes of Au nuclei implanted in PANI. Huang et al. demonstrated that the addition of PVP may significantly alter the morphology of PANI/Au hybrids.

The PANI/Au hybrid formed in the presence of the anionic surfactant SDS exhibited a core/shell structure, with solitary or multiple Au nanoparticles of 20 nm mean diameter encased by PANI with a well-defined tetrahedron shape and an average edge length of 150 nm. CTAB, a cationic surfactant, was also used. In contrast to SDS, the selective interaction between CTAB and Au nanocrystals promotes the production of anisotropic Au nanostructures. PANI nanotubes and Au nanoplates were created in distinct stages by combining aniline acidified with glacial acetic acid and chloroauric acid in the presence of CTAB.

In addition to functional doping acids and surfactants, deoxyribonucleic acid (DNA) used in the reduction of chloroauric acid adsorbed aniline, which resulted in highly branched dendritic Au nanoparticles with concomitant formation of PANI, as opposed to spherical Au nanoparticles in the absence of DNA.

4.1.4.2 Interfacial Polymerization

Monomers and noble metal salts are dissolved in immiscible solvents in the interfacial polymerization process, and the redox reaction occurs at the liquid/liquid interface. Monomer is typically dissolved in organic solvent, whereas noble metal salt is dissolved in aqueous solution. Monomer, on the other hand, may be dissolved in acidic aqueous solution, and noble metal salt can be dissolved in organic phase to achieve interfacial polymerization. When a solid membrane separates solutions containing monomer and noble metal salt, liquid/solid/liquid interfaces develop, and CPs and NMNPs form simultaneously in the membrane. Polycarbonate (PC) membranes have also been used in the same way. In addition to liquid/liquid interfacial polymerization, gas/liquid interface polymerization using chloroauric acid dissolved in aqueous solution and aniline introduced as a vapor has been proposed. The polymerization occurred at the gas/liquid interfaces, resulting in dendritic PANI/Au hybrids.

4.1.4.3 Template Polymerization

Another strategy to regulate the morphology of the hybrids is to use a hard template method to synthesize CP/NMNP hybrids. As a result, ternary hybrids with enhanced features will be created. The hard template is mostly responsible for the final morphology of the product. In other circumstances, rigid templates such as mesoporous silica SBA-15 and nickel foam have been employed to restrict the development of PANI/Ag and PPy/Ag hybrids. The nanofibrillar oligomers generated by the interaction of pyrrole monomers and V_2O_5 nanofibers would then react with the noble metal ions and reduce them to NMNPs, resulting in a PPy nanofiber network adorned with Pt, Au, or Ag nanoparticles in a nanofibrillar V_2O_5 seeding technique.

4.2 Conductive Polymers/Zeolite (Nano-)Composites

Zeolites are a vast class of hydrated aluminosilicates (AS) comprising group I and group II elements, namely sodium, potassium, magnesium, calcium, strontium, and barium. They feature a three-dimensional network of tetrahedrally connected $[AlO_4]^{5-}$ and $[SiO_4]^{4-}$ atoms that share all of the oxygen atoms. Their crystalline framework is organized in a distinct interconnected lattice form. $M_{x/n}[(AlO_2)_x(SiO_2)_y]\cdot wH_2O$, where M is an alkali or alkaline earth cation, n is the cation valence, w is the amount of water molecules per unit cell, x and y are the total number of tetrahedral per unit cell, and the ratio y/x normally ranges from 1 to 5. Zeolites have received a lot of attention due to their well-defined micro- and nanopore sizes for molecular shape selectivity, large specific surface areas, intrinsic acidity, high (hydro)thermal, chemical stability, and wide range of commercial and technological applications such as molecular sieves, ion-exchangers, adsorbers, catalysts, detergent builders, cement (natural zeolites), agricultural use, and selective removal of heavy metal ions from seawater and industrial wastewater. Conductive polymers/zeolite (nano-)composites have the potential to be useful hybrid materials in a wide range of commercial and technological applications, including sensors, cell cathodes, anticorrosive membranes, and selective removal of heavy metal ions from seawater and industrial wastewaters [7].

4.2.1 Preparation Methods of Conductive Polymers/Zeolite (Nano-)Composites

The insertion of functional conductive polymer chains into zeolites cavities increased their alignment and is intended to slow aging and boost electrical conductivity. Furthermore, the acidity of the zeolite surface provides good adherence of the two materials, and a conductive polymer/zeolite composite combines the polymer high electrical mobility with the zeolite capacity to absorb exchangeable cations into its structure.

Conductive polymers/zeolite (nano-)composites can be made in four ways: (i) the monomer is dissolved in an organic solvent and encapsulated into the cavities of zeolite by diffusion processes, and then polymeric chains are produced into the cavities of zeolite via oxidative polymerization, (ii) oxidant ion-containing zeolites, such as Fe(III) or Cu(II), are exposed to monomer and acid vapors, respectively, (iii) mechanical mixing of conductive polymer and zeolite powder, and (iv) via in situ polymerization of corresponding monomer in the presence of zeolite, polymer may occur within the zeolite channels as well as outside the zeolite cavities. Among the procedures mentioned for the synthesis of conductive polymers/zeolite (nano-) composites, two specific methods (i and ii) are more appealing because polymeric chains can be synthesized into zeolite cavities in the nanometer size domain and may improve their electronic, mechanical, chemical, and optical properties due to the nanometer scale organization of the polymeric chains.

4.2.2 Polyaniline/Zeolite (Nano-)Composites

Chuapradit et al. produced polyaniline/zeolite LTA composites and examined their electrical conductivity response to carbon monoxide (CO) in terms of dopant type, dopant concentration, zeolite LTA content, and zeolite pore size. Densakulprasert et al. produced PANI/zeolite composites and examined the effect of the composite ion exchange capacity on electrical conductivity reaction to carbon monoxide (CO).

Ma et al. described the manufacture of PANI/β-zeolite and PANI/FUYB-zeolite nanocomposites in HCl solution through oxidative polymerization of absorbed aniline monomers in the channels of the respective zeolite type. The gas sensitivity of the composite films to trimethylamine was substantially dependent on the type of zeolite used. They discovered that the gas sensitivity to trimethylamine was greatly reduced in the system including β-zeolite as compared to the polyaniline film, but significantly increased in the system having FUYB-zeolite. Furthermore, the results showed that, when compared to the polyaniline film produced by conventional in situ polymerization, not only was the shape of polyaniline composites noticeably different, but the thermal stability of polyaniline film was greatly enhanced by the addition of zeolite.

Safidine et al. described the manufacture of conducting polyaniline/zeolite (PANI/HY) composites using two distinct ways and evaluated their suitability as a carbon dioxide (CO_2) gas sensor. The first approach involved in situ polymerization of aniline in an aqueous media with ammonium peroxydisulfate as an oxidant and the addition of zeolite type HY-free acids. The second approach involves physically mixing different fractions of HY (5, 10, 15, 20, and 50 wt %) in a PANI-HCl matrix. They observed that when compared to PANI doped with HCl, the PANI/HY composite had a reduced conductivity by 7.74×10^{-3} to 10^{-2} S cm^{-1}. In fact, when exposed to CO_2, the conductivity of the produced composites rose; this behavior was well observed and explained as an interaction between PANI and acidic CO_2 molecules. Furthermore, the synthesized PANI/HY composite demonstrated a quick reaction time to CO_2. Milojevic-Rakic et al. produced polyaniline/ZMS-5 zeolite composites and tested them as glyphosate [N-(phosphonomethyl)glycine] adsorbents in aqueous solution. Freundlich and Langmuir–Freundlich models best suit the glyphosate adsorption isotherms on examined materials.

4.2.3 Polypyrrole/Zeolite (Nano-)Composites

Electrochemical polymerization can be used to create a polypyrrole/zeolite (nano-) composite. Via electrochemical polymerization, Trueba et al. created polypyrrole nanowires in the nanochannels of clinoptilolite (HNZ), a proton-modified natural zeolite, and Y-zeolite in acid form (HY). In comparison with the polymer formed in the hosts without prior proton modification, they discovered that the synthesis of PPy with electrochemical activity was preferred in a severely acidic nanoscaled environment. It has been postulated that the reduction/oxidation reactions are caused by polymer protonation/deprotonation, with the H^+ ions perhaps acting as charge transfer promoters. The well-defined redox signals of PPy contained in HNZ, compared to the HY, indicated that the polymeric units were held more closely to the channel walls due to its complex framework, according to a cyclic voltammetry investigation of the synthesized PPy/HNZ and PPy/HY. They concluded that the one electron oxidation process was not affected by first-order monomer concentration, but it was lower (0.38) for PPy development in HNZ, which was connected to the spatial restriction of the host.

Conductive polypyrrole/zeolite (nano-)composites have the potential to be used as sensitive sensors for detecting different chemical vapors and biological samples. Polypyrrole/zeolite (nano-)composites benefit from polypyrrole quick electronic mobility and zeolite capacity to integrate cations into its structure. Olad et al. created a polypyrrole/clinoptilolite (PPy/CL) nanocomposite and tested its effectiveness in batch studies for the removal of $Ni_{(II)}$ from aqueous solutions. The influence of several parameters such as solution pH, contact duration, and $Ni_{(II)}$ solution concentration on removal efficiency was examined. The best circumstances for high removal effectiveness were adjusting the pH of the solution to 4 and utilizing nanocomposite for around 24 h. Freundlich isotherm has a greater agreement with experimental data than Langmuir and the chelating and ion exchange mechanisms define the adsorption mechanism of nanocomposite.

4.2.4 Polythiophene/Zeolite (Nano-)Composites

Zeolites can be combined with polythiophene to create conducting (nano-) composites with enhanced physicochemical characteristics. Furthermore, polythiophene can be polymerized inside zeolite cavities to create a PTh/zeolite (nano-) composite with regulated structure and characteristics. Arvand et al. created a conductive polythiophene composite based on polythiophene/Y-zeolite by chemical oxidative polymerization of thiophene (Th) in the presence of a dispersion of Y-zeolite (powder) in chloroform solvent using anhydrous $FeCl_3$ as an oxidant and studied its reaction to sulfide ions. When compared to previous PTh-based systems, the synthesized PTh/Y-zeolite-$FeCl_3$ composite showed a much higher DC conductivity value (10^{-2} S cm^{-1}). They proposed that the emergence of high conductivity in PTh/Y-zeolite composites is a distinguishing feature of this polymerization. They used the manufactured PTh/Y-zeolite composite as ion-selective electrodes (ISE) to directly determine the sulfide ion in solution.

4.2.5 Other Conductive Polymers/Zeolite (Nano-)Composites

The techniques of synthesis, characteristics, and applications of PANI, PPy, and PTh (nano-)composites with synthetic or natural zeolites were addressed. Other conductive polymers, such as polyfuran, poly(N-vinylcarbazole), poly(p-phenylene vinylene), and poly(p-phenylene), have been used to make conductive polymers/zeolite (nano-)composites. Kamonsawas et al. synthesized an H_2SO_4-doped poly(para-phenylene vinylene) (dPPV)/Y-zeolite-based composite and examined its sensitive and selective reactions to ketone vapors. Incorporation of zeolites as inorganic fillers into polyfuran matrixes or in situ polymerization of furan inside zeolite cavities may result in a new class of organic–inorganic (nano-)composite materials with improved physicochemical properties for a variety of practical and technological applications.

4.3 Conductive Polymers/Graphene Composites

Graphene-based materials have been hybridized or mixed with CPs. When the materials are linked, the combined optical, electrical, thermal, mechanical, chemical, or electrochemical capabilities of each component can be exploited for chemical and biological sensing. Metal nanoparticles or biomolecules can be added into CP/GE composites for the requirement of certain new or better characteristics in the sensor fabrication procedure. The CP/GE composites serve as a loading platform for sensing material systems, harnessing or amplifying their adsorption, catalytic, redox, luminescence response, and transport characteristics. CPs can operate as conducting conduits that interact with graphene and analytes, or they can act as sensitizing agents, increasing sensitivity, lowering detection limits, or increasing selectivity of relatively innocuous substances and, in the end, give rise to high-performance sensors.

The preparation method is frequently linked to differences in the properties of CP/GE composites containing the same components. Because of the unique electrical, optical, chemical, and structural features of CPs and graphene-related materials, these types of composites have sparked growing interest in the senor field. A variety of methodologies, including chemical, electrochemical, and physical procedures, have been proven for the fabrication of CP/GE composites for chemical sensors and biosensors. A suitable combination of CP and graphene can result in nanostructured composites with a high surface area, providing a wide platform for analyte molecule adsorption.

In situ polymerization in a solution comprising monomer and graphene or GO is the most used chemical approach for the manufacture of CP/GE. In the case of GO as a precursor to graphene, post-reduction of CP/GO is required. Because both molecules have a conjugated orientation, the high affinity of these two components arises by π-π stacking during the in situ polymerization process. The matching aromatic monomer is typically added to an aqueous or organic solution containing GO or GE dispersion. In the polymerization reaction, an oxide agent or initiator, such as $FeCl_3$, $(NH_4)_2S_2O_8$ (APS), H_2O_2, and others, is typically utilized. Certain types of dopants, such as photon acids and salts like poly (styrenesulfonate) (PSS) and sodium dodecyl sulfate, are also used in the system as needed.

PANI/GO composites, for example, were created using the in situ polymerization approach using APS and HCl as initiator and acidic dopant, respectively. Furthermore, the GO is shown to behave as a macromolecular dopant in the meantime, and the resultant fiber-like composite exhibits improved electrochemical characteristics. This approach may produce a variety of nanostructured CP/GO composites or hybrids, such as PPy/GO, PEDOT/GO, and PANI/GO or their hydrogels, among others. A unique approach for producing PANI/GE composites employing GO as an oxidant and precursor and aniline as a reducing agent was disclosed, with no extra oxidant. RGO/MnO_2 hybrids, on the other hand, can operate as intermediate templates and oxidants for aniline monomer during the RGO/PANI polymerization process.

Owing to the ease of manipulation, the wet reduction strategy has been intensively researched for further in situ reduction of CP/GO. Hydrazine, glucose, $NaBH_4$, L-ascorbic acid, and other green agents are common reducing agents. Hot base treatment is also acceptable for achieving GO to RGO deoxygenation. To create the hybrid materials of graphene and PANI (GEP), a three-step procedure was established, in which the obtained PANI/GO was reduced under 8 M KOH at 90 °C for 5 h.

Liquid/liquid interfacial polymerization in an oil/water environment can be used to create CP/GE composites. The water phase comprises sulfonated graphene with H_2O_2 and $FeCl_3$, whereas the oil phase contains aniline and chloroform. The oil/water interface serves as a nice soft template for PANI polymerization on sulfonated graphene sheets. The sulfonic acid groups are responsible for the hydrophilic characteristics of the nanosheets and serve as the acid dopant required for the creation of PANI, leading to a PANI/sulfonated graphene composite. The distinct nanostructured composites of PANI/sulfonated graphene have been made with and without an extra dopant of hydrogen chloride.

Photochemical processes can also be used to create CP/GE composites. A one-step redox process under UV light including GO reduction and pyrrole polymerization via electron transfer at ambient temperature resulted in a quick and straightforward production of GE/PPy nanofiber composites. A NO_2 sensor was made from the highly conductive GE/PPy nanofiber combination.

Polymerization of a monomer in a solution containing GO or GE dispersion with or without additional dopants is a standard electrical polymerization technique for CP/GE. Via the electrochemical polymerization, a GO-doped PEDOT (PEDOT/GO) film was fabricated for the simultaneous measurement of hydroquinone (HQ) and catechol (CT). EDOT was electrochemically polymerized on a glassy carbon electrode (GCE) in the presence of GO (1 $mg{\cdot}mL^{-1}$) and EDOT (0.01 M) by CV with potential scanning between 0.2 and 1.2 V at 100 $mV{\cdot}s^{-1}$. PEDOT evenly combines with wrinkled GO sheets which generate many micro- or macroholes allowing the passage of molecules and ions in the solution. Similarly, other CP/GEs, such as PPy/GO and PEDOT/GO, PTh/RGO, and PEDOT/GE, have recently been electrochemically incorporated via potentiostatic polymerization of GO and monomer.

An interesting in situ electrochemical approach was described by two groups. The PANI/GE modified electrodes are directly manufactured employing aniline-GO modified electrodes by in situ electrochemical reduction of GO and its simultaneous inclusion with the in situ electrochemical polymerization of aniline in acidic electrolyte. The in situ electrochemical synthesis of uniform PANI/GE composites thin film in 0.5 M HCl was performed with the applied potential of -1.0 to $+1.0$ V at scan rate of 0.02 $V{\cdot}s^{-1}$. The well-attachment of PANI and graphene, caused by the interaction of $C{-}N^{+}$ of PANI and carboxylate of graphene, benefits the sensor sensitivity, reliability, and repeatability in the detection of hydrazine.

Another method of electrochemical preparation is the electrochemical polymerization of CPs on graphene or GO supports. A layer of GO or graphene is formed by dropping its suspension in a solvent (e.g., the commonly used DMF) over the surface of particular electrode substrates. Subsequently, CPs can be electropolymerized on the modified electrode with GO or graphene film. Here, the loose carbon layer functions as a crucial electrical substrate for loading nanostructured CPs.

Yang Bo et al. described a three-step electrochemical deposition procedure at constant current density to coat PANI nanowires on the GO-modified GCE in order to fabricate a novel electrochemical DNA sensor. In certain situations, graphene paper may be utilized directly as the working electrode for the manufacture of a flexible CP/GE electrode, which may be employed in numerous sectors. CPs may grow only on the surface of the electrode or intercalate into the exfoliated carbon sheets and then anchor on the surface of carbon sheets, depending on the experimental conditions and preparation procedure.

In the case of GO as the predecessor, post-reduction of CP/GO to CP/GE is required for a specific reason. It is possible to do this at a constant negative potential or by scanning potential. For example, PPy/RGO was created by electrochemically polymerizing PPy/GO at -0.1 to -1.0 V (vs. SCE) at 50 mV·s^{-1} for 5 cycles. It is worth noting that the 3D structure of the CP/GE composite hydrogel lends itself to use in chemical sensors and biosensors. PPy/GO hydrogels are typically generated by in situ chemical polymerization of monomers in aqueous dispersions of GO sheets followed by lyophilization. GO sheets serve as a 2D template during polymerization. PPy is wrapped around GO sheets and subsequently crosslinked through π-π interaction, electrostatic interaction, and hydrogen bonding, causing the GO mix solution to gel. The composite hydrogels exhibit low critical hydrogel concentrations (1% by weight), high storage moduli and electrical conductivity, and high electrochemical activity.

4.4 Conductive Polymer Nanocomposites

Conductive polymer nanocomposites, a distinct class of polymer nanocomposites that are made conductive by combining a polymer matrix (either conducting or insulating) with a nanofiller (either conducting or insulating), have enormous potential in a variety of fields due to the synergistic effects of the polymer matrix and the nanofiller. Polyaniline (PANI), polypyrrole (PPy), polyacetylene (PAc), polythiophene (PTh), and their variants are the most often investigated conductive polymers. Conducting nanostructures such as carbon nanotubes, graphene, metals, and insulating ceramic nanostructures are the most extensively used nanofillers. Furthermore, conductive polymers have been turned into nanostructures and served as nanofillers in insulating polymer matrixes, giving a variety of interesting structure/properties. This has resulted in the formation of a very interesting category of conductive polymer nanocomposites. Furthermore, non-conducting ceramic nanostructures can be coated with conductive polymers to create conducting hybrid nanostructures, which can then be used as conducting nanofillers to create more complicated conductive polymer nanocomposites with improved matrix-nanofiller compatibility. With the potential for increased demand for lightweight materials in a variety of practical applications ranging from transportation carriers (e.g., airplanes, automobiles) to construction structures (e.g., the flame-retardant insulating layer, rooftop lightweight solar panels), conductive polymer nanocomposites are viable options for replacing conventionally heavy components such as metals and ceramics while providing enhanced mechanical, thermal, and flame-retardant properties. As novel functionalities and synergistic performances are regularly developed, the huge array of nanofillers with vastly different structures/properties boosts the potential of conductive polymer nanocomposites.

In general, three primary techniques are used in the manufacture of conductive polymer nanocomposites: (1) direct mixing/blending: The conductive polymer is dissolved in a suitable solvent, and the well-wetted nanofiller is added to the polymer solution while vigorously swirling. (2) In situ polymerization in the presence of nanoparticles: Typically, the nanofiller is surface treated or modified first, followed by the monomer, oxidant, and/or doping agent being added under a chosen stirring technique to complete the polymerization and/or form a polymer network with nanostructures embedded relatively uniformly. This technique often leads to improved interactions between the polymer matrix and the nanofiller, which improves attributes such as mechanical strength. (3) Formation of nanoparticles in situ in the presence of a conductive polymer: In this technique, the polymer is typically dissolved in a suitable solvent before adding the nanofiller precursor, which is then treated thermally or electrochemically to generate the conductive polymer nanocomposite. This approach has resulted in highly interesting and homogenous nanostructures in a variety of morphologies inside the polymer matrix.

4.4.1 Application of Conductive Polymer Nanocomposite

4.4.1.1 Optoelectronic Devices

Conductive polymers—and their derivatives, conductive polymer nanocomposites—are widely used in various optoelectronic devices due to their outstanding electronic, optical, and mechanical properties, high structural stabilities, easy-processing features, and low synthesis costs. It is appropriate to improve their qualities, add new features, and increase their performance by creating, tailoring, and adjusting their structural compositions and morphologies.

Almost every component of an OLED may be made of conductive polymers. Poly(*N*-ethyl-2-vinylcarbazole), poly(2-vinylcarbazole), poly(9-vinylcarbazole), poly(1-vinylnaphthalene), and poly(2-vinylnaphthalene) are examples of typical polymer hole transporters. Conductive polymer nanocomposites with unique nanofillers were increased in characteristics and functioned as different device components for OLEDs, including emissive layer (EML), electrodes, and hole transport layer (HTL).

Emissive Layer (EML)

When compared to bulk-heterojunction PLEDs without ZeSe nanowires, those with ZnSe nanowires hybridized with poly[2-methoxy-5-((2- ethylhexyl)oxy)-1,4-phenylene vinylene] as the EML displayed a more than 330% increase in electroluminescence intensity and a substantially lower driving voltage. The PLEDs made with poly(9,9-dioctyl fluorene-*alt*-bithiophene)/reduced graphene oxide (rGO_2) nanocomposites as the active EML had a maximum luminance of 54.31 cd/m^2

and a maximum power efficiency of 0.084, which were approximately three and 4.7 times higher than the PLEDs fabricated with pure poly(9,9-dioctyl fluorene-*alt*-bithiophene). Furthermore, the turn-on voltage of both types of PLEDs was just 4 V. It was thought that when rGO_2 was introduced, the balance of hole and electron injection shifted toward the center of the EML, considerably improving the brightness and efficiency of PLEDs.

Electrode

In comparison with AgNW and PEDOT:PSS electrodes, the PLED employing AgNW/PEDOT:PSS composites as the electrode had a lower leakage current and a shorter working voltage. Without removing the supporting layer, the GR/PEDOT:PSS bilayer might be used directly. In the low working voltage range (4–7 V), the PLED with the GR/PEDOT:PSS electrode had a lower turn-on voltage and greater luminous efficiency than the PLED with the ITO electrode. As a consequence, Joule heating was minimized, and the operational lifespan of GR/PEDOT:PSS-based PLEDs was increased.

Hole Transport Layer (HTL)

Gupta et al. found that polymer nanocomposite HTL improved the performance of blue OLEDs. The emissive layer was polyfluorene, while the HTL was TiO_2 nanoparticle-doped PEDOT:PSS. The impacts of TiO_2 nanoparticle concentrations on the performance of OLED devices were explored, and increasing the concentration of TiO_2 nanoparticles dramatically improved blue emission at 430 nm. Furthermore, the TiO_2 nanoparticles improved the turn-on voltage. The optimal concentration of TiO_2 nanoparticles in the HTL was discovered to be 15 wt%, at which point the OLED devices performed optimally. The manufactured PLED with PANI·PAMPSA/POGr nanocomposites as the HTL outperformed the PANI·PAMPSA interpolymer complex in terms of current density, brightness, current, and luminous efficiency.

4.4.1.2 Photovoltaic Devices

Buffer Layer

Using hybrid poly(ethylene glycol)-titanium oxide (PEG-TiO_x) nanocomposite film as cathode buffer layers (CBLs), Yin et al. produced extremely efficient and stable inverted polymer solar cells (PSCs). Due to improved charge extraction/collection and lower interface energy barriers, PSCs with PEG-TiO_x CBLs outperformed those with pure TiO_x CBLs or without CBLs. Under AM 1.5G irradiation (100 Mw cm^{-2}) for PEG-TiO_x CBL-based PSCs, a power conversion efficiency of up to 9.05% was obtained, which was a record high figure among inverted PSCs with TiO_x-based interface materials.

Photoactive Layer

Chehata et al. investigated the impact of polystyrene functionalized multiwalled carbon nanotubes (PSMWCNTs) on PSCs. PSCs with an ITO/PEDOT:PSS/MEH-PPV:PSMWCNTs/Al structure were created, with MEH-PPV:PSMWCNTs serving as the active layer. Compared to 26% for MEH-PPV-based devices, the greatest performance with a fill factor of 36% was obtained at a PSMWCNT concentration of 0.15 wt %. Furthermore, the PSC with PSMWCNTs outperformed the PSC without CNTs in terms of power conversion efficiency.

4.4.1.3 Electrochromic Devices

Conductive polymers have attracted the interest of researchers working on electrochromic (EC) devices. Furthermore, conductive polymers offer greater coloring efficiency, faster response speed, higher color tailorability, and need lower redox switching potentials than metal oxides. Polymer nanocomposites, one type of organic–inorganic material, have been extensively researched for EC applications. Carbon nanotubes and graphene were mixed with conductive polymers to create EC conductive polymer nanocomposites. Furthermore, different metal oxides have been extensively investigated as nanofillers in electrochromic conductive polymer nanocomposites, such as TiO_2, NiO, WO_3, and IrO_2.

The optimum composition of Au-CdSe in the nanocomposite achieved a higher coloring efficiency (300 $cm^2\ C^{-1}$) and faster kinetic response (it colors in 4.5 s and bleaches in 1.5 s) for the PEDOT/Au/CdSe nanocomposite films, as compared to the neat PEDOT film, with a coloring efficiency of 191 $cm^2\ C^{-1}$ and kinetic response (colors in 5.2 s and bleaches in 3.1 s).

In general, nanocomposites comprising conductive polymer and inorganic materials outperform equivalent conductive polymers in terms of EC performance, such as greater color efficiency and faster reaction speed. Because CNT and graphene may minimize charge transfer resistance through the bridging effect of the conductive nanoparticles, they can increase coloring efficiency across crystalline domains of conductive polymers. The quick switching speed is ascribed to the shorter diffusion length in the very porous structure generated by inorganic nanofillers like metal oxides.

4.4.1.4 Other Devices Application

Conductive polymer nanocomposites not only have a high potential in optoelectronic devices due to the synergistic interaction of the polymer matrix and the nanofiller, but they also play essential roles in other types of devices due to their mechanical malleability and conductivity. If conductive polymer nanocomposites can replace conventional conductors, the possibilities are endless.

Stretchable Electronics

Stretchable/wearable electronics based on conductive polymer nanocomposites have made tremendous development. Due to the use of polymers, particularly elastomers, stretchy electronics are flexible and promising for applications such as virtual reality, robotics, and electronic skins. Amjadi et al. presented a strain sensor with great stretchability and sensitivity based on silver nanowire (AgNW)/PDMS elastomer nanocomposite. The sensor has a high piezoresistivity, with adjustable gauge factors (GFs) ranging from 2 to 14 depending on the loading of silver nanowires, and a maximum stretchability of 70%. The static and dynamic loading tests demonstrated good linearity with little hysteresis. They also disclosed a strain sensor made of carbon nanotube (CNT)/silicone rubber nanocomposite thin films. The strain sensors based on CNTs and silicone rubber demonstrated exceptional stretchability up to 500% strain.

Electrochemical Energy Storage Devices (EES)

Battery life has always been a hurdle to replacing current Li-ion batteries with innovative high-performance materials. Wu et al. described a stable Li-ion battery anode material in which silicon nanoparticles were integrated into a conductive polymer hydrogel network. Wu et al. demonstrated over 90% capacity retention after 5000 charging and discharging cycles using an electrically conducting polyaniline hydrogel as the network. The benefits of this one-of-a-kind hierarchical three-dimensional material include a continuous conductive channel and porous area for silicon volume growth. These benefits might be the explanation for the extended Li-ion battery life.

Electrochemical Biosensors

Conductive nanomaterials such as gold nanoparticles (GNPs), silver nanoparticles (SNPs), and carbon nanomaterials can also be coimmobilized in a polymer matrix with enzyme. These conductive nanoparticles have a large surface area and a high electric conductivity. Miao et al. developed an Au/polypyrrole (Au/PPy) biosensor to monitor the electrocatalytic reduction of oxygen by laccase. Colloids of Au/PPy nanocomposite tightly attach to the Au electrodes and have an excellent reduction capacity.

4.4.2 *Nanocomposites Based on Conductive Polymers and Carbon Nanotubes*

Carbon nanotubes (CNTs) have been demonstrated to exhibit remarkable electrical, mechanical, and thermal characteristics, making them appealing for a wide range of prospective applications spanning from nanoelectronics to healthcare devices. However, because CNTs are virtually insoluble and infusible, their use in practical applications has been severely constrained. CNTs are classified into two types: single-walled carbon nanotubes (SWCNTs) and multiwalled carbon nanotubes (MWCNTs). Aside from potential advantages in the mechanical and electrical characteristics of polymers, the production of CNTs/polymer composites has been and continues to be investigated as a viable strategy for effectively incorporating CNTs into practical devices. Composite materials based on the coupling of conducting polymers (CPs) and carbon nanotubes (CNTs) have demonstrated that they combine the characteristics of the separate components in a synergistic way. The combination of CNTs and CPs provides an appealing way for reinforcing the polymer while also introducing electronic features based on morphological alteration or electronic interaction between the two components. It has been proposed that in CP/CNTs composites, either the polymer functionalizes the CNTs or the CPs are doped with CNTs, resulting in a charge transfer between the two elements. CPs/CNTs composites are classified into two types: CPs functionalized CNTs and CPs doped with CNTs. To make CP/CNTs composites, three methods were used: (i) direct mixing of CP with CNTs, (ii) chemical polymerization of the matching monomer in the presence of CNTs, and (iii) electrochemical synthesis of CP on CNTs electrode.

4.4.2.1 Conductivity and Photoconductivity

The use of readily processable materials or coatings for electromagnetic interference shielding is a crucial application for many industries. This would need the use of materials or composites with conductivities close to 1 S/cm. Several studies have been conducted in this regard. Doping a poly (p-phenlyenevinlene-co-2,5-dicotoxy-m-phenylene vinylene) (PmPV) with CNTs results in a hybrid composite with 10 orders of magnitude enhanced conductivity owing to the introduction of conducting routes to the polymer. The behavior of conductivity as a function of doping is typical of percolation, with a mass fraction threshold of about 8.5wt %. Following an initial rise in conductivity, between 0 and 0.5 wt %, caused by the introduction of traps by the CNTs, effective mobility decreases below that of PmPV. The effective carrier density was seen to fall between 0 and 2% before continuously rising, demonstrating electron transport from the nanotube to the polymer. Mulazzi et al. have published some data on the photoconductive properties of CP/NC composites. The inclusion of nanotubes into the PPV precursor polymer solution, which is heated to perform conversion into PPV, results in substantial changes in both the structural aspects of the composite components and the electrical properties of the composites, according to these scientists. The insertion of nanotubes causes the PPV polymer matrix to become more disordered, resulting in a shortening of the polymer conjugated segments. Photoconductivity studies demonstrate that the percolation regime begins at around a 2% concentration of SWCNTs, indicating that a migratory network for photogenerated changes is developed above this threshold.

4.4.2.2 Applications

Electrochemical Capacitors

Because the redox sites in the polymer backbone are not sufficiently stable for many repeated redox reactions, CPs have a reduced cycle life and sluggish kinetics of ion transport. The incorporation of carbon nanotubes (CNTs) into a polymer matrix increases the electrical conductivity of original polymer matrix as well as its mechanical characteristics, while potentially offering an alternative active material for capacitive energy storage. The electrochemical capacitance of non-aligned composites of CPs/MWCNTs is twice as big as any of the component materials, according to an analysis of the electrochemical capacitance. The utilization of an aligned array of MWCNTs in these composites provides more control over the thickness of CPs deposited onto each nanotube as well as the size of the intertubular pores. These two variables are critical in influencing the electrolyte capacity to access the full composite structure and hence assist the ion transfer process required by redox pseudocapacitive materials like these. Aligned PPy/MWCNT composite films provide an interesting combination of remarkable charge storage capacities as big as 2.55 F cm^{-2} (many times that of MWCNTs or PPy) and enhanced device reaction times as compared to pure PPy films (an order of magnitude faster). The enhanced performance of the PPy/MWCNT composite over its component materials is attributed to the combination of electrolyte accessibility, shorter diffusion distances, and higher conductivity in the redox pseudocapacitive composite structure. These findings suggest that arrays of aligned CP/MWCNTs composites are not only suitable for energy storage applications such as supercapacitors and secondary batteries, but also for usage in devices including sensors that would benefit from this desired combination of features. PANI/SWCNTs composite, produced by chemical polymerization of aniline in the presence of SWCNTs, has higher specific capacitance, better power characteristics, and is more promising for use in capacitors than pure PANI electrode. PEDOT/MWCNTs composites appear to be more cyclable than any other forms of CPs because CNTs high resilience facilitates compensating volumetric variations in such composites during counterion insertion/deinsertion processes, particularly during cycle performance.

Sensors

Gas Sensors

The sensitivity of the gas sensor made with the SWCNT/PPy nanocomposite to NO_2, as measured by a direct voltage divider at room temperature, was about 10 times greater than that of PPy. The nanodispersing of the SWCNT bundles increased the specific surface area of the coated PPy, leading to an increase in sensitivity. It was also demonstrated that the recovery period might be decreased, particularly in nanocomposite, by utilizing the Joule-heating effect.

Biosensors

According to a recent impedance research, the addition of MWCNTs to a CP (e.g., POAS) matrix leads to the best reduction in charge transfer resistance and mass transfer impedance. This result shows that CNT-based materials are the most enhanced polymeric nanocomposites for biosensor applications because they give the fastest electron transfer and ensure quicker ion mass transfer.

Photovoltaic Cells and Photodiodes

SWCNT/poly(3-octylthiophene) (P3OT) composites were employed to create innovative solar devices. P3OT, as a photoexcited electron donor, is combined with SWCNTs, which operate as electron acceptors. The transferred electrons are conveyed in such devices via percolation channels supplied by the inclusion of SWCNTs. Diodes (Al/polymer–nanotube composite/ITO) with 1% nanotube concentration demonstrated photovoltaic behavior with an open circuit voltage of 0.7–0.9 V. When compared to pristine polymer diodes, the short-circuit current rose by two orders of magnitude, and the fill factor increased from 0.3 to 0.4 for the nanotube/polymer cells. The scientists hypothesized a photoinduced electron transfer at the polymer/nanotube interface as the primary cause of this rise. It was discovered that the internal polymer/nanotube junctions operate as dissociation centers, capable of splitting apart the excitons while simultaneously forming a continuous channel for electrons to be effectively transferred to the negative electrode. This increases electron mobility and, as a result, balances charge carrier transit to the electrodes. Furthermore, the composite conductivity is boosted by a factor of ten, showing percolation channels inside the materials. These findings suggest that the CP/SWCNTs composite provides an alternate type of hybrid organic semiconducting material that is potential for increased performance in organic photovoltaic cells.

Optical Limiting Devices

The dispersion of SWCNTs in a host polymer (PmPV, hole conducting) traps the holes in an organic light emitting diode depending on the applied voltage (DE-OLED). This hole trapping causes the radiative recombination zones in the DE-OLED to change. Nile Red-doped tris-(8-hydrozyquinolinolato) aluminum (Alq3) was employed as an emissive material between the polymer composite and cathode. The device without SWCNTs distributed in the PmPV exhibited a dominating emission around red at 600 nm, which is in the region of the typical emission of Nile Red-doped Alq3, whereas the addition of a minor amount of SWCNTs enhanced a green emission. The polymer/SWCNTs composite-fabricated devices demonstrated an increase in the oscillator strength of the green emission, with a dominating emission peak near 500 nm, the characteristic emission of PmPV. This was reported for SWCNT concentrations as low as 0.1 wt%. The change in emission suggests that the SWCNTs in the PmPV matrix function as a hole-blocking material, causing the recombination zone to move from the Nile Red-doped Alq3 layer to the PmPV composite layer. Following that, Ago et al. postulated hole collecting characteristics of MWCNTs from a CP at the polymer composite area, while Romero et al. investigated CNT heterojunction with an organic semiconductor. This polymer/CNT composite is projected to form a polymer–nanostructure matrix with cooperative behavior between the host and additive, hence altering the electrical characteristics of the polymer composite.

Solar Cells

Because the diffusion lengths for excitons in CPs, such as PPV, have been reported at 10 nm, the need for a substantial percolation network of electron-accepting dopants in the polymer composite is supported. Electrical conductivity measurements have confirmed that SWCNTs-doped polymer composites have an extraordinarily low percolation threshold. Polymer composites doped with as low as 1 wt % SWCNTs showed a 70% improvement in heat conductivity at 40 K. The feasibility of embedding SWCNTs into a CP for photovoltaic devices was demonstrated using SWCNT/poly(3-octylthiophene) (P3OT) composites. Their findings demonstrated a diode response for devices built in a sandwich configuration with the composite film sandwiched between an indium tin oxide (ITO) front contact and an aluminum back contact. Under AM 1.5 illumination, both the pristine P3OT device and a 1% SWCNT/P3OT composite blend showed a photoresponse, though the composite exhibits current densities that are several orders of magnitude higher. The photoresponse of a 1% doped SWCNT/P3OT composite was reported to have an open circuit voltage V_{oc} of 0.75 V and a short-circuit current density I_{SC} of 0.12 mA cm^{-2}, compared to $V_{oc} = 0.35$ V and $I_{SC} = 0.7$ μA cm^{-2} for the pristine P3OT device. The inclusion of nanostructure–SWCNT complexes into the polymer has been recommended as a future optimization of carrier transport and exciton dissociation using these innovative solar systems.

Schottky Diodes

The nano-engineering of composites incorporating CP and CNTs for the production of Schottky diodes would improve the thermal and mechanical stability as well as the overall conductivity of such devices. It has been proposed that evenly dispersed nanotubes in the polymer operate as nanometric heat sinks, avoiding the accumulation of substantial thermal effects and, as a result, decreasing material and device deterioration. Composites containing 1 wt% CNTs, in particular, have been discovered to exhibit superior mechanical characteristics, as well as minimal surface roughness and increased DC conductivity. Schottky diodes made from composites of high molecular weight PANI and MWCNTs achieve current values that are much greater than pure PANI devices. The lack of a single linear zone on semi-log I-V curves for these devices was discovered, which contradicted thermoionic emission. This behavior was explained by linear regions with two different slopes observed on a log–log scale, where at lower voltages the charge transport mechanism is consistent with Ohm's law, but at higher voltages the charge transport mechanism is consistent with Child's law of space charge-limited emission. Localized defect states have been observed to have a considerable effect on this non-ideal diode behavior.

High-Resolution Printable Conductor

Blanchet et al. demonstrated that the addition of SWCNTs to PANI doped with dinonylnaphtalene sulfonic acid (DNNSA) forms a highly conducting three-dimensional percolating network, revealed by the linear temperature dependency above percolation, a fact that confirms their metallic nature. These composites were created as a high-resolution printed conductor for organic electronics applications. Thin composite sheets were printed with high resolution using laser ablation while keeping adequate conductivity. Printing structures that might act as a source and drain with a 7 μm channel and 2 S/cm conductivity for usage in plastic transistors demonstrate the applicability of these results. Although its intrinsic conductivity of 10^{-4} S/cm renders it unsuitable for use as a printed conductor, it has been claimed that SWCNTs can be used to enhance conductivity.

Electromagnetic Absorbers

On the basis of a polyurethane composite including carbon nanotubes, carbon fibers, and microballoons, as well as a polypyrrole fabric with varied surface resistances, EM absorbers with different dielectric characteristics and thicknesses were tested. It has been demonstrated that the surface resistance of the PPy fabric, as well as the sequence in which the composite layers are placed, is crucial for the sample reflection characteristic. In the 4–18 GHz frequency band, a combination of the PPy fabric and the composite resulted in higher than 15 dB reflection loss. Using a mix of PPy fabric and composite layer stacks, the desired bandwidth and performance may be reached with correct configuration.

Transistors

Qi et al. revealed that SWCNT-contacted P3HT FETs have three orders of magnitude greater current modulation (I_{max}:I_{min}) than metal-contacted devices across the same gate voltage V_{gs} equivalent to −2 to 2 V gate range [1].

Detection of Heavy Metal Ions

When compared to granular or ceramic composite materials, CPs can give more desired qualities such as increased tensile strength, elastic modulus, and improved flexibility. Due to its numerous favorable qualities such as environmental stability, flexibility, lightweight, high conductivity, facile synthesis via chemical and electrochemical processes, and superior processability, CPs are commonly employed in the composite construction with CNTs. Because of their high surface-to-volume ratio as compared to other carbon materials, CNTs/CPs can operate as efficient filler materials. However, a CNT/CP composite can provide synergistic effects suited for advanced applications. CNTs/CPs composites are formed as a result of π-π interactions between both components, which frequently result in collaborative effects in boosting electrochemical performances. CNTs and CPs are easily bindable and exhibit high interfacial coupling via donor–acceptor binding and π-π interaction because CNT π-electrons may interact strongly with the quinoid ring of CPs. Many research communities have planned to use the ultimate characteristics of these two organic materials, organic conductive polymers, and carbon nanotubes, for heavy metal ion detection by overcoming their limits and downsides. However, the inclusion of CNTs into polymeric matrix may extensively harness most of the advanced and excellent features of CNTs, and the resulting composite materials can broaden the range of more beneficial and attractive properties. The combination of conductive polymers with carbon nanotubes is useful for a wide range of technological solutions.

The performance of glassy carbon electrodes (GCEs) covered with polyaniline/multiwalled carbon nanotubes (MWCNTs/PANI) nanocomposite was studied to determine Pb(II) ions. It was discovered that the MWCNTs/PANI-coated electrodes performed better than the naked GCEs. The ultimate mechanism behind the CPs and CNTs composite is that prepared CNTs conduct. The fundamental drawback of CNTs is that they are functionalized with restricted functional groups, such as –COOH, –C=O, and –OH. Furthermore, CNTs are quite hydrophobic in nature. Owing to the lack of functional groups on their surfaces, unaltered CNTs have no special affinity for metal ions and cannot be used as an electrode material for metal ion sensing applications. However, CPs have high environmental stability and chelating capabilities, as well as an innate affinity for heavy metal ions. As a result, the combination of CPs and CNTs became fantastic, with CNTs enabling quicker passage of electrical signals due to their high conductivity and CPs providing advanced affinity toward metal ions for the reasons stated above. Therefore, the combination of CNTs and CPs proved suited for the construction of extremely sensitive and selective metal ion sensors [2].

4.4.3 Conductive Polymer/Clay Nanocomposites

Natural clay may be divided into four groups: kaolinite, montmorillonite(MMT)-smectite, illite, and chlorite. The MMT clays are the most commonly employed for intercalation or exfoliation of polymer/clay nanocomposites. Polymer nanocomposites, which are composed of polymer and clay in a combination of nanoparticle building blocks, have a type of nanoscaled structure that improves the macroscopic properties of products. The nanocomposites are multiphase materials with two or more different compounds organized in the nanometer regime and can exhibit electronic, mechanical, magnetic, and optical properties not present in the individual phases. Creating conductive polymer/clay nanocomposites is a novel way to build organic/inorganic hybrid systems and provide new synergistic properties that cannot be obtained from individual materials, such as improved conductivity and mechanical or thermal stability through the synthesis of the nanocomposites.

Several ways for producing conductive polymer/clay nanocomposites have been described. In a manner similar to how general polymer/clay nanocomposites are made, conductive polymer/clay nanocomposites can be made by intercalating the monomer of conductive polymer into the gallery of clay layers, followed by emulsion polymerization and in situ polymerization, which contributes to the maximization of the affinity between the hydrophilic host of clay and the hydrophobic guest of monomer. Furthermore, a process known as solution compounding is suggested, in which ready-made polymer chains, rather than monomers, are injected directly into the layers of clay by dissolving the two phases in polar solvent in the case of soluble conductive polymers. It should be noted that, while the electrochemical polymerization process is frequently used for the synthesis of various conductive polymer species, its application in the construction of nanocomposite is uncommon.

After the conductive polymer/clay nanocomposites are synthesized, various physical characterizations such as morphology are examined to determine whether the conductive polymer is successfully inserted into the layers of clay, whether the nanocomposites synthesized using different methods indicate morphology differently, and whether all clays are exfoliated or intercalated with increased *d*-spacing between clay layers. Images from scanning electron microscopy (SEM) and transmission electron microscopy (TEM) can be used to solve all of these suspicions. SEM micrographs of PANI/MMT nanocomposites with varying PANI concentration, as well as pristine MMT clay, reveal that the PANI content in the nanocomposites has a considerable influence on the surface morphologies of the nanocomposites. A pure MMT clay is composed of multilayer structures with multiple sheets piled in nanoscale thickness. The minute granular particles of PANI deposited on the surface of the MMT platelets rose steadily as the synthesized PANI content increased, until the MMT platelets were almost coated with granular powders of PANI. Excess aniline molecules that could not be absorbed into clay layers were polymerized outside them with increasing aniline loading, and the resultant PANI particles fully covered the surface of clay platelets.

With a more ordered system, increasing the degree of coherent layer stacking leads in a relative drop in the "fwhm" of the basal reflections upon hybrid formation. A decrease in the degree of coherent layer stacking with a more disordered system, on the other hand, results in peak widening and intensity loss. Due to the successful intercalation of conductive polymer fillers, the *d*-spacing of these conductive polymer/clay nanocomposite becomes wider than that of pure clay. The nanoscopic integration of conductive polymers with clay alters the thermal properties of polymer/clay nanocomposites owing to considerable increases in the interfacial adhesion between the polymer and the clay gallery.

The clay layers in nanocomposite systems disrupt the effective doping process and produce poor interchain contact. Furthermore, the weakening interchain interaction in the PANI chains results in charge carrier localization. In the PANI/clay nanocomposites, this results in relatively low $\sigma_{dc}(T)$ and a considerable temperature dependency of DC. Yeh et al. created a unique method for improving corrosion protection coatings using soluble electrically conductive PPy/clay nanocomposite materials. The PPy-coated CRS coupon had a higher E_{corr} value than the uncoated CRS. Its E_{corr} value, however, was lower than that of the specimen covered with polymer/clay nanocomposite materials. Furthermore, visual inspection of the corrosion products indicated that the polymer/clay nanocomposite sample demonstrating corrosion protection is a gray oxide layer formed over the naked exposed CRS surface, similar to what Wessling saw on steel with PANI dispersion coatings. The superior corrosion resistance of PPy/clay nanocomposite materials over pure PPy might be attributed to the dispersion of silicate nanolayers of clay in the PPy matrix, which increased the tortuosity of the diffusion pathway of oxygen gas and H_2O vapor. Both the PPy/clay and PANI/clay systems pass the corrosion protection test. Through a series of electrochemical measurements including corrosion potential, polarization resistance, and corrosion current on CRS in aqueous NaCl electrolyte, it was discovered that electronically conductive polymer/clay nanocomposite materials with low clay loading showed an advanced corrosion protection effect compared to pristine PPy.

In the absence of an applied electric field, shear stress grows monotonically with shear rate. In the presence of applied electric fields, the PANI/clay nanocomposite-based ER fluids had larger shear stresses. The ER behaviors of the PPy/clay nanocomposite-based ER fluids differed significantly. In contrast to the PANI/clay nanocomposite system, when shear rate rose, shear stress reduced initially and subsequently increased above a critical shear rate [3].

4.4.4 Conductive Polymer/Nanodiamond Nanocomposites

Nanodiamond (ND), a zero-dimensional nanoparticle, possesses desired and tailorable features such as nanosize, low dimensionality, high conductivity, high strength, and low aspect ratio. Because of their nanosize and higher surface-to-volume ratio, ND particles may aid in the formation of a distinct interphase, which can have a major influence on polymer characteristics even at modest loading levels. Fine ND dispersion in a polymer matrix is also required for ND-matrix compatibility. Furthermore, NDs increase thermal conductivity, luminescence, and semiconductivity of a material.

4.4.4.1 Polyaniline/Nanodiamond Nanocomposites

Gopalan et al. electrochemically polymerized aniline on amine-functionalized ND to generate a PANI-g-ND nanocomposite by grafting PANI chains on ND. Tamburri et al. used an electrochemical polymerization approach to create PANI and DND-based nanocomposite. They investigated the influence of DND nanoparticles on the conductive polymer nucleation process. A template-free technique and chemical oxidative precipitation polymerization were used to create the PANI-DND nanocomposite. The ND filler improved the electrical characteristics and thermal stability of PANI matrix. The increased conductivity was caused by the polymer chain being oriented perpendicular to the nanofiber axis. SEM examination revealed the presence of PANI-DND nanofibrils. Many crystallites were discovered to align and organize themselves neatly in the direction perpendicular to the polymer chain. Irrera et al. created a nanocomposite from self-assembled PANI micrometrical fibers doped with ND. The self-assembled and ND-doped nanofibrils were discovered using energy dispersive X-ray (EDX) and Raman analyses. The study efforts revealed that the interactions within the PANI/ND nanocomposite might result in remarkable characteristics and morphological organization in these materials. Using functional ND, the crystalline properties and charge transport capabilities of the nanocomposite may be changed. Modified nanoparticles can also improve the nanoscale conductivity and electrical characteristics of PANI fiber.

4.4.4.2 Polypyrrole/Nanodiamond Nanocomposites

PPy/ND nanocomposites transmit electricity, absorb visible light, and create a free charge, making them ideal for light-harvesting components in photovoltaic systems. Furthermore, covalent and non-covalent bonding interactions between the PPy and the ND may generate nanocomposites with stability. PPy/ND nanocomposites outperformed pure PPy in terms of charge transfer and physical properties.

4.4.4.3 Polythiophene/Nanodiamond Nanocomposites

Because of its surface functionalization, nontoxicity, and excellent chemical and physical stability, ND is also utilized as a significant filler to adjust the features of nanocomposite materials. The impacts of ND surface and size on solar cell performance have been investigated. The addition of ND improves the photoconductivity of P3HT. The ND-filled PTh and P3HT have higher short-circuit current, fill factor, and power conversion efficiency than the plain polymers. The characteristics of the ND reinforced regioregular P3HT nanocomposite were reported by Xu et al. The incorporation of NDs increased the length and crystallinity of the P3HT conjugation. The photoinduced charge transfer efficiency was shown to be higher between P3HT and ND. The morphology of the nanocomposite is an interconnected network of P3HT fibrils and ND domains. The ND energy band gap was wider than the P3HT band gap. The crystallinity and conjugation length of PTh and its derivative polymers were enhanced by the addition of ND.

4.4.4.4 Technical Applications of Conductive Polymer/Nanodiamond Nanocomposites

Corrosion Protection

PANI and PANI/ND coatings on mild steel and iron provided excellent corrosion resistance against corrosive conditions such as sulfuric acid and its halide solutions. The mechanism includes the formation of a protective barrier layer by the adsorption of organic molecules. The corrosion potential was altered to higher positive values due to the charge transfer of conductive polymers. Using electrochemical polymerization, Ashassi-Sorkhabi and Eshaghi created a polyaniline/nanodiamond (PANI/ND) nanocomposite covering for mild steel (cyclic voltammetry technique). The coatings exhibited excellent adhesion, minimal porosity, and fine dispersion. In a 0.5 M H_2SO_4 corrosive media, the corrosion protection efficacy of PANI sheets with additional ND was enhanced. According to the electrochemical modification, the ND included PPy increased coating corrosion resistance. The protective film efficiency increased when operative factors including ND concentration, current density, and surface shape were changed. Similarly, various conductive polymer/ND nanocomposite depositions can be investigated by altering parameters in order to get high-performance coatings.

Supercapacitors

The hydrogen bonding between the polymer and the ND filler in PANI/ND nanocomposites is critical for effective electrode formation. As a result, a PANI and functional ND nanocomposite-based electrode with specific capacitance has been shown (640 Fg^{-1} per 10,000 potentiodynamic cycles). The performance and cycle stability of the

Table 4.1 Device photovoltaic performance at 1.5 wt% nanodiamond from 5 to 250 nm, under AM 1.5 G at 100 mW/cm^2 illumination. Reprinted from ref. [6], copyright 2012, with permission from Elsevier

Device	V_{oc}	Jsc (mA cm^{-2})	FF (%)	PCE (%)
P3HT:PCBM	0.559	4.91	65.6	1.80
Nanocomposite ND5	0.579	5.61	65.7	2.14
Nanocomposite ND50	0.579	6.62	62.3	2.39
Nanocomposite ND250	0.539	4.89	45.0	1.21

neat ND and activated carbon composites were much greater. Furthermore, the supercapacitor behavior of PANI/ND nanocomposite minimized the volume fluctuations in PANI nanoparticles during cycling and increased the electrode durability.

Photovoltaics

Polymer/ND nanocomposites are utilized to create photovoltaic devices that are large-area, flexible, mechanically stable, and low cost. Poly-(3-hexylthiophene) and [6]-penyl-C_{61} butyric acid methyl ester (P3HT: PCBM) are the most often utilized copolymers for thin film production in organic solar cells. This polymer has the potential to function as both an electron donor and acceptor in organic solar cell systems. P3HT:PCBM with nanofillers may have energy conversion efficiency of more than 5%. As p-type semiconductors, nanodiamond-based solar cells and metal phthalocyanine derivatives have been created. Hsiao et al. used 1.5 wt% NDs of various sizes (5, 50, and 250 nm) in P3HT:PCBM films. The film root mean square roughness increased as the ND concentration rose. Furthermore, the absorption intensity of P3HT:PCBM and ND films was greater than that of the clean polymer films. The P3HT:PCBM with 50 nm ND increased the power conversion efficiency of the device from 1.8 to 2.4%. The photovoltaic performance of devices with ND sizes ranging from 5 to 250 nm is shown in Table 4.1. The additional ND enhanced the short-circuit current (J_{sc}) as well. With 50 nm ND particles, the highest current density of 6.6 mA/cm^2 was achieved. However, at the 250 nm ND size, the efficiency reduced. Better designs for devices with improved power conversion efficiency are needed in the fascinating field of polymer/ND-based solar cells [8].

4.4.5 *Electrode Materials Based on Conductive Polymers/Metal Nanocomposites*

Conductive polymer/metal nanocomposites, which combine the electrocatalytic capabilities of metal nanoparticles (NP) with conductive polymers, are a potential family of electrode materials. Owing to the porous nature of conductive polymers, they may be used as matrices loaded with metal particles that serve as reaction centers. Because they integrate the qualities of both components, conductive polymer/metal composites created by loading the conductive polymers with metal nanoparticles are of particular interest. Polymer/incorporated noble metal particles have greater activity than compact metals due to their increased surface area-to-volume ratio and unique surface characteristics. The ability of nitrogen-, oxygen-, or sulfur-containing groups in polymer molecules to stabilize nanosized clusters by interacting with metal atoms and thus restrict cluster aggregation is an appealing property of materials based on conductive polymers. As a result, in addition to acting as a support for metal particles and enabling electric integration, the polymer matrix also works as a stabilizing medium, preventing metal particle agglomeration during catalyst manufacture and operation.

Due to the porosity and strong electrical conductivity of the conductive polymer, the composite materials have good electron–ion transport characteristics in the material bulk. Metal particle integration into polymers results in nanostructured polymer/metal compounds with features promising for use in a variety of scientific and technological domains. Polymer matrices have pores of varying sizes, which are defined by the technique of synthesis and the physical state of the polymer. The chemical composition of the matrix and the manner of metal loading influence the stability and size parameters of metal dispersions. The incorporation of noble metals (platinum, palladium, gold, etc.) recognized for their high catalytic activity is frequently used in the creation of novel electrode materials based on conductive polymers.

Methods for creating conductive polymer/metal composite films on electrodes may be classified into four major groups: electrochemical or electroless deposition of metal NP into pre-synthesized films of conductive polymers; incorporation of pre-synthesized metal NP during electrochemical or electroless synthesis of conductive polymers; synthesis of polymer/metal composite films from colloid dispersions of polymers and metal NP; one-step synthesis of polymer/metal composite films. Each of the above synthetic techniques has benefits and disadvantages, and the choice of one or more approaches is determined by whether the chosen process produces nanocomposite materials that are compatible with the solution of the specific scientific or practical issue.

4.4.5.1 Electrochemical Reactions on Conductive Polymer/Metal Nanocomposite Electrodes

The basic scheme of electrode processes seen on a conductive polymer/metal composite sheet is depicted in Fig. 4.5. The localization of electrochemical reactions involving species from an external redox system is a topic worthy of special attention. In theory, electrode operations involving such electrolyte solution species can take place at the following interfaces: (1) conducting substrate/polymer film; (2) polymer film/electrolyte solution; and (3) metal cluster/electrolyte solution. The first instance is likely only for films with sufficiently high porosity, i.e., holes of varying

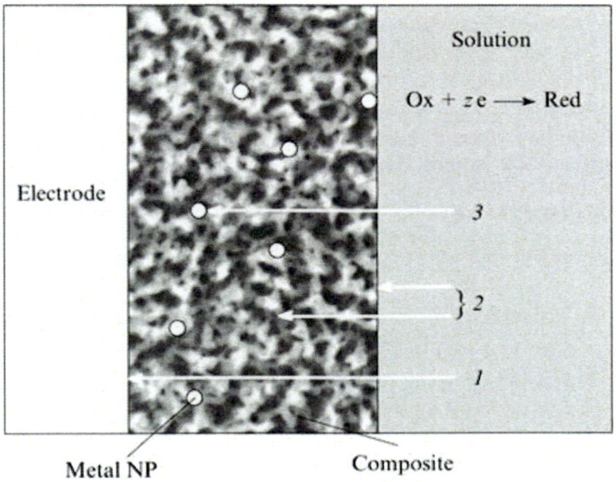

Fig. 4.5 Scheme of the electrode processes on a composite film. (1 ± 3) Interfaces (see the text). Ox is the oxidant, Red is the reductant. Reprinted from ref. [10], copyright 2016, with permission from IOP

sizes that provide reacting species access to the substrate through electrolyte. Films porous structure also allows for operations at interfaces 2 and 3. They can exist on the apparent exterior polymer surface and metal clusters on it, as well as on the interior surface of film pores and metal clusters within them. Indeed, examining the structure of such films revealed the existence of metal particle inclusions on the surface as well as within the pores. Furthermore, the position of a metal cluster in contact with electrolyte solution, either inside the film or near the film/electrolyte interface, has a significant effect on the process kinetics: In the former case, an additional stage of ion transport in the film appears, whereas in the latter case, this stage is absent.

Scientists continue to focus on electrocatalytic processes on various conductive polymer/metal composite electrodes, which is related with the development of low-temperature fuel cells and sensors. The electrocatalysis of hydrogen, formic acid, hydrazine, methanol, and ethanol oxidation and oxygen reduction processes is particularly important. The oxidation of methanol and other small molecules is known to take place in a series of parallel processes, particularly those that produce intermediates that can poison the catalytic metal particles. These intermediates include carbon monoxide and formaldehyde, which can be heavily adsorbed on the catalyst and impede its operation; hence, attention is focused on the choices for further oxidation of these products.

The vast majority of research paid little attention to determining the real size impact, that is, assessing the catalytic activity per unit real surface area of the catalytic metal. This is attributed, in part, to the difficulty in determining the true surface area of metal nanostructures embedded in a polymer, as well as its probable alterations during electrochemical studies under different circumstances.

The electrochemical adsorption of oxygen is a promising approach for determining the true surface area of platinum group metal and gold particles. This approach was found to be inaccurate for polymers that are quickly oxidized at relatively high anodic potentials of oxygen adsorption. However, several conductive polymers, such

Table 4.2 Experimental values of the surface area of palladium metal incorporated into a PEDOT film at different times of deposition. Reprinted from ref. [10], copyright 2016, with permission from IOP

Pd deposition time/s	$Q_{Pd} \times 10^{-5}$/C	S_{pd}/cm^2
400	13.95	0.332
240	7.575	0.180
120	5.467	0.130
60	4.116	0.098
30	2.176	0.052

as polythiophenes, have relatively strong oxidation resistance in the stated potential range. The PEDOT/Pd composite data can be used as an example of such estimations. The oxygen desorption peaks in voltammograms are highly repeatable in dilute acid solutions (0.1 M H_2SO_4, 0.1 M $HClO_4$), indicating a strong reliance on the timing of film loading with palladium particles. The charge (Q_{Pd}) used in the reduction of palladium surface oxide may be used to calculate the palladium surface area. The Q_{Pd} value was calculated by integrating the current beneath the cathodic peak of oxygen desorption from the palladium surface after adjusting for the charging current.

The charging currents of the PEDOT film are easily isolated from the currents of palladium oxide reduction on the PEDOT/Pd film. This allows for accurate estimations of the true surface area of palladium particles based on the proximity of the specific charge density consumed for oxygen desorption from palladium particles and the equivalent value for the polycrystalline palladium electrode. The experimental surface areas of palladium particles (S_{Pd}) integrated into the PEDOT film at various loading durations are shown in Table 4.2.

The impedance spectra of PEDOT/Pd films in 0.1 M H_2SO_4 in the potential regions of electrochemical sorption/desorption of hydrogen on the surface of incorporated palladium particles showed clear-cut semicircles with the diameter (determining the resistance to charge transfer and inversely proportional to the surface area of these particles) decreasing as the palladium amount in the film increased. Under the parameters mentioned, the latter appears to be connected with an increase in the surface area of palladium particles exposed to hydrogen sorption/desorption. The true surface area of palladium particles within the composite film has been determined using the assumption that the specific resistance to charge transfer is the same for smooth palladium and palladium particles embedded in a film. Based on the true surface area of Pd particles, the specific surface area was determined to be 36 $m^2 g^{-1}$ for the palladium loading employed, but the average size of palladium particles computed in terms of the model of spheres was found to be close to 14 nm. Under the same circumstances, a comparable size of palladium NP (10–20 nm) was discovered using the SEM approach.

The majority of studies base the electrocatalytic activity of such composites solely on the gain in the rate of some process occurring in these objects when compared to the rate (current) observed on the electrode chosen for comparison (polymer film, electrode substrate, etc.), i.e., in reality, the potential changes in the actual surface area of the catalytic metal while switching from test to reference systems are largely overlooked. The most consistently established and significant advantages of dispersing the catalytic metal throughout the polymer matrix include an increase in both the active surface area of the catalytic metal and the stability of its particles in the nano-sized state, which provides long-term stability of electrode characteristics. These findings were reached in the majority of investigations, but with varying degrees of reliability. However, more profound generalization of results from various studies, even those obtained for similar systems, is frequently complicated by differences in conductive polymer/metal composite synthesis conditions, dissimilar metal inclusion characteristics, and ambiguities in electrochemical response acquisition.

4.4.6 Nanocomposites Based on Graphene Analogous Materials and Conductive Polymers

Because of their graphene-like two-dimensional structure, graphene analogous materials (GAMs) are described as "analogous to graphene." GAMs, like graphene, are made up of individual atomic layers that can be separated from a three-dimensional bulk material made up of stacked two-dimensional layers. This results in extraordinary characteristics as a result of low dimensionality and edge states. They have been used in business, research, and electrical gadgets. GAMs are van der Waals materials that can be easily created by exfoliating their three-dimensional equivalent layered structures along the z-axis. These exfoliated individual layers may often be disseminated into a conductive polymer matrix, resulting in nanocomposites. Transition metal dichalcogenides (TMDs), boron nitride (BN), and MXenes are the most common graphene analogue materials (GAMs).

GAMs/polymer nanocomposites are created using procedures similar to graphene-based polymer nanocomposites. Solution mixing, melt mixing, and in situ polymerization are the most common and practical techniques. Non-covalent grafting of polymers onto the filler material by π-π stacking is one of these additional ways. Other techniques include in situ polymerization, melt mixing, and solution mixing. Emulsion polymerization, phase transfer procedures, lyophilization methods, shear pulverization, layer-by-layer assembly, and mechanochemical treatment are a few examples.

4.4.6.1 Transition Metal Dichalcogenides/Conductive Polymer Nanocomposite

Transition metal dichalcogenides (TMDs) are a class of compounds with the general formula MX_2, where M stands for Mo, W, V, Nb, Ta, Ti, Zr, Hf, and so on, and X stands for S, Se, or Te. Molybdenum disulfide (MoS_2) and tungsten disulfide (WS_2) are two of the most frequent TMDs utilized in polymer-based nanocomposites.

4.4.6.2 MoS_2/Conductive Polymer Nanocomposite

Several researches have been conducted in an attempt to entrap various polymers in the layered structure of MoS_2. Kanatzidis et al. were the first to describe the intercalation of polyaniline (PANI) into MoS_2. Pre-intercalation of Li^+ into MoS_2 was followed by a reaction with water to create exfoliated MoS_2 nanosheets. When an NMP solution of PANI in the emeraldine base form was added to the separated layers, an intercalated nanocomposite was formed. Powder X-ray diffraction revealed that the PANI/MoS_2 nanocomposites produced by Kanatzidis et al. and Bissessur et al. were intercalated.

Zhang et al. created MoS_2/PANI nanocomposites with three distinct morphologies: aggregated particles (MoS_2/PANI-1), nanowires (MoS_2/PANI-2), and nanotubes (MoS_2/PANI-3). The pH employed in the second phase of the synthesis, as shown in Fig. 4.6, determined the morphology of the final product. Because of its hollow structure, MoS_2/PANI-3 had the maximum specific capacitance of 375 Fg^{-1} at a current density of 50 Ag^{-1} and also displayed good charge/discharge cycle stability. These nanocomposites have been shown to be useful as electrodes in supercapacitor applications, a sensing platform for simultaneous determination of adenine and guanine, a modified electrode for detection of bisphenol A, an electrode with applications in DNA detection, as biosensor for detection of chronic myelogenous leukemia, and a photocatalyst for the degradation of organic pollutants such as methylene blue (MB) and 4-chlorophenol.

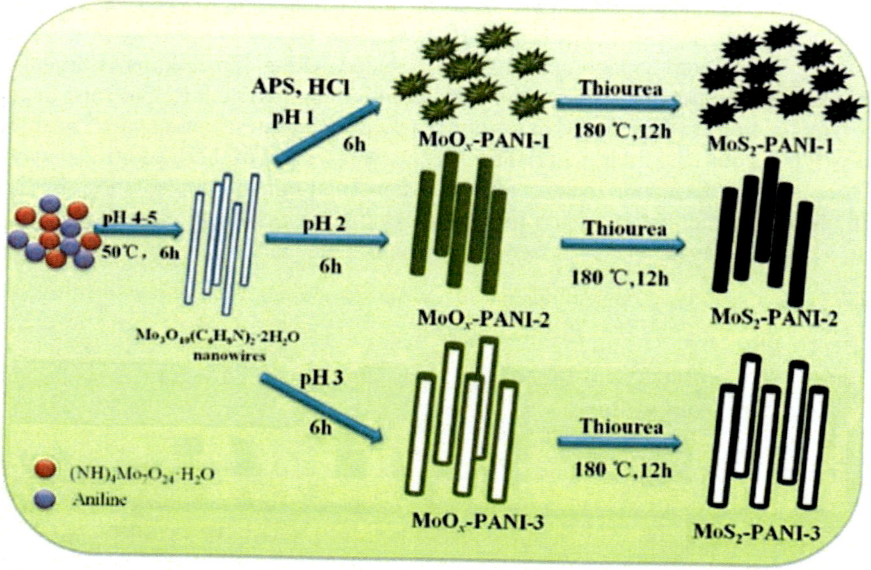

Fig. 4.6 Preparation of MoS_2-PANI-1, MoS_2-PANI-2, and MoS_2-PANI-3. APS:$(NH_4)_2S_2O_8$. Reprinted from ref. [17], copyright 2019, with permission from Elsevier

Wang et al. were the first to demonstrate polypyrrole intercalation into MoS_2. In the intercalation procedure, pyrrole was used instead of aniline, and iron(III) chloride was used as the oxidizing agent instead of ammonium peroxydisulfate. Powder X-ray diffraction tests gave clear evidence of full intercalation, according to the scientists. To accommodate a single layer of the polymer chain, the *d*-spacing of the host material increased by about 4.5 A°.

Bissessur and Liu successfully intercalated polypyrrole (PPy) into MoS_2. Polypyrrole was first synthesized and then ultrasonically disseminated in NMF. The polymer colloidal solution was then applied to exfoliated layers of MoS_2 in water. This technique resulted in the polymer being inserted into the restacked MoS_2. The interlayer expansion value of the sandwiched compound in comparison with pristine layered MoS_2 was typically around 3.5 A°. This is consistent with the polymer chain having a single layer that is parallel to the dichalcogenide sheet. In 0.1 M H_2SO_4, the corrosion rates of various coatings were measured. With a corrosion protection efficacy of 99.38%, polypyrrole coating with 25 ppm MoS_2 provided the best corrosion protection for brass, outperforming polypyrrole coating with a corrosion protection efficiency of 77.68%.

MoS_2 has also been intercalated with polythiophene (PTh). Lin et al. were successful in creating PTh/MoS_2 intercalated nanocomposites with thiophene as the monomer. The electrical conductivity of the PTh/MoS_2 nanocomposite was found to be 16 times better than that of pure stacked MoS_2. The coating of polythiophene onto MoS_2 nanotubes has piqued the researchers' curiosity. MoS_2 has also been intercalated with PEDOT. The monomer 3,4-ethylenedioxythiophene was added to MoS_2 single layers. After refluxing the reaction mixture, an aqueous solution of iron(III) chloride was added. This technique resulted in the creation of PEDOT/MoS_2 intercalated material. Powder X-ray diffraction revealed evidence of intercalation. The conductivity of the PEDOT/MoS_2 was discovered to be one order of magnitude greater than that of pristine MoS_2. When employed as a cathode in a lithium cell, the better conductivity of the PEDOT/MoS_2 and its greater interlayer spacing of 13.76 A° contributed to a considerably increased discharge capacity.

4.4.6.3 WS_2/Conductive Polymer Nanocomposite

Conductive polymers, albeit to a lesser extent than MoS_2, have been intercalated into multilayer WS_2. Polypyrrole encapsulation in multilayer WS_2 resulted in a polypyrrole intercalation compound within WS_2, as was demonstrated by XRD. The measured interlayer expansion of 3.43–3.76 A° corresponded to a polypyrrole monolayer placed parallel between the WS_2 sheets. Lane et al. described the synthesis and characterization of PANI and WS_2 exfoliated nanocomposites. They found an eightfold increase in conductivity (24.5 S cm^{-1}) when 15 wt% by mass of exfoliated WS_2 was disseminated. Exfoliated WS_2/polythiophene nanocomposites may be produced, as demonstrated by Arsenault et al. They discovered that a nanocomposite containing 5% by mass WS_2 had an order of magnitude better conductivity than pure polythiophene.

4.4.6.4 Boron Nitride/Conductive Polymer Nanocomposite

Boron nitride (BN) can exist as nanosheets (BNNS), which are hexagonal arrangements of boron and nitrogen atoms covalently bound to produce a strong two-dimensional material. BNNS is similar to graphene. Nanocomposites made of BN and conductive polymers have been investigated. Mixing and sonicating the two components in N,N-dimethylformamide resulted in self-organized BNNTs/PANI nanocomposite films. The positive interaction of EB with BNNT resulted in its ordering inside the composite material, as shown by a sharpening of the wide peak corresponding to the EB ($\theta = 19°$) in the composite XRD pattern. The composite Raman shifts were similarly compatible with interaction between the EB and BNNT. EB Raman spectra had two main peaks at 1350 and 1579 cm^{-1}. The EB peak at 1350 cm^{-1} could not be seen in the composite Raman spectra due to substantial overlap with the BNNT peak, and the BNNT itself moved from 1365 to 1352 cm^{-1}. The EB peak at 1579 cm^{-1}, however, was relocated to 1589 cm^{-1}.

An in situ polymerization approach was also employed to generate BNNTs/PANI nanocomposites, and the presence of BNNTs was observed to cause the production of crystalline phases inside the PANI. The nanocomposites were made by first absorbing aniline on BNNTs and then immersing them in an aqueous solution of ammonium peroxydisulfate (APS) for 24 h. The typical thickness of the PANI layer on the surface of the BNNT was 2–5 nm as measured by TEM, and the thickness could be increased to 20 nm by applying extra aniline. As aniline molecules and APS solution were sucked within the nanotubes where the polymerization reaction took place, the encapsulated PANI engaging with the inner surface of the BNNTs (Fig. 4.7), TEM indicated that part of the BNNTs is packed with PANI chains.

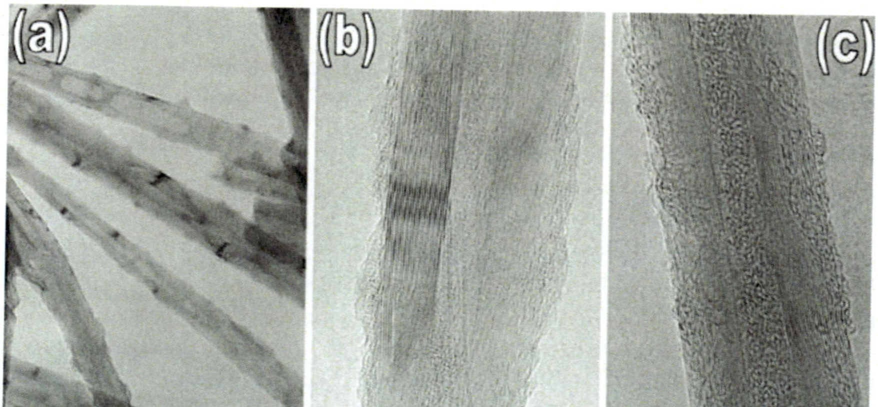

Fig. 4.7 **a** Low magnification TEM image of PANI/BNNTs, **b** high-resolution TEM image of PANI/BNNTs and **c** high-resolution TEM image showing BNNT filled with PANI. Reprinted from ref. [12], copyright 2008, with permission from the American Chemical Society

Powder X-ray diffraction revealed that the crystallinity of the PANI in the PANI/BNNTs nanocomposites increased. The XRD pattern of pure PANI revealed two wide peaks at 21° and 25°, indicating the polymer amorphous nature. The peak corresponding to PANI at 25° in the nanocomposite XRD pattern could not be recognized due to considerable overlap with the peak from the BNNT component. The PANI peak at 21°, on the other hand, was divided into three distinct peaks at 19.9°, 21.3°, and 22.1°, indicating a structural ordering of the PANI when embedded inside the BNNTs.

Polypyrrole (PPy)/*h*-BN nanocomposites have also been reported. Madakbas et al., for example, developed PPy/*h*-BN nanocomposites comprising 1, 2, 3, and 4 wt% h-BN with the goal of enhancing the thermal stability of PPy as well as fine-tuning its dielectric characteristics. The inclusion of *h*-BN in the range of 1–4 wt% boosted the thermal stability of the polypyrrole substantially. Furthermore, the dielectric constant of the nanocomposites dropped as the *h*-BN concentration increased. By non-covalently binding regiorandom poly(3-hexylthiophene) (rra-P3HT) to the surface of BNNTs, Martinez-Rubi et al. were able to produce rra-P3HT/BNNT nanocomposite (Fig. 4.8a). Rra-P3HT in $CHCl_3$ displayed a wide envelope (λ_{max} = 432 nm), which was attributed to an intrachain π-π* transition caused by polymer chain twisting, culminating in a coil-like conformation. Surprisingly, the absorption band of the rra-P3HT/BNNT nanocomposite was redshifted and partially resolved, with faint peaks at 520, 555, and 603 nm (Fig. 4.8b). The observed redshift was ascribed to a favorable and strong π-π stacking interaction between P3HT and BNNT, which discouraged rra-P3HT chain twisting. As a result, the polymer chain is planarized with extended conjugation on the surface of the BNNT. The scientists performed structural characterization of the P3HT chains on the surface of BNNT using polarized excitation fluorescence microscopy and determined that the chains were aligned along the BNNT axis.

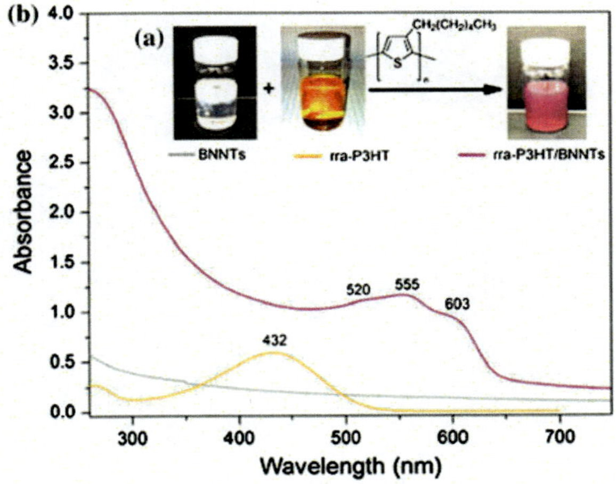

Fig. 4.8 **a** Photographs of rra-P3HT/BNNT, **b** absorption spectra of BNNT (0.2 mg/mL), rra-P3HT solution, and rra-P3HT/BNNT suspensions in $CHCl_3$ (rra-P3HT to BNNT weight ratio = 0.08). Reprinted from ref. [13], copyright 2015, with permission from the American Chemical Society

4.4.6.5 MXenes/Conductive Polymer Nanocomposite

MXenes are a class of layered ternary carbides, carbonitrides, and nitrides with the common chemical formula $M_{n+1}AX_4$, where M is an early transition metal and A is the main group element, often from groups 13 or 14. Similar to MoS_2, WS_2, and BN, nanocomposites based on conductive polymers and MXenes have been produced, albeit to a limited degree due to the relatively new synthesis of this sort of 2D structure.

Ren et al. created a PANI/Ti_3C_2 nanocomposite for use in high-performance supercapacitors. PANI particles were found to be adsorbed on the surface and edges of Ti_3C_2 nanosheets using electron microscopy (Fig. 4.9). Powder X-ray diffraction observations agreed with electron microscope findings. PANI/Ti_3C_2 nanocomposite had good electrochemical performance, with a maximum specific capacitance of 164 Fg^{-1}, 1.25 times that of Ti_3C_2. Furthermore, after 3000 cycles, 96% of its initial capacity was preserved, proving its remarkable cycling stability.

Zhang et al. created composite films from $Ti_3C_2T_x$ and polyaniline that were doped with dodecylbenzenesulfonic acid (DBSA) and hydrochloric acid. Figure 4.10 depicts the process utilized to create these films. Vacuum-assisted filtering was used to create $Ti_3C_2T_x$/cPANI films with weight ratios of 1:1, 3:1, 5:1, and 7:1. The scientists discovered that increasing the mass percentage of $Ti_3C_2T_x$ in the composite material boosted its electrical conductivity, electromagnetic shielding efficacy, and tensile strength. $Ti_3C_2T_x$/cPANI film (7:1) had a greater electrical conductivity of 24.5 S cm^{-1}, a better electromagnetic shielding effective of 36 dB, and a higher tensile strength of 19.9 MPa than cPANI, which had values of 0.3 S/cm, 16 dB, and 2.6 MPa, respectively.

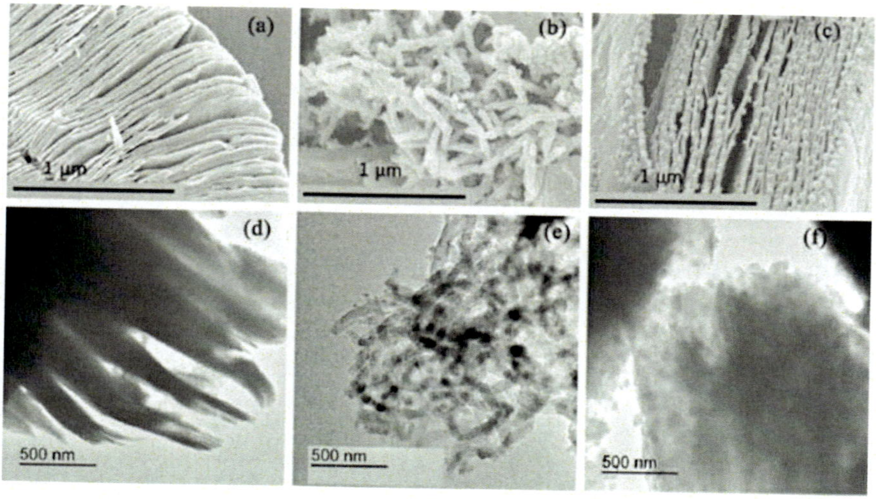

Fig. 4.9 SEM and TEM images of Ti_3C_2 (**a**), (**d**), PANI (**b**), (**e**), PANI/Ti_3C_2 (**c**), (**f**). Reprinted from ref. [16], copyright 2018, with permission from Elsevier

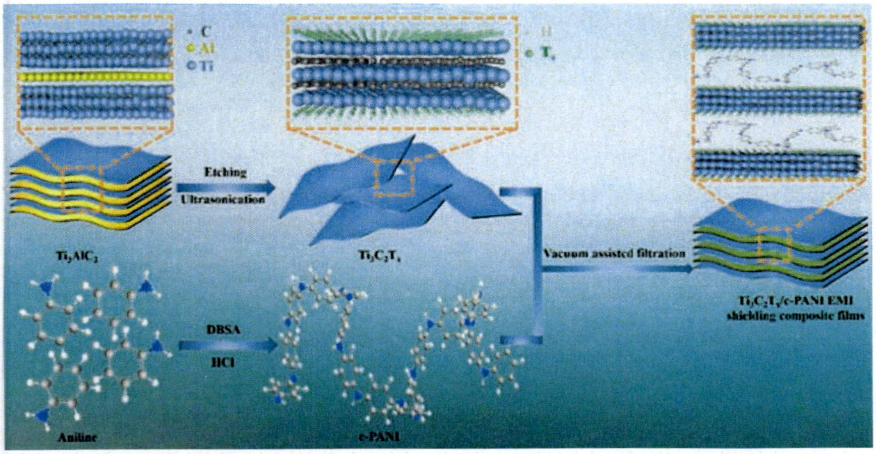

Fig. 4.10 Steps leading to the formation of $Ti_3C_2T_x$/cPANI film. Reprinted from ref. [9], copyright 2019, with permission from Elsevier

Wei et al. also created PANI/Ti_3C_2 composites that absorbed microwave radiation well. The nanocomposite prepared by Wei et al. demonstrated excellent microwave absorption potential, with a maximum reflection coefficient of -56.30 dB at 13.80 GHz for a sample with a thickness of 1.8 mm. Furthermore, with an adjustable thickness of 1.5–2.6 mm, the effective absorption bandwidth (90%) spanned from the X-band (8–12.4 GHz) to the Ku-band (12.41–18 GHz).

PANI/$Ti_3C_2T_x$ nanocomposite was created by Zhao et al. At 200 ppm, the PANI/$Ti_3C_2T_x$ nanocomposite demonstrated a quick detection response for volatile gases such as ethanol, methanol, and ammonia at ambient temperature. Among the gases examined, the nanocomposite was the most selective for detecting ethanol, with a greater sensitivity than bare MXene.

Boota et al. created a polypyrrole/$Ti_3C_2T_x$ nanocomposite. The fluorine atom from the $Ti_3C_2T_x$ was used to dope the polypyrrole. Raman and FTIR spectroscopy provided evidence for the production of doped polypyrrole. The $Ti_3C_2T_x$/PPy (2:1) powder pattern revealed that $Ti_3C_2T_x$ layers extended by 10 A°, indicating the creation of an intercalated nanocomposite. The TEM data was supported by the XRD data. The nanocomposite had a high volumetric capacitance of around 1000 F cm^{-3} and preserved 92% of its capacitance after 25,000 cycles.

Yan et al. electroplated polypyrrole (PPy) onto the surface of a $Ti_3C_2T_x$-coated cloth. The material was coated with $Ti_3C_2T_x$ using a simple dipping–drying method. The electrochemical deposition of PPy on the MXene/textile inhibited the development of TiO_2 on the MXene edge. The PPy/Mxene/textile combination was used as a flexible electrode in a supercapacitor application, with a specific capacitance of 343.20 Fg^{-1}.

Wu et al. intercalated polypyrrole into Ti_3C_2. FTIR and Raman spectroscopy gave evidence for polypyrrole production. The polypyrrole was found in the Ti_3C_2

gallery area, as evidenced by XRD. The PPy/Ti_3C_2 nanocomposite had a specific capacitance of 184.36 Fg^{-1} at a scan rate of 2 mV s^{-1}, which was 37% greater than Ti_3C_2. After 4000 charge/discharge cycles at a current density of 1 Ag^{-1}, the nanocomposite preserved 83.33% of its capacitance.

4.5 Conductive Polymer Reinforced Polyurethane Composites

Commercial uses of CPs have been realized. However, disadvantages such as poor processability, low solubility, low yield, and poor mechanical qualities have limited their direct uses in a variety of disciplines such as coatings, optical and solar devices, and so on. To overcome these drawbacks, researchers have proposed a number of strategies in the form of new synthesis methods, in which these polymers are added to conventional polymers to form copolymers, blends, IPNs, and composites and nanocomposites. These investigations indicate that there is a lot of room for improvement in their features and applications. However, in order to disperse CPs, a proper insulating polymer matrix must be used, since the matrix provides processability and stability to the CP nanoparticles dispersed composites. The characteristics of such composites are primarily determined by the nature of the CP dispersion inside the insulating polymer matrices, as well as the interactions between the CP fillers and the insulating polymer matrices.

Among many insulating polymer matrices such as epoxy, alkyl, polyacrylates, PU, and others, PU is the most significant, frequently used, and adaptable, consisting of hard (isocyanate) and soft (polyols) segments. It possesses distinct qualities such as strong elasticity, elongation, high impact and tensile strength, high abrasion resistance, good weathering resistance, outstanding gloss, color retention, and corrosion resistance. The synergistic impact of CPs (filler) and insulating PUs (matrices) has been demonstrated to improve their processability, stability, solubility, thermal, mechanical, electrical, and optical characteristics. Even under extreme conditions, these upgraded CP/PU composite materials are capable of providing increased service life in the aforementioned regions. As a result, there is plenty of room for research and development into other forms of CP/PU composites that may potentially have value-added features. Because of their versatility and breadth of applications, CP/PU composites have received a lot of interest. Melt mixing, solution blending, and in situ polymerization are three commonly used procedures for manufacturing CP/PU composites. Prior to the production of a CP/PU composite, either both components (CP and PU) were synthesized or one of these two components (CP or PU) was purchased or synthesized and then distributed together. It has been discovered that polyurethane of varying molecular weights, ranging from 23,000 g/mol to 108,430 g/mol, has been employed in the manufacturing of CP/PU composites.

PANI-based polyurethane composites have been proven to be more stable and processable than other conductive polymer-based polyurethane composites. This is due to the development of H-bonding between the –NH group of PANI and the –NHCOO group of PU, which results in improved miscibility, conductivity, and tensile strength. The influence of H-bonding between –NH of PANI and –NHCOO of PU/PMMA on the physicomechanical, electrical, and thermal characteristics of IPN was shown by Siddaramaiah et al.

Robila et al. created a conductive composite of PPy and sulfonated PU anionomers (PUA/PPy). The structure, characteristics, and concentration of the ionomer, as well as the influence of the synthesis method on performance, were thoroughly investigated. According to a thermogravimetric investigation, PU anionomer containing PPy has superior thermal stability than pure PU anionomer. According to a thermogravimetric investigation, PU anionomer containing PPy has superior thermal stability than pure PU anionomer. They created a carboxylated PU anionomer/PPy composite in their subsequent research by soaking the PU in an aqueous solution of Py monomer and $FeCl_3$. In this study, the mechanical and electrical characteristics of the composite were compared to those of PU anionomer, and it was discovered that the inclusion of PPy boosts electrical conductivity while decreasing mechanical qualities.

Yalinmaz et al. used ceric ammonium nitrate to create the PU-PPy nanofiber composite. The addition of PPy (molecular weight = 93,000 g/mol) to the PU matrix improves the dielectric, electrical, mechanical, and thermal characteristics of the composite. Wang et al. created a series of waterborne cationic PPy/PU composites by incorporating an oxidant and Py into a cationic PU dispersion. The effects of Py content, reaction duration, temperature, and oxidant-to-Py molar ratio on the surface resistivity of conductive films were thoroughly examined. In parallel, they created a waterborne PU/PPy dispersion to increase the compatibility of PPy and PU.

When utilized in particular applications, PTh and its derivatives, such as PEDOT and P(3HT), demonstrated the best characteristics with an insulating polyurethane matrix. Razal et al. reported the wet-spinning process for producing homogenous PU/PEDOT:PSS composite fiber. It was discovered that when the loading of PEDOT:PSS in PU fiber composites increased, Young's modulus grew exponentially and the yield stress increased linearly.

Excellent work has been reported on the dispersion of conductive polymer in the polyurethane matrix, which is used in the processing of CP/PU composites, which have a wide range of applications in the fields of electromagnetic interference (EMI) shielding, sensors, biomedical, shape memory polymers (SMP), membranes, anticorrosive coatings, films, and foams.

4.5.1 EMI Shielding

Among different polymer composites, conductive polymer-based polyurethane composites are determined to be the best choice for meeting the EMI shielding qualities of electronic equipment. Mathew et al. investigated the EMI shielding capabilities of a PANI/PU composite at *S* and *X* frequencies. The composite was discovered to be a possible option for EMI applications, and their shielding efficacy rose as the sample thickness grew. Furthermore, electrically conductive castor oil polyurethane composites containing polypyrrole-coated peach palm fiber (PPy/PPF) were created by dispersing varying quantities of PPy/PPF in PU matrix. They discovered that a composite having 25 wt% of PPy/PPF has strong EMI shielding properties. Gunjan Gupta et al. have created EMI shielding polyurethane composite coatings based on PANI/PPy. The polyurethane matrix was infused with varying concentrations of PANI/PPy. It was discovered that coatings containing 6% PANI and 6% PPy have significant EMI shielding characteristics.

4.5.2 Sensors

The conductive composite demonstrated exceptional chemoresistive capabilities for chloroform vapor detection, which were determined to be superior to that of plain PANI. This was ascribed to a shift in the conducting channel of PANI layers caused by the exceptional and reversible solvent-swelling capabilities of the PU fiber matrix. They explored the high sensitivity (4×10^{-4} ppm), rapid reaction time (>30 s), and low detection limit (30 ppm) of this sensor, which may be accomplished using composite fibers of 55 μm diameter PU and 550 nm thickness PANI layer for chloroform gas detection.

4.5.3 Biomedical

Because of their high physicomechanical qualities and strong cytocompatibility, conductive polyurethane composites have piqued the interest of researchers in exploring their uses in biomedical fields. The smart and next-generation conductive composite material has been studied for its potential uses in medication delivery and tissue engineering. Broda et al. created a PPy/PU composite for use in biological tissue engineering. The composite with a 1:5 PPy/PU ratio had the maximum conductivity (2.3×10^{-6} S cm^{-1}), whereas the composite with a 1:100 ratio had the lowest conductivity (1.0×10^{-10} S cm^{-1}). They conducted a cytocompatibility test on human dermal cells (C_2C_{12} myoblast cells) and found no evidence of cytotoxicity. Madrigal et al. created nanomembranes by spin coating a polythiophene derivative (P3TMA) and thermoplastic polyurethane (TPU) combination in various weight ratios. They explored the biomedical applications of TPU:P3TMA nanomembranes and discovered that increasing the concentration of P3TMA enhanced membrane swelling and enzymatic breakdown. TPU:P3TMA nanomembranes function as a biodegradable and bioactive material that promotes cell viability.

4.5.4 Shape Memory Polymer

The researchers created conductive shape memory polyurethane by combining conductive polymers. Sahoo et al. used a chemical oxidation approach to create an electroactive shape memory PU/PPy conductive composite. The shape recovery test using bending mode and a 40 V electric field yielded an excellent shape recovery of 85–90% in 25 s.

4.5.5 Membrane

Membranes are mostly utilized in processes such as microfiltration, ultrafiltration, reverse osmosis, electrodialysis, and others. M.A. Shehzad et al. used solution casting to create cation-exchange membranes of thermoplastic polyurethane (TPU) and polyaniline (PANI) doped with camphorsulfonic acid (CSA). Almeida et al. provided a comparative analysis of castor oil polyurethane (CAPU) and TPU-modified membranes with PANI. These membranes were employed in electrodialysis testing, and it was discovered that they had a greater mechanical property.

4.5.6 Anticorrosive Coatings

A vast number of publications have been written on the use of various forms of conductive protective coatings to shield metal surfaces from severe weather conditions. Among them, CP/PU-based coatings have demonstrated promising protection capabilities, giving twofold protection to coating materials via redox and barrier mechanisms. PU serves as a barrier covering, preventing penetration of corrosive ions. CP, on the other hand, produces a passive oxide layer between the metal and coating interfaces and serves as a protective barrier until the CP is capable of continually undergoing charge transfer at the metal-coating interface. Furthermore, the high contact of CP/PU with the metal surface is an essential element in enhancing metal adhesion. The anticorrosive performance of poly(urethane-co-pyrrole)s (CPUPYs) coatings was compared to that of bare steel, parent PU, and neat PPy-coated stainless steel, and it was discovered that the CPUPYs coating, prepared with 0.0023 mol of pyrrole and 0.0050 mol of CAN, had the highest conductivity and R_{ct} values. These findings demonstrated the strong corrosion protection capacity of the coatings by generating a passive metal oxide layer at the metal-coating contact.

4.5.7 *Films*

Organic molecular and polymeric composite conductors in the form of thin to thick films have a wide range of industrial uses. Among these, CP/PU composite films have seen extensive application in the electrical industry. Malmonge et al. created flexible and free-standing PANI dispersed castor oil polyurethane (PANI/CPU) films in various mediums (DMF and aqueous). The DMF-synthesized film had a better conductivity (10^{-2} S/cm) than the aqueous medium-synthesized film (10^{-4} S/cm). Zeghina et al. created a lightweight film out of pTSA and HCl-doped PANI/PU composite. The influence of dopant type, PANI concentration, and film thickness on morphological, dielectric, and microwave absorption characteristics was examined. PANI-PTSA/PU films exhibit increased permittivity and microwave absorption characteristics than PANI-HCl/PU for the same weight fraction of PANI.

4.5.8 *Foams*

Owing to the numerous applications of conductive polyurethane foams, several researchers and manufacturers have created conductive polyurethane foams by adding CPs into PU foams. Vapor phase oxidative polymerization was used to create a conductive composite of PPy/PU Foam. Harry et al. created electrically conductive PANI/PU foams. The DC conductivity PANI in open cell PU foams was also measured. They proposed that the created PANI/PU foams be employed as electronic component packaging materials.

References

1. Baibarac M, Pedro G-R (2006) Nanocomposites based on conducting polymers and carbon nanotubes: from fancy materials to functional applications. J Nanosci Nanotechnol 6(2):289–302
2. Deshmukh MA, Shirsat MD, Ramanaviciene A, Ramanavicius A (2018) Composites based on conducting polymers and carbon nanomaterials for heavy metal ion sensing (Review). Crit Rev Anal Chem 48(4):293–304
3. Fang FF, Choi HJ, Joo J (2008) Conducting polymer/clay nanocomposites and their applications. J Nanosci Nanotechnol 8(4):1559–1581
4. Han J, Li L, Guo R (2016) Novel Approach to controllable synthesis of gold nanoparticles supported on polyaniline nanofibers.Macromolecules 43(24):10636–10644
5. Han J, Wang M, Hu Y, Zhou C, Guo R (2017) Conducting polymer-noble metal nanoparticle hybrids: Synthesis mechanism application. Prog Polymer Sci 52–91
6. Hsiao YJ, Fang TH, Ji LW et al (2012) Size effect of nanodiamonds on P3HT: PCBM heterojunction solar cells. Electrochem Commun 18:4–7
7. Jaymand M (2014) Conductive polymers/Zeolite (nano-)composites: Under-exploited materials. RSC Adv 4(64):33935–33954
8. Kausar A (2019) Review on conducting polymer/nanodiamond nanocomposites: Essences and functional performance. J Plast Film Sheeting 35(4):331–353

9. Khatoon H, Ahmad S (2017) A review on conducting polymer reinforced polyurethane composites. J Ind Eng Chem 53:1–22
10. Kondratiev V, Maleev V, Eliseeva S (2016) Composite electrode materials based on conductive polymers loaded with metal nanostructures. Russ Chem Rev Rev Curr Topics Chem 64(8):1919–1925
11. Liu Z, Ma Z (2013) Fabrication of an ultrasensitive electrochemical immunosensor for CEA based on conducting long-chain polythiols. Biosens Bioelectron 46:1–7
12. Martinez-Rubi Y, Jakubek ZJ, Jakubinek MB, Kim KS, Cheng F, Couillard M et al (2015) Self-assembly and visualization of Poly(3-hexyl-thiophene) chain alignment along Boron Nitride nanotubes. J Phys Chem C 119(47): 26605–26610
13. Ren Y, Zhu J, Wang L, Liu H, Liu Y, Wu W et al (2018) Synthesis of polyaniline nanoparticles deposited on two-dimensional titanium carbide for high-performance supercapacitors. 214:84–87
14. Shan J, Ma Z (2017) A review on amperometric immunoassays for tumor markers based on the use of hybrid materials consisting of conducting polymers and noble metal nanomaterials. Mikrochimica Acta Int J Phys Chem Methods Anal 184:969–979
15. Xu JJ, Hu JC, Quan BG, Wei ZX (2009) Decorating polypyrrole nanotubes with Au nanoparticles by an in situ reduction process. Macromol Rapid Commun 30:936–940
16. Zhang Y, Wang L, Zhang J, Song P, Gu JJCS (2019) Technology. Fabrication and investigation on the ultra-thin and flexible Ti3C2Tx/co-doped polyaniline electromagnetic interference shielding composite films. 183:107833
17. Zhi C, Zhang L, Bando Y, Terao T, Tang C, Kuwahara H et al (2008) New Crystalline phase induced by Boron Nitride nanotubes in polyaniline. 112(45):17592–17595

Chapter 5
Conductive Polymers and Their Composites for Biological Application

Conductive polymers (CPs) have gotten a lot of interest since the discovery of metallic polyacetylene in the late 1970s, when it was discovered that CPs may display electrical conductivity in a wide range by utilizing a suitable dopant in a doping procedure. They not only have electrical and optical properties similar to metals and semiconductors, but they also have the above-mentioned benefits of traditional polymers. Furthermore, the degree and duration of electrical stimulation may be adjusted externally using CPs, which is advantageous for biological applications. The chemical structures and biomedical applications of the aromatic CPs are presented in Table 5.1.

As compared to typical electronic materials such as metals and inorganic semiconductors, conductive polymers (CPs) provide many major benefits for biological interactions such as programmable physiochemical characteristics, changeable form factors, and mixed conductivity (ionic and electronic). Organic conductive polymers, because of their soft nature, have superior mechanical compatibility and structural tunability with cells and organs than typical electronic inorganic and metal materials. Polyaniline (PANI), polypyrrole (PPy), and polythiophene, as well as their derivatives and composites, are appealing biomaterials due to their biocompatibility, ease of synthesis, simplicity of modification, and ability to electronically control a range of physical and chemical properties via (i) surface functionalization techniques and (ii) the use of a diverse range of molecules that can be entrapped or used as dopants.

Because of these advantages, they are appealing in a wide range of biomedical applications, including drug delivery systems, artificial muscles, bioactuators, biosensors, brain recording, and tissue engineering. Not only are CPs biocompatible, but they may also induce biological processes such as cell adhesion, migration, proliferation, differentiation, and protein secretion at the polymer tissue interface with or without electrical stimulation.

CP-based biomaterials are particularly effective in the engineering of electrically sensitive tissues including skeletal muscles, cardiac muscles, neurons, skins, and bones. Biomaterials containing CPs were found to dramatically improve cell adhesion and proliferation in a variety of cells, including L929 fibroblasts, C2C12 myoblasts,

Y. Wang and W. Feng, *Conductive Polymers and Their Composites*,
https://doi.org/10.1007/978-981-19-5363-7_5

Table 5.1 Chemical structures and biomedical applications of common aromatic conductive polymers (CPs): polypyrrole (PPy), polyaniline (PANI), polythiophene (PTh), and poly(3,4-ethyelenedioxythiophene) (PEDOT)

Polymer	Chemical Structure	Applications
Polypyrrole (PPy)	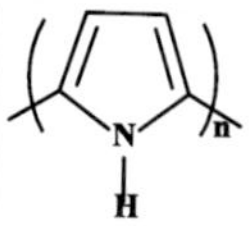	Fuel cell, corrosion protection, computer displays, microsurgical tools, biosensors
Polyaniline (PANI)	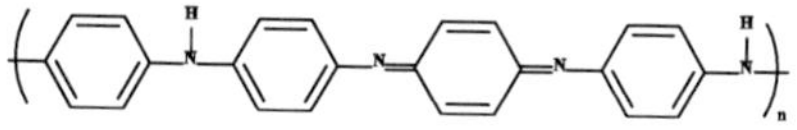	Biosensors, neural probes, drug delivery, tissue engineering
Polythiophene (PTh)	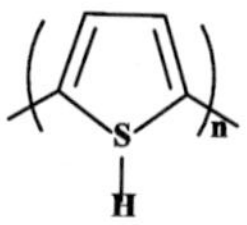	Biosensors, solar cells, tissue engineering, photosensitizers, supercapacitors
Poly(3,4-ethyelenedioxythiophene) (PEDOT)	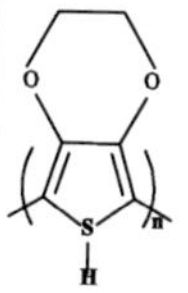	Neural electrodes, nerve grafts, heart muscle patches

PC12 cells, RSC96 Schwann cells, H9c2 cardiac cells, primary cardiomyocytes, MC3T3-E1 cells, and mesenchymal stem cells.

5.1 Biocompatibility of Conductive Polymers

The biocompatibility of conductive polymers in vivo and in vitro is another significant advantage for their use in biomedical engineering. Their polymer compositions, which include chemical structures, functional groups, morphologies, and synthesis methods, influence their biocompatibility. Diverse conductive polymers, such as polypyrrole (PPy) and polyaniline (PANI), have demonstrated excellent cellular response and growth support for a variety of cell types, which is an important feature of biocompatibility in biomedical applications. Furthermore, the conductive polymer biocompatibility may be easily increased by incorporating biocompatible molecules, segments, and side chains into the polymers.

Ramanaviciene et al. found that chemically produced PPy particles were highly biocompatible in mice after six weeks of therapy, with no deleterious impacts on cell survival and proliferation [22]. Although the biocompatibility of PPy has been called into question in some cases, it has been shown to support the adhesion, growth, and differentiation of a diverse range of cell types in vitro, including bone, neural, glial, rat pheochromocytoma and endothelial cells, fibroblasts, keratinocytes, and mesenchymal stem cells. Acute toxicity, mutagenesis, pyretogen, hemolysis, or allergy reactions were not observed in Schwann cells treated with a PPy powder solution. PPy-PLGA composite meshes showed to be biocompatible with embryonic hippocampal neurons and PC12 cells. The inclusion of a PPy coating has no effect on the biocompatibility of polyester textiles. PPy superior biocompatibility was also demonstrated in animal models: the results reveal that PPy has no substantial long-term effect in vivo or generates only a little tissue response.

Humpolicek et al. have observed that PANI had excellent biocompatibility in terms of cutaneous irritation and sensitization. PANI biocompatibility findings are mixed: it was stated to enhance neural cell development, offer appropriate proliferation and adhesion, preserve sufficient biocompatibility, and not induce substantial inflammation. PANI emeraldine base and emeraldine salt forms were reported to be cytocompatible with H9c2 cardiac myoblasts and did not cause inflammation in rodent models. During a 90-day period, none of the emeraldine, nigraniline, or leucoemeraldine variants of PANI were observed to elicit an inflammatory response in rats in a comparable study. It was blatantly non-cytotoxic, but that surface modifications were required to improve its biocompatibility.

PEDOT has demonstrated high biocompatibility with epithelial, neuronal, and neuroblastoma cells, as well as L929 and NIH3T3 fibroblasts. PSS and tosylate anion-doped PEDOT films, for example, have been shown to promote fibroblast adhesion and proliferation. PEDOT was doped with NGF in an attempt to improve its compatibility with neural tissue, and it did indeed boost the proliferation of PC12 cells. Similarly, PEDOT nanotube-coated electrodes implanted into the barrel cortex of rats were found to provide a better tissue response than uncoated electrodes.

5.2 Conductive Polymers for Tissue Engineering

Tissue engineering has made significant progress in recent decades as an interdisciplinary field that integrates engineering and biological science ideas to the production of tissue replacements. The primary objective of tissue engineering is to repair or replace damaged, sick, or missing tissues and organs. Conductive polymers are utilized solely in tissue engineering applications. Conductive polymers, such as polyaniline, polypyrrole, and polythiophene, and/or their derivatives and composites, provide compatible substrates that promote cell growth, adhesion, and proliferation at the polymer-issue interface via electrical stimulation due to their ability to electronically control a variety of physical and chemical properties.

Specific cell responses are determined by polymer surface attributes including roughness, surface free energy, topography, chemistry, charge, and other features such as electrical conductivity or mechanical actuation, which are determined by the synthesis conditions used. Biomaterials with regulated nanometer-scale organization in the case of conductive polymers and electrical stimulation can significantly improve cell biological processes. Cell interactions and responses are vitally dependent on polymer chemical structures (polyaniline, polypyrrole, or polythiophene), macroscopic forms (films, powder, membranes, nanofibers), and geometric cues (2D, 3D).

CPs were created as bioactive biomaterials for tissue engineering applications using various approaches, and their applications in tissue engineering include bone, skeletal muscle, neural, cardiac, and wound healing, as shown in Fig. 5.1.

5.2.1 Polyaniline in Tissue Engineering

PANI powders and films in the forms of emeraldine (EM), nigraniline (NA), and leucoemeraldine (LM) were implanted beneath the dorsal skin of male Sprague Dawley rats for an extended length of time to assess the in vivo tissue response. After 50 and 90 weeks, the inflammation associated with the various kinds of PANI was shown to be negligible. To assess tissue development, interstitial pressure was measured around implantation. For nearly two years after implantation, no strong positive pressures were recorded in the animals during this assessment. This demonstrated PANI biocompatibility in the dorsal portion of the skin. The adhesion and proliferation of H9c2 rat cardiac myoblasts were used to test the biocompatibility of PANI thin films in both non-conductive EM base and its conductive salts (E-PANI). Cell attachment tests revealed that H9c2 cardiac myoblasts connected to both conductive and non-conductive versions of PANI quickly, within 15 min, and

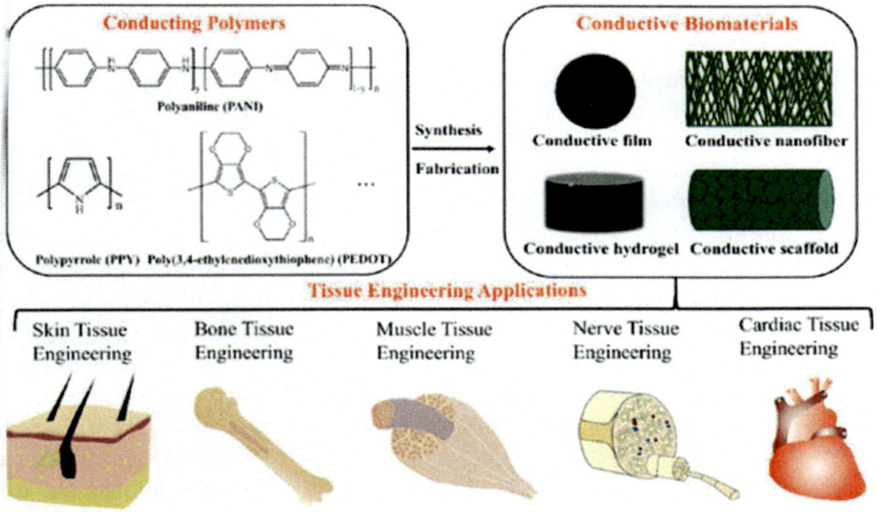

Fig. 5.1 Conductive polymers and conductive biomaterials and their tissue engineering applications. Reprinted from ref. [7], copyright 2018, with permission from American Chemical Society

in similar numbers, within 7% of controls. Following cell proliferation assays, cells seeded on both the native and electroactive forms of PANI remained viable and proliferated at comparable rates, although with a minor delay, when compared to tissue-culture polystyrene (TCP) (used as a control).

5.2.2 *Polypyrrole in Tissue Engineering*

Polypyrrole (PPy) is the most thoroughly researched conductive polymer for biological applications due to its high electrical conductivity, flexible method of preparation, ease of surface modification, excellent environmental stability, ion exchange capacity, in vitro and in vivo biocompatibility, and ability to support cell adhesion and growth of a variety of cell types.

The majority of research has been on modifying PPy to maximize its interaction with certain cell types and other features that are required for its application in vivo. Several investigations have shown that PPy is compatible with cells and tissues in vitro and in vivo. Electropolymerization was the most often used technology for producing PPy films on various substrates. A systematic study linking the methodologies used for PPy synthesis to its basic polymeric properties (e.g., hydrophilicity, surface roughness) and the biological effects these properties have on cells emphasized the relationships between synthesis parameters, polymeric properties, and biological compatibility.

5.2.2.1 Polypyrrole Films and Powders

Williams and Doherty established in vitro and in vivo cytocompatibility of PPy films with mouse fibroblast and neuroblastoma cells, as well as the ability to provide electric fields to neuroblastoma cells cultivated on PPy. Positive electrical stimulation of polystyrene sulfonate (PSS)-loaded PPy films during the growth of rat PC12 cells and primary chicken sciatic nerve explants was also found to improve attachment and neurite extension. In the case of PC12 cells, there was no significant difference in neurite extension between cells grown on TCP plates and cells grown on unstimulated PPy. There was, however, a considerable increase in neurite length for the electrically stimulated PPy substrate. On mouse peritoneum cells, chemically produced PPy particles had no obvious cytotoxic impact. They did not cause an allergic reaction, nor did they have an effect on spleen, kidney, or liver indices. The cell attachment, proliferation, and differentiation of mesenchymal stem cells (MSC) from the bone marrow stroma were studied using PPy thin films coated on TCP plates and generated by admicellar polymerization. The effect of PPy film thickness on MSC in vitro behavior was also examined. Increased monomer concentration resulted in thicker PPy films and rougher surfaces, which had a significant impact on MSC capacity to attach to these surfaces.

5.2.2.2 Doping and Entrapment of Biomolecules in PPy Films

Surface roughness, surface energy, conductivity, mechanical actuation, and dopant retention are all features of conductive polymers that impact their biocompatibility. Furthermore, when considering tissue engineering applications, the anionic dopant and any other existing excipients must be addressed in addition to the conductive polymers. During the synthesis process, several biomolecules can be introduced into PPy as biodopants. When conductive polymers are generated chemically or electrochemically by oxidation of the monomer, negatively charged dopant molecules are simultaneously incorporated. By altering the biodopant, PPy can be tailored to encourage the proliferation of different cell types or to induce certain elements of wound healing after the incorporation of biomolecules (proteins, peptides, or ECM components). The type of biomolecular dopants influences both the electrical and biological characteristics of the resultant surfaces. Several research examined PPy modulation using various dopants, such as dermatan sulfate for enhancing keratinocyte survival, heparin (HE) for boosting endothelial cell proliferation, and laminin-derived peptides for controlling neuron and astrocyte adhesion. Such biomolecules can be entrapped in PPy films alone or in combination with other dopants. PPy was used to entrap adenosine 50-triphosphate (ATP) and NGF for tissue engineering or medication administration. The coentrapment of NGF and dextran sulfate (DS), followed by the regulated release of NGF upon PPy reduction, was reported to induce PC12 cell neurite extension.

5.2.2.3 Covalent Grafting of Biomolecules on PPy Films

The most promising method for avoiding biomolecule denaturation is covalent attachment of the biomolecules to a polymer film containing pendant-reactive groups such as $-NH_2$ or $-COOH$. Because of the lower physisorption, the chances of the trapped biomolecules being leached are similarly reduced. Doping PPy with polyglutamic acid (PGlu), for example, supplies carboxylic acid groups for future functionalization. Because the PPy-PGlu matrix is conductive, it can be employed to induce cell adhesion. This technology may be used to easily include a variety of different biomolecules, and it can also be used to build multilayers of biomolecules, which may lead to three-dimensional substrates appropriate for tissue engineering applications. One disadvantage of this strategy is that glutamic acid is a neurotransmitter that may cause excitotoxicity. The similar method was used to immobilize nerve growth factor (NGF) on PPy films. In this scenario, an intermediate photocrosslinker composed of polyallylamine conjugated to an arylazido functional group was utilized. As a result, the development of covalent bonds between bioactive compounds and the surface of conductive polymer films might be considered as an alternate method for designing tissue engineering scaffolds. When compared to other procedures (dopant or entrapment), which generate a drop in conductivity by orders of magnitude, these covalent techniques utilized in the modification of PPy surfaces give both electrical and biological stimulation.

5.2.3 Polythiophenes in Tissue Engineering

In comparison with research on PPy and PANI, studies on the application of polythiophene (PTh) and its derivatives in tissue engineering are scarce and new. The polythiophenes were discovered to have characteristics similar to, and in some cases superior to, those of PPy. The cytocompatibility of high-quality films of PTh substituted with methyl and dodecyl side chains produced by in situ chemical deposition in the presence of dodecylbenzenesulfonate (SDBS) was tested by growing PC12 pheochromocytoma cells and NIH3T3 fibroblasts on these surfaces. The in vitro testing demonstrated that these films were not cytotoxic and may enable cell adhesion and proliferation. By altering the length of the alkyl side chains of the replacements present at the 3-position, the hydrophobicity of the films could be altered. The increased hydrophobicity of poly(3-dodecylthiophene) explained its superior ability for protein adsorption and proliferation of PC12 cells over poly(3-methylthiophene). Organo-soluble poly[2-(3-thienyl) ethyl octanoate] (POTE) and poly(3-(2-hydroxyethyl)thiophene) (PHET) films were also tested for their capacity to support the adhesion, proliferation, and differentiation of skeletal myoblasts.

Because of its features (greater electrical conductivity and chemical stability), poly(3,4-ethylenedioxythiophene) (PEDOT) is regarded the most effective PTh derivative, allowing its practical usage in many disciplines such as biomedicine and biotechnology. Del Valle et al. studied the surface of anodic polymerized PEDOT films in terms of epithelial cell adhesion and proliferation. The findings revealed that epithelial cells Hep-2 had high activity on the surface of PEDOT electrodeposited on stainless steel electrodes, and no cytotoxicity was identified as a result of this polymer. Using cyclic voltammetry, the electrochemical properties of PEDOT coated with cells were determined in various biological media. It was discovered that the connected cellular monolayer considerably improves the electroactivity of PEDOT film. The comparison of PEDOT coated with cells to PEDOT with adsorbed BSA indicated that the increased electroactivity of the former system originates from the cell reaction to anodic scanning, which stimulates the diffusion of anions through the cellular membrane. These findings demonstrated PEDOT biocompatibility with Hep-2 cells as well as electrical biocompatibility with live cells [3].

5.2.4 *Fabrication of Conductive Biomaterials for Tissue Engineering*

5.2.4.1 Pure Conductive Polymer Films for Tissue Engineering

Conductive polymers (e.g., PANI, PPy, and polythiophene) promote the adhesion, proliferation, and differentiation of a wide range of cell types in vitro, showing that they are cytocompatible. Furthermore, the biocompatibility of CPs was demonstrated in animal models. Several findings revealed that CPs had no long-term effect in vivo or only caused a minor tissue response. Chemical or electrochemical synthesis can be used to create CPs. Electrochemically produced conductive polymers are often in the form of sheets on the electrode that may be utilized for cell culture. The PPy membrane, for example, was electronically formed on an indium tin oxide glass plate at the nanoscale. The nanoscaled PPy particles were scattered equally, with an average diameter of 62 nm. Preosteoblast MC3T3-E1 cells were cultured on the PPy membrane with both electrical and mechanical stimulation, and the results in terms of cell proliferation and collagen-I gene expression showed that combined electrical and mechanical stimulation greatly promoted the proliferation and differentiation of MC3T3-E1 cells compared to single stimulation, indicating that the nano-PPy membrane may offer a way to stimulate bone tissue repair.

5.2.4.2 Conducting Blends or Composite Films for Tissue Engineering

Because CPs are extremely brittle, producing pure conductive polymer films from them is extremely challenging. As a result, combining CPs with other biodegradable polymers is commonly employed to create conductive biomaterials for tissue engineering. PLA, PLGA, PCL, chitosan, and silk fibroin were blended with CPs such as PANI and PPy. A conductive PPy/PCL film, for example, was created by first immersing a PCL film in a polystyrene sulfonic acid and pyrrole DI water combination. The mixture was then oxidized with ferric chloride, resulting in an interpenetrating network of PPy-coated PCL conductive film. The resistivity of the film was 1.0 ± 0.4 kΩ cm, which was comparable to that of native heart tissue. Using emulsion polymerization and dip-coating processes, conductive PPy/poly(D,L-lactic acid) (PDLLA) composite films and nerve conduits were created.

5.2.4.3 Conducting Copolymer Films for Tissue Engineering

CPs have been widely employed as biomaterials in tissue engineering; however, their in vivo uses are limited because of their non-degradability. Although CPs are not naturally biodegradable, the usage of aniline/pyrrole-based copolymers functionalized with hydrolyzable groups resulted in materials with similar electroactivity as CPs as well as the added benefit of being erodible (biodegradable).

5.2.4.4 Conducting Nanofibers for Tissue Engineering

Biomaterials should be designed to resemble the structure of the extracellular matrix (ECM). Collagen, the most abundant protein in ECM, takes the shape of nanofibers. Nanofibrous polymeric scaffolds have been produced to mimic the natural ECM utilizing a variety of methods such as electrospinning, phase separation, and molecular assembly. Electrospinning is the most often used technology for creating nanofibers. A range of biodegradable polymers, both natural and synthetic, were combined with CPs and electrospun into nanofibers for tissue engineering.

Electrospun fiber has a large surface area and porosity, and its diameter may be adjusted between few nanometers and several micrometers. PANI and PLA electrospun conductive nanofibrous scaffolds were created. The scaffolds were created by electrospinning a PANI-CSA and PLA mix in hexafluoroisopropanol. When the PANI level in the PLA polymer was increased from 0 to 3 wt%, the scaffolds had equal fiber diameter but steadily increased conductivity, which might be employed for cardiac tissue engineering.

Designated fibrous structures have been proven to direct myoblasts and neurons. It was also shown that aligned structures can boost cell proliferation and direct cell differentiation. As a result, the aligned conductive nanofibers were created further. Well-ordered conductive PCL/PANI nanofibers, for example, were created. Using the magnetic field-assisted electrospinning method, pure PCL fiber demonstrated non-detectable electrical conductivity, but PCL fibers containing 3 wt% PANI had a conductivity of 63.66.6 mS/cm.

Electrospinning a block copolymer comprising aniline tetramer and PCL resulted in fibrous non-woven scaffolds. The copolymer outperformed the PANI-conductive polymer in terms of processability and biodegradability. The oxidative fibrous scaffold demonstrated fibronectin (Fn) adherence, but the doped fibrous scaffold demonstrated decreased Fn unfolding length on the membranes.

5.2.5 Conducting Hydrogels for Tissue Engineering

Hydrogels are an important type of biomaterials because of their rubbery nature, adjustable characteristics, and great biocompatibility. Conductive hydrogels of various architectures and compositions have been created. Free radical polymerization of acrylic acid, double bond-modified PEDOT, and poly(ethylene glycol) diacrylate resulted in an electroactive poly(ethylene dioxythiophene) (PEDOT)/poly(acrylic acid) (PAA) hydrogel. A hydrophilic PAA network was covalently linked to the double bond-functionalized PEDOT. Furthermore, the covalently crosslinked PEDOT is stable inside the hydrophilic polymer network, and such a conductive hydrogel endows the scaffolds with customized features such as a significant swelling ratio, suitable mechanical properties, and electroactivity.

Injectable hydrogels are attractive biomaterials for cell encapsulation and tissue engineering applications due to their ability to successfully encapsulate cells. Although chemical gelation of injectable conducting hydrogels is routinely utilized to make biomaterials, the hydrogel system frequently contains leftover monomers, initiators, or crosslinkers. Some residual agents are hazardous to cells, such as glutaraldehyde and isocyanate. Physically crosslinked hydrogels, on the other hand, have an advantage over chemically crosslinked hydrogels since they often go through light gelation processing and do not require crosslinking agents. Physical conductive hydrogels were created using a variety of non-covalent interactions. Thermosensitive gelation is the most often utilized method for creating injectable conductive hydrogels. By chemically grafting an electroactive tetraaniline segment onto the chain end of the thermosensitive Pluronic F127 copolymer, an injectable electroactive hydrogel was created. The hydrogel demonstrated excellent electroactivity and cytocompatibility. The PF127-tetraaniline hydrogel also showed improved mechanical capabilities for hydrophobic contacts, stacking, and hydrogen bonding between tetraaniline molecules.

5.2.6 Conductive Polymer Scaffolds for Tissue Engineering

A scaffold is a tissue engineering structural element that acts as a template for the regeneration of functioning tissues and organs. Materials for tissue engineering scaffolds must have the following features: hydrophilicity, biocompatibility, biodegradability, and mechanical properties similar to those of tissues. As a result, CPs are potential scaffolds for tissue engineering to heal or replace injured organs such as skin, tissue, and the spinal cord.

Pure CPs with cytocompatibility, such as PPy, PANI, and PTh, were used as scaffolds because they are capable of adhesion, proliferation, and differentiation of many cell types. These polymers can be chemically or electrochemically produced as films on the surface of an electrode. To improve mechanical qualities, CPs are frequently blended with other biodegradable and more flexible polymers such as poly(lactic acid) (PLA), poly(lactic-co-glycolic acid) (PLGA), and polycaprolactone (PCL).

Another sort of modification that allows CPs to be employed more broadly in tissue engineering is the use of conducting copolymer films. To imitate the natural extracellular matrix (ECM), nanofibrous polymeric scaffolds have been widely investigated. Electrospinning, phase separation, and molecular assembly can all be used to create CP nanofibers. The CPs are combined with biodegradable polymers and electrospun into nanofibers during the electrospinning process. They have a large surface area, porosity, and adjustable sizes in the nanometer range. Scaffolds made by electrospinning conductive nanofibers with PANI combined with PLA or gelatin were produced and showed an electrical conductivity of 4.2×10^{-3} S cm^{-1}, which is significant enough to be used in cardiac tissue engineering. Electroconductive hydrogels are hybrid materials made up of CPs mixed with traditional polymers to form aqueous gel networks. These hydrogels have rubber-like flexibility, tunability, and biocompatibility, all of which are required for usage in scaffolds.

5.2.6.1 CP Scaffolds for Skin Tissue Engineering

Chitin/PANI electrospun nanofibrous scaffolds were discovered to stimulate cell development and dispersion in a bipolar, rather than multipolar, manner. Electrospun nanofibrous scaffolds are composed of camphorsulfonic acid (CPSA)-doped PANI blended with poly(L-lactide-co-Ɛ-caprolactone) (PLCL). PANI electrical conductivity is increased by doping with CPSA. PLCL, on the other hand, is a bioresorbable, elastic polymer that gave the composite nanofibers elastic characteristics. In comparison with the PLCL-only control, the scaffolds were observed to increase cell adhesion of human dermal fibroblasts, murine NIH3T3 fibroblasts, and murine C2C12 myoblasts. Furthermore, the application of external electrical stimulation aided the development of NIH3T3 cells.

By far, the most often reported CP put into skin tissue engineering scaffolds or membranes was PPy. PPy was widely employed in conjunction with lactic acid polymers or its cyclic di-ester, lactide, which has proven popular in tissue engineering because of their bioresorbability, biocompatibility, thermoplasticity, and favorable mechanical characteristics. Under electrical stimulation, a membrane made of PPy and poly(D,L-lactide) (PDLLA) promotes the proliferation of human dermal fibroblasts. Furthermore, after 1000 h of applying a 100 mV DC voltage, this composite material was shown to retain its electrical conductivity at clinically significant levels. Other researchers investigated the impact of functionalizing PPy with diverse elements as well as the introduction of other compounds into PPy films. Collier et al. investigated the capacity of HA-doped PPy films to promote the development of PC12 rat adrenal gland pheochromocytoma cells in vitro [18].

5.2.6.2 CP-Based Nanofiber Scaffolds for Tissue Engineering Applications

Conductive nanofibers were manufactured and evaluated as cell culture scaffolds employing a variety of cell types. Electrically excitable cells such as neuronal cells, cardiomyocytes, and myoblasts have been cultivated on CP nanofibers in several circumstances. The goal of developing electrically conducting nanofibers for tissue engineering applications is to create effective scaffolds capable of supplying topographical and electrical cues at the same time, enabling the regeneration of wounded tissues. As a result, in addition to the manufacture of electrically conducting nanofibers, experimental research into the interacting functions of electrical and topographical signals in cell behaviors must be proven using specific target applications. Surprisingly, increased differentiation of certain cells, such as cardiomyocytes and myoblasts, on electrically conducting substrates was seen in some circumstances even in the absence of external electrical stimulation. These findings might be explained by the induction of electrical connection between cells via conducting substrates, changes in protein adsorption to substrates, and changes in the hydrophilicity/hydrophobicity of the substrate surface. The primary uses of conducting nanofiber-based scaffolds are neural, muscle, and stem cell culturing.

Neural Tissue Engineering

CPs have gained favor as artificial nerve guidance conduits (NGCs) for peripheral nerve regeneration. Several attempts have been undertaken to investigate the applicability of conducting nanofibers to neuronal cells. Cell growth with contact guiding has been shown using CP nanofibers. When developing neural cells atop aligned conducting nanofibers, neurites and axons typically align. On aligned conducting nanofibers, increased neurite outgrowth has also been reported. These findings are obvious because topographical signals are provided by scaffold characteristics rather than chemical properties of the material. PC12 cells grown on PPy-coated PLGA nanofibers showed greater neurite outgrowth on aligned fibers than on random fibers. Electrical stimulation (10 mV/cm) of cells via random and aligned conducting nanofibers increased neurite formation and extension by 40–50% and 40–90%, respectively. For the first time, electrical stimulation has been shown to improve neurite outgrowth on three-dimensional nanofibrous scaffolds. Cells on the aligned fibers also extended further and had longer neurites in the presence of electrical stimulation than cells on the random fibers, indicating that electrical stimulation may be paired with topographical signals. Dorsal root ganglia cultivated on aligned PPy-PCL nanofibers enhanced maximal neurite length by 82% over random fibers. When compared to unstimulated controls with random and aligned fibers, electrical stimulation (10 V) through random and aligned PPy-PCL fibers increased maximal neurite length by 83 and 47%, respectively.

Cardiac Tissue Engineering

Cardiac tissue engineering scaffolds have been created with a variety of material properties, including electrical activity, nanofibrous features, and flexibility. Two-dimensional materials are incapable of providing the matrix porosity essential for medium exchanges and cell infiltration. In the case of nanofibrous scaffolds, fiber orientation, in addition to porosity and large surface area, can create anisotropic cardiomyocyte shape comparable to natural myocardium. For cardiac tissue applications, PPy-contained PCL/gelatin nanofibers containing 15% PPy demonstrated a balance of conductivity, mechanical characteristics, and degradability. These nanofibers benefited primary cardiomyocyte adhesion, proliferation, and differentiation. Cardiomyocytes cultivated on aligned nanofibers of PANI and PLGA produced gap junction proteins (i.e., connexin 4/3) and demonstrated synchronous beatings inside individual cell clusters, indicating the formation of electrical connection among cells. External electrical stimulation might be used to modify the heartbeat frequency. As a result, conducting nanofibers can help in cardiac tissue engineering by facilitating differentiation and synchronous electrical connection of myocardial constructs.

Musculoskeletal Tissue Engineering Applications

Because electrical stimulation of multiple types of cells has been shown to modify cell behaviors, conducting nanofibers can be used for additional tissue regeneration applications (e.g., skeletal muscle, articular cartilage, bone) in addition to neural and cardiac tissue engineering. Increased numbers of sarcomeric myosin-positive cells on PANI/PLCL nanofibers compared to plain PLCL nanofibers in the absence of external electrical stimuli revealed that the material accelerated myoblast differentiation into myotubes, most likely due to greater electrical coupling between cells. Myoblast culture on PCL/PANI nanofiber scaffolds revealed that aligned nanofibers (both conductive and non-conductive) favored myotube development over random nanofibers. Myoblasts on aligned PCL/PANI (conductive) nanofibers displayed higher myogenic differentiation than those on aligned PCL (non-conductive) nanofibers or random PCL/PANI nanofibers, demonstrating that electrical and guiding signals work together to influence myotube formation.

Stem Cell Applications

Nanofibers with particular geometries (i.e., diameters and alignment) can drive stem cell development into specialized lineages. Furthermore, electrical stimulation and/or culture of stem cells on conductive substrates have been shown to boost stem cell differentiation capacity. Electrically conductive fibers have been shown to encourage and enhance stem cell development. When compared to non-conductive PLA scaffolds, PPy-PLA scaffolds increased hASC proliferation and alkaline phosphatase activity (indicative of osteogenic differentiation). In another work, mouse nerve stem cells (C17.2) were cultivated on PANI-PLGA nanofiber scaffolds and electrically stimulated via the scaffolds to promote neurite production and elongation.

5.2.7 Conductive Biomaterials for Various Tissue Engineering Applications

Biomaterials that are developed to physically increase tissue development and certain cell activities are becoming increasingly essential. Conductive matrix materials, which may stimulate electrically responsive cell types such as bone, muscle, nerve, heart, and skin, have been widely employed in the regeneration of various electrical signal-sensitive tissues.

5.2.7.1 Bone Tissue Engineering

Conducting biomaterials can improve MC3T3-E1 cell adhesion and proliferation, as well as osteoblast-like SaOS-2 cells, C2C12 cells, and mesenchymal stem cells. Ice-templated porous and conductive PEDOT:PSS scaffold with median pore diameter greater than 50 μm permitted MC3T3-E1 cell infiltration and matrix deposition inside the void space. Furthermore, the scaffold may boost ALP, COL1, and Runx2 gene expression, improve extracellular matrix mineralization, and promote osteocalcin deposition in MC3T3-E1 cells. These findings show that the conductive PEDOT:PSS scaffold may promote osteogenic precursor cells (MC3T3-E1) to differentiate into osteocalcin-positive osteoblasts.

The cytocompatibility of 3D conductive PLA/PANI composite nanofibrous scaffolds for bone marrow-derived mesenchymal stem cells was good (BMSCs). The conductive nanofibrous scaffolds improved BMSC osteogenic differentiation, which elevated the production of alkaline phosphatase (ALP), osteocalcin (Ocn), and runt-related transcription factor 2 (Runx2), as well as BMSC mineralization.

One advantage of using conductive polymers as scaffolds is the ability to provide electrical stimulation to the substrates. A heparin-doped PPy/PLLA conductive sheet might aid in the adhesion, proliferation, and differentiation of osteoblast-like SaOS-2 cells. Electrical stimulation (ES) increased osteoblast adhesion and development, resulting in considerably greater calcium and phosphate content in mineral deposition compared to the control group. Furthermore, the ES dramatically increased the expression of osteoblast-specific markers (ALP, BMP2, and Runx2).

5.2.7.2 Skeletal Muscle Tissue Engineering

Skeletal muscle tissue engineering frequently depends on in vitro muscle tissue prefabrication by differentiation and maturation of muscle precursor cells or stem cells on a functional scaffold. Poly(L-lactide-co-caprolactone)/PANI fibers were cytocompatible and had a higher percentage of cells positive for sarcomeric myosin than the PLCL fiber group. Furthermore, PLCL/PANI fibers increased myogenin expression above PLCL fibers and significantly improved the expression of troponin T and myosin heavy chain genes, demonstrating that electrically conductive substrates controlled the induction of myoblasts into myotube formation without further electrical stimulation.

The scaffold aligned construction can imitate the anisotropic shape of elongated myofibers in skeletal muscle. Electrospun scaffolds made of aligned nanofibers were discovered to be topographical cues that improve skeletal muscle cell alignment and elongation. As a result, aligned conductive nanofibers for skeletal muscle tissue engineering were created. Conductive PCL/PANI nanofibers having a highly aligned structure, for example, may be able to direct myoblast orientation and promote myotube production better than random fibers. Furthermore, conductive and aligned PCL/PANI nanofibers had much higher myotube maturation than non-conductive and aligned PCL fibers and random conductive PCL/PANI fibers, suggesting that the combined impact of both guiding cues was more effective than an individual cue. To improve the efficiency of skeletal muscle engineering, a number of critical scaffold qualities, such as conductivity, degradability, and elasticity, must be adjusted.

5.2.7.3 Nerve Tissue Engineering

Neurons in the nervous system are electrically excitable cells that rapidly convey messages. Conductive polymers such as PPy and PANI have been created as conductive scaffolds to aid in the formation and regeneration of nerve tissue. For the existence of neurite outgrowth from PC12 cells on the conductive nanofibers, the PCL/PPy conductive nanofibers with high cytocompatibility facilitate PC12 cell differentiation. Furthermore, the PC12 cells covered a much bigger area than the PCL without the PPy covering. These findings showed that conductive nanofiber scaffolds may be used to construct nerve tissue.

Because of their high surface area and ability to increase efficient ion exchange between recording locations and surrounding tissue, CPs are appealing options for brain probes and implanted electrodes. To increase the function of brain probes, PANI was employed as an electrode coating material. When compared to the uncoated Pt electrode, the PANI-coated Pt electrode surface tended to consolidate rather than distribute retinal fragments, indicating a high potential for reducing inflammation and scar formation in long-term implantation. Under electrical stimulation, the PANI coating displayed long-term stability for 6 months. Lacour's group announced the creation of a microfabricated conformable electrode array with tiny (100 μm diameter) electrode sites covered with PEDOT:PSS to provide high charge injection qualities and to safely activate the auditory system with small stimulation sites. These investigations revealed the feasibility of utilizing conducting PEDOT:PSS coatings on tiny electrode locations for electrochemically safe and efficient central auditory system stimulation.

Scaffolds suitable for tissue engineering should be as structurally and biologically similar to natural ECM as feasible. Therefore, nanofibrous scaffolds with nanometer-scale fibers, rather than typical scaffolds, might serve as a possible substrate for cell attachment, function, and proliferation. One of the most essential ways for producing nanofibers is electrospinning. However, the main impediment to the development of electrically conductive polymers has been the difficulties in manufacturing these materials. To address this issue, most researchers have electrospun conductive polymers by combining conductive polymers with other spinnable polymers, degrading the composite fibers conductivity. The characteristics of the resulting nanofibers are affected by the mixing of conductive polymers with other polymers. PANI was combined with a PCL-gelatin solution to create conductive nanofibrous scaffolds appropriate for nerve tissue creation.

5.2.7.4 Cardiac Tissue Engineering

The well-known feature of excitation–contraction coupling of the heart is induced by the coordinated passage of electrical impulses across the cardiac cells. When compared to PCL film, PPy/PCL conductive film could improve the velocity of calcium wave propagation and shorten the calcium transient duration of cardiomyocyte monolayers. They also supported cardiomyocyte adhesion, and more cardiomyocytes on the PPy/PCL film showed peripheral localization of the gap junction protein connexin-43 (CX43), as compared to pure PCL. The gene expression level of CX43, on the other hand, did not change between the two materials.

5.2.7.5 Skin Tissue Engineering

Antibacterial properties were also demonstrated by conductive polymers such as PANI. Because of these benefits, conductive polymers are attractive biomaterials for wound healing applications. The wound healing effectiveness of conductive nanofiber composites based on poly(aniline-co-aminobenzenesulfonic acid), poly(vinyl alcohol), and chitosan oligosaccharide, for example, was investigated using a double full-thickness skin wound model on the dorsum of SD rats. Throughout the test, the conductive nanofiber composite dressing showed less provoking reactions than the control group. After 15 days of treatment, the conductive dressing showed virtually full healing and increased collagen and granulation compared to the control group, indicating that conductive nanofiber composites have a promising use for wound healing.

5.2.8 *Modification of Conductive Polymers for Tissue Engineering Applications*

Despite the fact that conductive polymers have several benefits over other materials due to their electrical characteristics, further optimizing these materials for tissue engineering applications remains difficult. The design of bioactive surfaces is of particular interest in the field of biomaterials because biological systems interact with biomaterials via the interface, and a variety of surface modification and immobilization methods have been developed to create surfaces with bioactive ligands to interact with biomolecules and cells. Several attempts have been made to combine conductive polymer conductivity and biocompatibility, such as doping with biological dopants, incorporation of bioactive molecules for increasing cell adhesion and proliferation, patterning of conductive scaffolds for improving surface topography, and surface modification of conductive polymeric materials with biological moieties.

5.2.8.1 Modification of Conductive Polymers with Bioactive Molecules

The natural extracellular matrix (ECM) molecules are mostly used as dopants with the expectation that the resultant polymer will have a better affinity for cell adhesion molecules. Polymers can be changed for specialized functionality by including proteins, peptides, or ECM components using bioactive compounds as dopants. Conductive polymers are being modified by doping them with bioactive molecules such as heparin, dextran sulfate, hyaluronic acid, chitosan, collagen, growth factors, oligodeoxyguanylic acids, and ATP. However, there are significant limits to biomolecule doping, such as low loading and decreased conductivity, and in induced release scenarios, with polymer reduction, supply becomes limited and release happens faster.

The incorporation of bioactive substances into conductive polymers is regarded as a key method for enhancing cell-tissue interactions. Electrically conductive and physiologically active scaffolds are useful for improving cell adhesion, proliferation, and differentiation, particularly that of neurons. The integration of bioactive molecules into conductive scaffolds, such as neuroactive molecules, has also been used to modify conductive scaffolds. In the electrochemical deposition of PPy and PEDOT, Kim et al. used nerve growth factor (NGF) as a codopant. An in vitro investigation utilizing nerve cells revealed that the addition of NGF can alter the biological interactions of the conductive polymers. Direct covalent attachment of moieties to the backbone of a conductive polymer, on the other hand, generally has a negative influence on the polymer electrical characteristics.

5.2.8.2 Modification of Conductive Materials with Biopolymers

Hybrid material fabrication has also been researched as a means for changing the characteristics of conductive polymers. Blending electroactive polymers with natural polymers typically improves their biocompatibility. Chitosan-PPy hybrids, collagen-PPy hybrids, and a combination of poly(-caprolactone) (PCL), gelatin, and PANI were used to create conductive nanofibrous scaffolds for nerve tissue creation. This scaffold benefited from the use of both a synthetic polymer with a PCL domain that gives mechanical strength and a natural polymer (gelatin) that promotes cell adhesion and proliferation.

5.2.8.3 Topographical Modification of Conductive Scaffolds

The topographical characteristics of a scaffold have a substantial influence on tissue engineering, and surface roughness of scaffolds has been shown to alter cell behavior, including cell attachment, proliferation, and differentiation. Furthermore, topographical characteristics are characterized as having a favorable influence on axonal orientation, where physical guiding of axons is an important component of nerve healing. Gomez et al. also electrochemically synthesized PPy microchannels to create electro-conductive, topographical substrates for neural interfacing and discovered that PPy microchannels aided axonal development of rat embryonic hippocampus neurons [5].

5.2.9 Biomimetic Conductive Polymer-Based Tissue Scaffolds

It is currently difficult to produce therapeutically appropriate CP-based tissue scaffolds with biomimetic chemical, mechanical, and topological features (as shown in Fig. 5.2).

5.2.9.1 CP-Based Tissue Scaffolds with Biomimetic Chemical Properties

ECM, a protein-polysaccharide mixture, exhibits biochemical signals that regulate cell activity and control how easily cells adhere to them via interactions with glycoproteins on the cell surface. Integrins are a kind of cell adhesive glycoprotein that recognizes certain peptide sequences in ECM proteins such as collagen, fibronectin, laminin, and vitronectin. Cell adhesive proteins and peptides are widely found in biomimetic biomaterials intended for use as tissue scaffolds.

A number of methods may be used to create CP-based materials having ECM-mimetic chemical characteristics. It is possible to non-covalently incorporate both

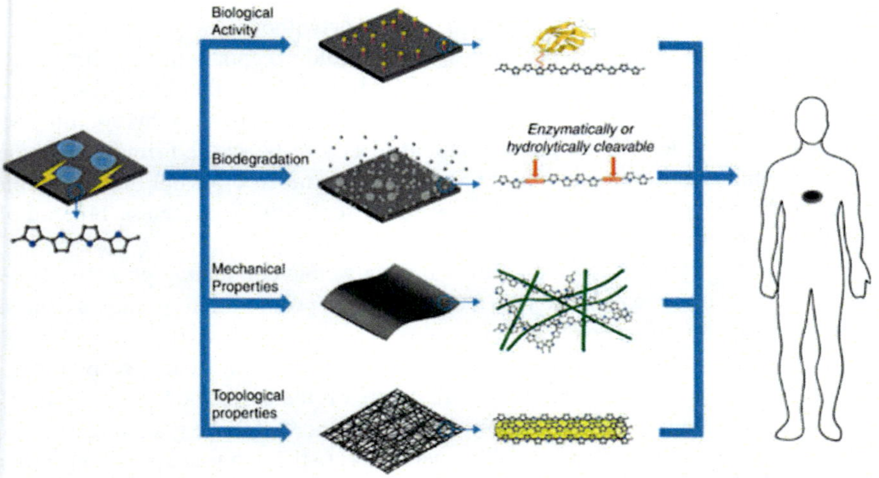

Fig. 5.2 Schematic illustrating the most important aspects considered when designing biomimetic conductive polymer-based materials. Reprinted from ref. [8], copyright 2013, with permission from Elsevier

high and low molecular weight ECM components/derivatives as dopants during electropolymerization reactions. For example, doping with collagen or low molecular weight peptides derived from laminin improved PC12 cell adhesion to poly(3,4-ethylenedioxythiophene) (PEDOT) films. By polymerizing monomer functionalized ECM derivatives, electropolymerization also has the potential to covalently integrate ECM-derived dopants (e.g., collagen or hyaluronic acid). Although electropolymerization is commonly employed to create thin 2D films of CPs, it is also applicable to 3D substrates such as interpenetrating networks of PEDOT and ECM protein-based foams (produced from decellularized tissues) and, more excitingly, the in vivo creation of PEDOT hydrogels.

ECM-mimetic properties can also be imparted to materials by covalently modifying their surfaces with ECM derivatives, most commonly using carbodiimide chemistry, though there are a variety of other methodologies available, some of which are capable of producing surfaces that can spatially control cellular interactions via ECM derivative functionalization (i.e., printing patterned surfaces). Furthermore, non-covalent interactions can be utilized to change surfaces, often depending on protein adsorption via non-specific interactions. However, phage display can be used to find peptides that engage specifically with CP substrates (e.g., THRTSTLDYFVI with polypyrrole). These peptides can be changed to produce physiologically active peptides, such as the IKVAV peptide produced from laminin.

Because the ECM is intrinsically biodegradable and undergoes substantial modification throughout the normal wound healing process, degradable CP-based tissue scaffolds are preferable because they allow for ultimate replacement with natural functioning tissue. Bioerodible materials are those that degrade by the solubilization of an originally water insoluble polymer (with or without changes in their chemical structure) under the molecular weight threshold acceptable for renal filtration (<50 kDa). Biodegradable polymer-based materials, on the other hand, disintegrate owing to scission of the polymer backbone (e.g., enzymatic or hydrolytic bond cleavage). The deterioration process should ideally result in the formation of non-toxic components. Commonly used CPs are neither bioerodible nor biodegradable, making them unsuitable for clinical use as tissue scaffolds. Nonetheless, bioerodible or biodegradable CP-based scaffolds may be prepared.

The Schmidt group reported entirely biodegradable conducting polyester-based scaffolds that are acceptable for the attachment and proliferation of human neuroblastoma cells in vitro, as well as being non-immunogenic in vivo in rats (although in the undoped condition). Biodegradable CP-based scaffolds can also be created by grafting water-soluble conducting oligomers on the surface of a biodegradable scaffold. Several groups, most notably Albertsson and Wei, have grafted conducting aniline oligomers (most often an aniline pentamer) to the surfaces of a wide range of biodegradable polymers of natural and synthetic origin. Although studies have clearly proven that the aniline oligomers generated during biodegradation are hazardous, these systems tend to exhibit modest levels of cytotoxicity in vitro. Electrical stimulation of PC12 cells attached to oligoaniline-modified polylactide films enhanced neurite extension in vitro, and electrical stimulation of preosteoblastic MC3T3-E1 cells attached to oligoaniline-modified poly(ester amide) copolymer films increased levels of intracellular free Ca^{2+} ions and alkaline phosphatase activity in vitro, which are early markers of osteogenic differentiation. The Langer group has revealed the construction of scaffolds made of naturally occurring melanins, which is an elegant alternative (a class of electrically conducting biopolymers). They were shown to be appropriate for the adhesion and proliferation of Schwann cells and PC12 cells in vitro. Surprisingly, melanin films implanted in rats were slowly bioresorbable (>2 months), and the authors hypothesized that the film fragility led to breakage after implantation, followed by absorption by macrophages and giant cells.

The Langer group also published the first bioerodible CP-based scaffolds, which were made of a sparsely water-soluble self-doped CP (pyrrole-4-butyric acid). The scaffolds were demonstrated to erode over weeks at physiological pH levels due to slow polymer breakdown and to be appropriate for the attachment and proliferation of human mesenchymal progenitor cells in vitro. Following that, the Wallace group published erodible multilayer films made of self-doped polyanionic polythiophene and polycationic polyethyleneimine. The films were demonstrated to degrade over a three-month period and to be appropriate for in vitro adhesion and proliferation of L929 and C2C12 cells.

5.2.9.2 CP-Based Tissue Scaffolds with Biomimetic Mechanical Properties

Biological tissues have distinct mechanical characteristics, and mechanical inputs are known to impact cellular activity via a range of mechanisms collectively referred to as mechanotransduction. Mismatches in the mechanical characteristics of a tissue scaffold and the tissue in which it is implanted can cause inflammation of the surrounding tissue, followed by encapsulation of the implanted scaffold inside an avascular network of fibrous tissue. As a result, the development of CP-based materials with biomimetic mechanical characteristics is of particular interest.

Because polymers have limited conformational flexibility in 3D, materials formed only of CPs are very inelastic. As a result, electropolymerized films tear readily. This is an important issue since the handling qualities of biomedical products are critical for their effective translation from the laboratory to the clinic. Flexible CP-based tissue scaffolds are particularly appealing for muscular (and maybe cardiac) tissue engineering. By spreading a significant amount of conductive filler (e.g., CP nanoparticles) inside an elastomeric matrix, such as polycaprolactone or polyurethane, C2C12 myoblasts were found to adhere, proliferate, and differentiate into myotubes in vitro. Multiblock copolymers made of alternating blocks of conducting and elastomeric blocks can also be used to create flexible CP-based tissue scaffolds. Polypyrrole and polycaprolactone, for example, have been demonstrated to adhere and proliferate PC12 cells, and electrical stimulation has been proven to promote neurite extension in vitro.

Because of their high water content, porosity, and mechanical qualities similar to soft tissues, electrically conductive hydrogels are particularly appealing as tissue engineering scaffolds (typically ranging from sub-kPa to hundreds of kPa). Guiseppi-Elie, Martin, Poole-Warren, Wallace, and Yaszemski's groups have produced notable electroconductive hydrogels. Conducting hydrogels composed of polypyrrole and photocrosslinked oligo(-polyethylene glycol) fumarate were shown to be suitable for PC12 cells to adhere and extend neurites into the scaffold in vitro, and biodegradable conducting hydrogels formed by crosslinking poly(3-thiophene acetic acid) with carbonyldiimidazole were shown to be suitable for the adhesion and proliferation of C2C12 myoblast cells in vitro.

5.2.9.3 CP-Based Tissue Scaffolds with Biomimetic Topological Properties

Natural tissues are three-dimensional (3D) composite materials having distinct topological features that are critical for their function. Anisotropic properties are frequently detected in functional tissues (including cardiac, ligament, musculoskeletal, neurological, and vascular tissues), generally in the form of anisotropically dispersed extracellular matrix components that impact the alignment, shape, and behavior of resident cells. In recent years, there has been a great deal of interest in the creation of tissue scaffolds that imitate such finely organized natural tissues, and a variety of CP-based tissue scaffolds with biomimetic topological features (e.g., aligned nanofibers) have been studied.

Electrospinning is a prominent technique for creating nanofibrous tissue scaffolds with variable fiber orientation. Electrospinning CP-based composites is a straightforward method for producing electrically conductive nanofibrous tissue. Nanofibers made of polyaniline and gelatin, for example, have been found to aid in the adhesion and proliferation of cardiomyocytes in vitro. The addition of polycaprolactone into analogously spun fibers improved their mechanical qualities considerably, and electrical stimulation of the scaffolds improved the proliferation and neurite outgrowth of neural stem cells cultivated on them in vitro. Similarly, nanofibers made of polyaniline and either poly(L-lactide-co-ecaprolactone) or polycaprolactone were shown to be appropriate for fibroblast and myoblast attachment and proliferation in vitro.

Another straightforward way for producing electrically conducting nanofibrous tissue scaffolds is to chemically alter the surface of non-conductive fibers using CPs. For example, it is possible to coat the surface of non-conductive polymer (e.g., poly(lactic acid-co-glycolic acid)) nanofibers deposited on the surface of an electrode with CPs (e.g., PEDOT) via electropolymerization or vapor phase polymerization of CPs from fibers containing an initiator (e.g., iron(III) p-toluenesulfonate) when exposed to a suitable monomer (e.g., pyrrole), or indeed, the bulk polymerization of CPs (e.g., polypyrrole) in solution in the presence of nanofibers. Interestingly, electrical stimulation of PC12 cells cultivated on such scaffolds was shown to boost neurite outgrowth in vitro, while immobilization of nerve growth factor on the surface of conducting nanofibers improves neurite outgrowth even more. It is widely assumed that 2- and 3-dimensional printing will play an increasingly essential role in the future development of CP-based tissue scaffolds with biomimetic topological features. Excitingly, it has already been demonstrated that CP-based composites comprising biopolymers (e.g., chitosan, hyaluronic acid, or collagen) can be printed by extrusion or ink-jet printing and that a submicrometer level pattern can be applied to them using a low-energy infrared laser. Printed polypyrrole and collagen composites were found to be suitable for PC12 cell attachment and proliferation, and electrical stimulation of cells cultured on micrometer-scale lines was found to enhance neurite outgrowth and orientation (preferentially along the long axis of the printed lines) in vitro.

5.3 Artificial Muscles: State of the Art

Electric pulses, polymeric chains, aqueous solutions, volume fluctuations, and strain and stress changes must all be present in artificial devices attempting to replicate muscles. Artificial muscle growth necessitates volume or dimensional differences in films. During oxidation/reduction processes caused by subjecting the film to potential sweeps, square potential steps, or square current steps, length fluctuations of films were observed under continuous mechanical stress. Polymeric actuators are macroscopic or microscopic devices, or polymer chains (molecular motors), that interact with polymer films to convert energy from various sources (electric fields, electrical currents, light, temperature, pressure, and so on) into macroscopic mechanical energy. It is worth noting that the majority of polymeric actuators (electromechanical, chemomechanical, photomechanical, thermomechanical, and baromechanical) are referred to as artificial muscles in the literature.

5.3.1 *Electrochemomechanical and Electromechanical Muscles: Electroactive Polymer Actuators*

Electromechanical actuators are artificial capacitive muscles that respond to electric fields, E, and are composed of electroactive and electronic insulator polymers, polymer solvated salts (salt + polymer), or gels (polymer + solvent + salt) that respond to high electric fields (potential gradients). The dimension variations are proportional to E^2 in electrostrictive actuators or to E in piezoelectric actuators, ferroelectric actuators, electrostatic actuators (including ions), and electrokinetic (electroosmotic) actuators (including ions and solvent). During actuation, the polymer chains do not engage in chemical processes.

Electrochemomechanical actuators are dimensional changes controlled by an electrochemical reaction (low potential gradient, high current) and proportionate to the spent charge Q. Electrochemomechanical actuators may be made from any redox material, organic or inorganic, that can be laminated or incorporated in electrode films and has redox sites available for oxidation/reduction processes producing ionic exchange for charge balancing. Electromechanical and electrochemomechanical actuators can also be distinguished by the presence of reactive or non-reactive electrodes in the actuator construction. Natural muscles are outperformed by polymeric actuators in terms of elasticity, strain and stress generation, controllability, and molecular interactions. Natural muscles are now the champions in terms of energy conversion efficiency, hierarchical structure (from molecular to macroscopic three dimensional), and durability.

5.3.2 *Bending and Linear Electrochemomechanical Artificial Muscles*

Electrochemomechanical artificial muscles serve as actuators, converting electrical energy into macroscopic mechanical motions and mechanical work via electrochemical reactions in polymer films. The volume change of the CPs as a result of the electrochemomechanical reactions paves the path for their use as actuators in artificial muscle applications. Tuning the applied voltage controls the volumetric expansion of a CP in electrolyte. Because of the ionic crosslink at the polaron site and the delocalization of the π-electron, the oxidized state is stiffer than the reduced state. The size of the anions utilized in the process influences whether the electrochemomechanical reaction is anion or cation driven. Small anions, such as ClO_4^- and bis-(trifluoromethanesulfonyl)imide (TSFI), emerge from the bulk CPs during reduction, shrinking the reduced state, a process known as anion drive. When big anions, such as dodecylbenzenesulfonic (DBS) acid, are employed, the lengthy alkyl chains of the anions create tangling and immobilization in the network. The CPs are subsequently reduced by the cation drive, in which cations supplied from the electrolyte solution cause a swelling polymer network due to the volume of extra cations.

5.3.2.1 Bending Artificial Muscles

The volumetric change of the CPs in actuators can be used to generate a linear or rotational movement. A bilayer structure made of a CP film grown on a metallic electrode and a non-conductive coating is one of the most often employed designs. The gadget shows macroscopic bending behaviors as a result of swelling or shrinking caused by an electrochemical stress gradient across their contact. Specifically, following oxidation, the anion-driven device bends toward the CP-convex shape due to swelling of the CP. The cation drive, on the other hand, causes the device to bend in a CP-concave fashion when they decrease due to oxidation. The employment of a metallic counter electrode in a bilayer actuator is still inevitable. However, it frequently results in actuating film degeneration due to corrosion of the metal electrode exposed to a hostile environment (i.e., a sudden change in pH and chemical).

To facilitate current flow, all bilayer devices require a metallic counter electrode. A significant portion of the consumed electrical energy is squandered in order to create the solvent discharge, which necessitates a high over-potential and hence a significant portion of the used electrical energy. Furthermore, these interactions cause pH changes and the formation of new chemicals, which move to the muscle, supporting the continuous degeneration of the actuating layer.

From the bilayer, a three-layer arrangement CP/tape/CP arises, attempting to eliminate the metallic counter electrode (Fig. 5.3), gaining greater efficiencies by employing the same current twice to create opposing electrochemical reactions and volume changes in CP films. By separating the two CP films by an ionic conducting membrane, three-layer devices may move outside of a liquid electrolyte medium. This membrane may be formed by solvent evaporation and UV irradiation, trated networks. The two CP films are generated by chemical polymerization on the exterior surface of the membrane. Multilayer devices have also been built and researched.

In order to eliminate the metallic counter electrode, a trilayer actuator was created by attaching two CP films to double-sided tape. When the trilayer structure is submerged in an electrolyte, two opposing volume changes occur as one of the CP films suffers anion-driven swelling and the other experiences cation-driven shrinkage. The fundamental benefit of using CPs as actuators for artificial muscles is that they have a low operating voltage (as low as 2 ~ 10 V). Furthermore, they produce greater stresses at a lower cost than carbon nanotubes. However, the electromechanical coupling, or the efficiency of converting electrical energy to mechanical energy, is not as great as in carbon nanotubes. As a result, the use of the CPs in a large-scale actuator has been limited.

5.3.2.2 Artificial Muscles Giving Linear Movements

The simplest linear actuator is a free-standing CP film or threads. Depositing platinum on the CP was proposed to boost conductivity but generated brittle films. The strategy of Prof. Kaneto's group was to synthesize CP in the presence of large anions in order to produce large volume and length changes in multilayered devices, in which several

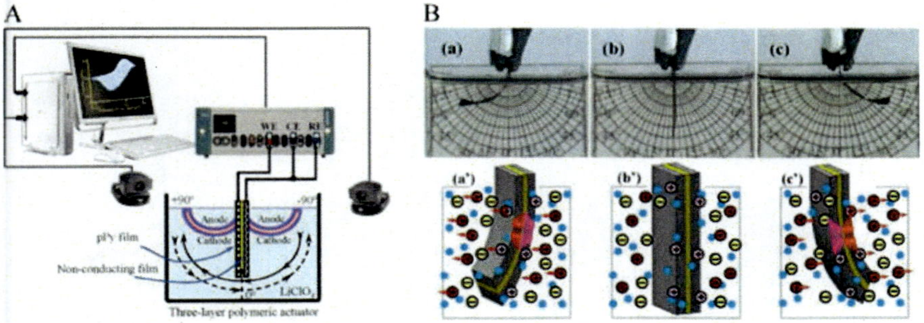

Fig. 5.3 **A** Scheme of the electrochemical cell and configuration of electrodes (WE, working electrode; CE, counter electrode; RE, reference electrode) used to study the electrochemical actuation and sensing behavior of the triple-layer actuators (conductive polymer film/tape/conductive polymer film). A vision system was used for film recording and to quantify the angular movement. **B** Anticlockwise, from **a** to **c**, or clockwise, from **c** to **a**, bending movements and concomitant ionic exchanges and stress gradients, **c′** and **a′** from a triple-layer PPy-DBS/Tape/PPy-DBS in 0.1 M $LiClO_4$ aqueous electrolyte. Reprinted from ref. [4], copyright 2011, with permission from Royal Society of Chemistry

thin laminations of CP films and an electrolyte (ionic liquid-soaked paper) are used to develop a compact and scalable linear CP actuator with a high work output. Linear motions are also provided by folded films with Origami forms. The electrolytic medium separated the CP films, allowing them to move in air. CP microrods and nanorods have also been tested. Weight vertical displacements were produced using bundles of films or fibers.

Another method for obtaining a linear displacement is to combine various bending structures, such as bilayers or trilayers, to achieve longitudinal displacements greater than 60% of their original length. These combinations necessitate that the bending angles of the individual layers be less than 20%, resulting in minimal mechanical fatigue and increased longevity. While operating, reactive artificial muscles will detect any change in a physical or chemical variable acting on the reaction rate (electrolyte concentration, working temperature, driving currents, or masses linked to the muscle) (Fig. 5.4). The sensing magnitudes are the muscle potential or the consumed energy.

5.3.2.3 Tactile Muscles

The length increment in linear muscles or the angle (rad) represented by bending muscle is proportional to the spent charge Q, according to the electrochemical basis of such changes (C or As). The constant k (rad s^{-1} A^{-1}) is a property of the CP, and given the same CP, it is a property of the type of the exchanged ion (electrolyte).

$$\alpha(\text{rad}) = k \cdot Q \quad (5.1)$$

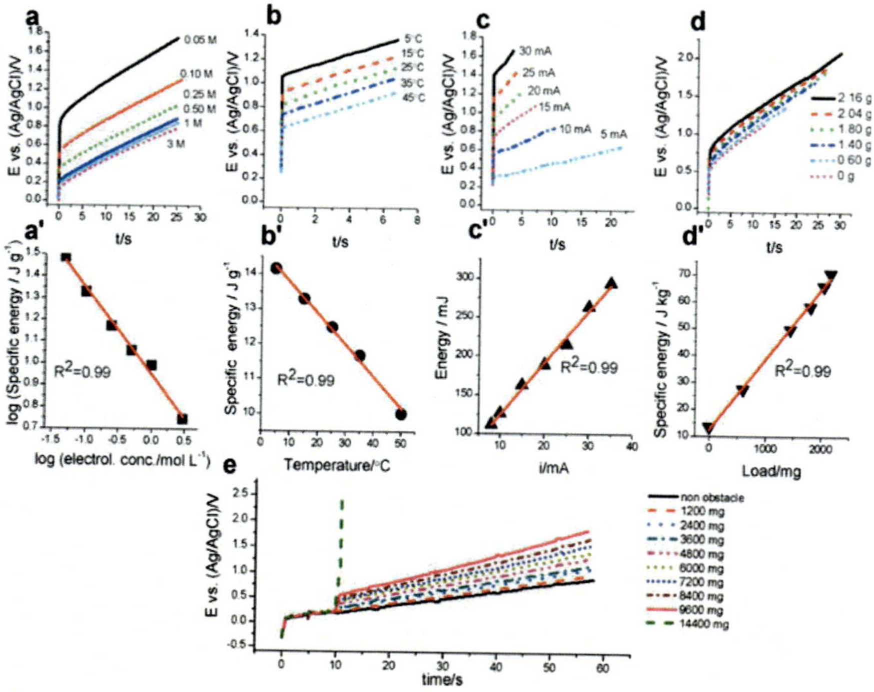

Fig. 5.4 Artificial muscles describing the same angular movement sense working conditions: chronopotentiometric responses, **a** from different concentrations of the electrolyte, **b** at different temperatures, **c** by flow of different currents, and **d** shifting different steel masses (the device contains 12 mg of PPy). Consumed electrical energy by the device as a function of the different studied variables: **a′** electrolyte concentration, **b′** temperature, **c′** current, and **d′** shifted masses. Tactile muscle (e): the muscle moves freely, meeting an obstacle after 10 s and sliding it. Obstacles weighing 1.2, 2.4, 3.6, 4.8, 6.0, 7.2, 8.4, and 9.6 mg were slided. The muscle was unable to push and slide an obstacle weighing 14.4 g. Reprinted from ref. [14], copyright 2012, with permission from Elsevier

The rate of movement is a linear function of (controlled by) the applied current, I; similarly, the angular rate $\omega(\text{rad s}^{-1})$ is a linear function of (controlled by) the applied current, I.

$$\omega\left(\text{rad } s^{-1}\right) = \frac{\alpha}{t} = \frac{kQ}{t} = kI \tag{5.2}$$

By changing the current, I, by the specific current, i (Ag^{-1}), from Eqs. (5.1) and (5.2), where the new dynamic constant, k' (rad $g\ s^{-1}\ A^{-1}$) is a general constant of the system CP/electrolyte. When the reactive polymer consumes one unit of charge per unit of CP mass, this angle is specified.

$$\omega\left(\text{rad } s^{-1} g^{-1}\right) = k'i \tag{5.3}$$

The muscle potential evolves at increasing potentials for a muscle operating under the flow of a constant current, defining the same angle, and following increasing weights. If the muscle comes into touch with an object at a given angle, it senses mechanical resistance at the point of contact and the potential increases in order to create the additional energy ($I \cdot \Delta E$) necessary to push and shift the impediment. The potential step is proportional to the weight of the obstacle: this is a tactile muscle that provides information about the contact moment, the extra energy necessary to shift the impediment, or whether the muscle is unable to move it (vertical step of the potential for numerous obstacles). Artificial muscles detecting working circumstances and the faradic equations (5.1–5.3) provide a solution to the unresolved technical challenges of soft robot activeness and deformability. Deformable and reactive muscle gels serve as both acting and sensing devices in this case. Uncertainties associated with the effect of the environment on gel actuators and gel robots do not present for CP sensing motors.

5.4 Computer/Neuron Dialog: Artificial Synapses

Artificial synapses are mostly used in the literature to refer to simulations of natural neural pulses using dry semiconductor circuits and mathematical models. We are referring to genuine moist synaptic space between neurons as well as any attempt to replicate its physical and chemical features. When one of the cations or anions involved in neuron pulse transmission (Cl^-, Na^+, K^+, Ca^{2+}, glutamate, aspartate, D-serine, γ-aminobutyric acid (GABA), or other neurotransmitters) is stored in a polymeric film, direct contact with neurons is possible. The CP, which receives an electric pulse and liberates ions in the synaptic region, replaces the axon of the talking neuron. The rise in ionic potential acts on the electromechanical protein of this ion channel, which is situated at the neuron dendrite, and the ionic pulse enters the neuron, resulting in a neural pulse. A nervous pulse was produced as a result of the electric stimulus.

In this approach, the biocompatibility of most conductive polymers allows for the development, adhesion, and proliferation of glial, neurite, or neuron cells in vitro. Coated electrode in vivo activity for recording cortical activity, retinal or cochlear implants, or neuron stimulation and differentiation has become standard clinical practice. Some preliminary research on ionic calcium transducers (or calcium pumps) has recently been published. There are high hopes for medical and biological breakthroughs based on conductive polymers in a variety of new neurological fields.

The culture of neurons around microelectrodes coated with conductive polymers that store different neurotransmitters can allow for the liberation of this neurotransmitter under faradic control. Cl^-, Na^+, or Ca^{2+} ions are the most basic neurotransmitters. The liberation of those ions close to a dendrite end promotes their potential increase in solution (Nerst Eq.), and above the threshold potential, the electromechanical protein of the concomitant ion channel opens, allowing ionic flow inside the cell and connecting the computer to the cell.

5.5 Conductive Polymers in Bioelectronics

Bioelectronics is the integration of electrical devices with biological systems for the monitoring and control of biological processes. Bioelectronics includes cardiac pacemakers and brain implants used to treat arrhythmia, epilepsy, and Parkinson's disease, as well as devices that stimulate the vagus nerve to treat inflammation. Retinal and cochlear implants are used to activate the retina and restore hearing capabilities. Bioelectronic devices, as opposed to traditional pharmacological approaches, can enable tailored therapeutic therapy and localized intervention [10].

5.5.1 Molecular Bioelectronics

The combined electronic-ionic transport of CPs is used in organic bioelectronic electrodes and devices to transform ionic processes happening in aqueous solutions, cells, and tissues to electronic transport in human-made technology and vice versa. Researchers intentionally created synthetic electrogenic molecules, such as CPs, to interact or integrate into the cell membrane in order to influence existing cellular processes (such as electronic and ionic transport) or to endow biomembranes with novel capabilities. The chemical structure of CPs determines their specific function and ability to integrate into membranes.

5.5.2 Selected Application of CPs in Bioelectronics

Electrodes covered with doped CPs are a popular biointerfacing application. They are mostly employed to record local changes in potential and are hence non-specific bioelectronic sensors, while they have been used to record potential dependent oxidation/reduction of biomolecular species (i.e., fast-scan cyclic voltammetry). Their usefulness, ease of deposition, facile functionalization, and low impedance have enabled high-quality recordings of electrophysiological signals with high signal-to-noise as microelectrodes, as well as small-footprint, high current injection efficiency for electrical stimulation. Unlike electrodes, which are considered as passive components in biosensors, transistors have been targeted as a technique to locally amplify and transduce biological signals. These devices are made up of a channel (either intrinsic or doped semiconductor) through which a source-drain current flows.

When the channel comes into direct touch with an externally gated sensing environment, the transistor is said to be "electrolyte-gated." While inspired by typical ion-selective or biosensing inorganic transistors, these transistors with an organic channel are divided into two categories: electrolyte-gated organic field-effect transistors (EGOFETs) and organic field-effect transistors (OECTs). EGOFETs are typically made from more traditional organic electronic polymers and small molecules, and they are functionalized either at the gate electrode–electrolyte interface, the channel-electrolyte interface, or buried beneath the channel material, resulting in multiple modes of functionalization and sensitivity. While EGOFETs rely on charge buildup at the electrolyte channel interface, OECTs work by using bulk variations in channel conductivity.

5.6 Conductive Polymers for Drug Delivery

Drug distribution is a critical aspect of the healthcare industry. The three major aims of drug delivery strategy research are to enhance patient compliance, boost treatment efficacy, and reduce undesirable side effects of drugs delivered. The most recent medication delivery research has focused on the creation of micro- and nanoparticles for passive and active targeting when administered, most commonly intravenously. Alternatively, drug delivery incorporated with conductive polymers and CP-based electronic devices opens up several exciting possibilities, including the ability to deliver drugs without the use of a solvent ("dry" delivery), to deliver drugs locally with implantable devices, to electronically control release by adjustment outside the body, to release in response to a stimulus (via a coupled electrode or sensor), and to provide personalized and autonomous treatment (as part of closed loop systems).

When conductive polymer films oxidize, their positive charge is connected with counter ion flow from solution into polymer to compensate for the charge surplus, resulting in an increase in film volume. The counter ions are released from the film during the reduction state, causing the film to shrink. These features can be used to medication delivery systems, in which medicines are integrated into the polymer film during oxidation and released when the film is reduced.

Polymers can be generated chemically or electrochemically from an aqueous solution containing monomers, and the medicine can be included during the polymerization process. Various types or numbers of pharmaceuticals and compounds, whether anionic, cationic, or neutral ions, can be included into the polymer backbone. By combining conductive polymers with other materials and nanostructures, such as titanium and carbon nanotubes, the surface area of the films for storage may be increased. Surfaces and compositions with variable mechanical and electrical properties can be created by adjusting the electropolymerization settings. Polymer device miniaturization can also be utilized to include pharmaceuticals.

Conductive polymers go through a reversible redox process that causes ion movement in and out of the polymer bulk. To release or capture ions, a potential difference of less than 1 V is typically required between the polymer film and the electrolyte, depending on the ambient circumstances. Conductive polymers can function in a wide temperature range in a liquid electrolyte or in air using a polymer electrolyte. Polypyrrole, polythiophene, polyaniline, and its derivatives are widely considered in the literature to have some biocompatibility with live body tissues and fluids. Long-term (90-day) in vitro and in vivo tests revealed no signs of toxicity or immunological issues. However, it has been stated in the literature that the polyaniline exhibited considerable cytotoxicity during in vivo experiments.

A conductive polymer release system may be generally categorized into numerous categories based on the parameters that govern drug release. The drug is released chemically in the first kind by utilizing a redox reagent that is thermodynamically capable of reducing the oxidized conductive polymer or by elevating the pH. Theoretically, the conductive polymers can selectively sense the redox reagent in the solution; concurrently, the redox reagent initiates drug release from the conductive polymer, and the drug is released as a function of the concentration of the detected redox reagent where ionic exchange happened. Pernat et al. employed hydrazine N_2H_4 as a powerful reducing agent at pH = 12 to liberate ATP from the PPy film. The amount released was 70 nmol cm^{-2}, which is ≈80% less than the amount produced by electrochemical stimulation of the same film at N_2H_4 concentrations up to 10 M. The authors speculate that the low porosity of the film prevents the hydrazine molecule from penetrating and reducing the PPy film. Although hydrazine is employed as a pharmaceutical intermediate, it is poisonous and unstable at ambient temperature. Another example is dithiothreitol (DTT), as a powerful reducing agent, which failed to release measurable ATP from the PPy film after an hour of exposure. Furthermore, by lowering the pH, the integrated medicine might be freed. This results in chemical deprotonation of the conductive polymer and drug diffusion out of the film in the treated areas of the body. Pernaut et al., for example, used an alkaline NaOH solution to release the ATP drug anions from the PPy. The alkaline solution deprotonated and decreased the conductive polymer, causing ATP ions to be ejected and hydrated sodium cations incorporated. The pH change, on the other hand, has a greater release rate but produces ≈60% less than electrochemical stimulation.

The molecules are supplied in the second type of release mechanism by providing an external electrical potential to oxidize and/or decrease the films. In general, two types of stimulation protocols, step potential and cyclic potential, can be utilized. By introducing a negative potential, the polymer film is reduced and its cationic charge is neutralized, resulting in the electrostatic ejection of the anionic drug synchronized with the entry of hydrated cations into the polymer bulk. This causes the film to bulge. However, exposing the film to a negative stimulus potential for an extended amount of time might cause the film to lose electrical conductivity, which is not always recoverable. Cycling the potential causes hydrated ions to move in and out of the conductive polymer, generating expansion and contraction that forces medicines out of the film.

When compared to step potential, cyclic stimulation is possibly more effective and capable of releasing a greater amount of medication. Nevertheless, cyclic potential exposes the film to physical stress as a consequence of swelling and contraction, producing delamination, cracks, and breakdowns. It has been observed, for example, that when cyclic potential stimulation was utilized to produce neurotrophin-3 (NT-3), the PPy film began to delaminate after 12 min.

Despite the greater flexibility and control given by the electrochemical release approach over the chemical one, the necessity for an external power supply prohibits its widespread application in in vivo release systems. Some drug delivery systems do not require external power sources and may release the medication through chemical, pH, and temperature changes, although such systems are not without flaws. The amount of medicine released, for example, is less than that released by the electrochemical approach.

One issue with these systems is that the medicine is released spontaneously, with little or no control over the process. To address this issue, a layered system of conductive polymers has been developed. The multilayer polymer films are made up of independent conducting layers, each with its own redox potential. For example, Fig. 5.5 depicts a bilayer conductive polymer system in which the first layer of conductive polymer (CP1) is electrodeposited on the electrode surface from a solution containing the monomer and the anionic drug A. On top of the CP1 surface, the second layer CP2, with a greater redox potential than the CP1 layer, is electrodeposited. When this bilayer system is employed, there is no spontaneous drug release from the system. Between the release medium and the CP1 layer, the unstimulated CP2 layer functions as a protective layer. The integrated A drug is released into the medium once the bilayer system has been totally reduced. Furthermore, the CP2 may be doped with different medications, transforming the device into a dual-drug delivery system. It is important to note that the oxidation potential of the first layer should be lower than that of the next layer; otherwise, the CP1 would function as an insulator before it can oxidize, preventing electropolymerization of the second layer. Furthermore, the oxidation potential of the second layer should be in a range that does not promote over-oxidation of the first layer, which might result in a loss of conductivity. If the system has more than two layers, this condition also applies to the other layers [2].

In drug delivery applications, CPs are electrically stimulated to release a variety of therapeutic proteins and medicines. Because of the controlled and reversible redox processes that enable drug delivery, PPy is one of the most intensively investigated CPs for drug delivery. The rate of drug release in CPs may be accurately regulated by changing their redox state, which results in changes in polymer charge, conductivity, and volume. In particular, as shown in Fig. 5.6, the process of drug release by CPs is similar to that of bioactuators. Anionic drugs prepared with small anions leach out as the volume of the CP decreases during reduction, whereas cationic drugs prepared with immobile anions are released by cation-driven actuation when the CPs are oxidized.

5.6.1 Neuromodulatory Ions and Neurotransmitters

As a supplementary dopant, physiologically active molecules in their ionic form, such as neurotransmitters, can be electrostatically integrated into CP electrodes. When electrical stimulation is provided, the CP transitions from a doped to a neutral

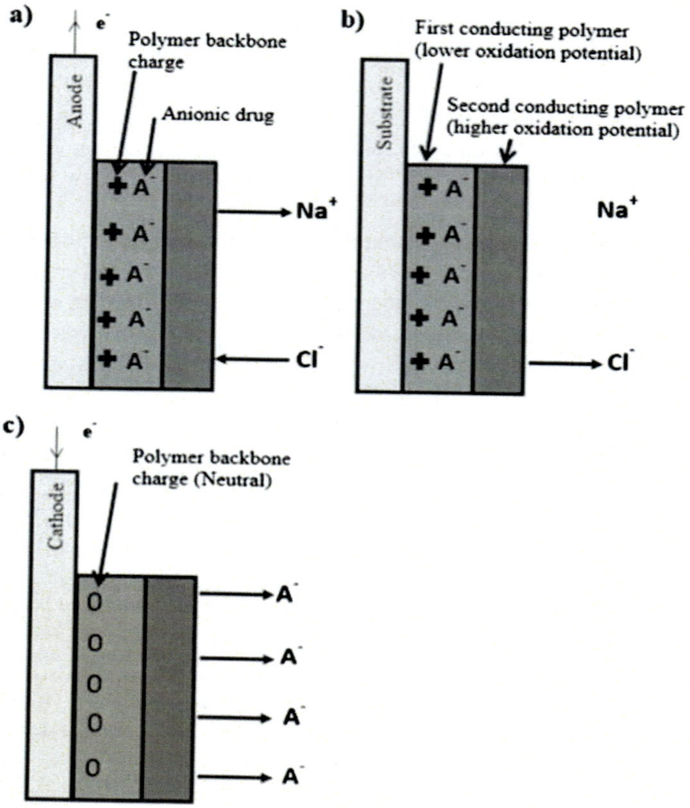

Fig. 5.5 Film consistent of bilayer conductive polymers for drug delivery system (dual ions transport). Reprinted from ref. [16], copyright 2015, with permission from Springer Nature

state, releasing these ions. This method enables neural recording and stimulation electrodes to also transport drugs.

In addition to doping with the desired substance, CPs can be loaded with the drug via entrapment. Supercritical carbon dioxide treatment of electropolymerized PEDOT:PSS films was utilized to imbue the film with the cation acetylcholine. Calcium responses in SH-SY5Y cells demonstrated that acetylcholine produced from electrical stimulation maintained its activity. Polymer swelling and hydration during oxidation and reduction were responsible for the release mechanism.

Beyond doping and trapping, the mixed conductivity of PEDOT:PSS has been used for ion delivery using electrophoretic ion pumps. The ion pump device is made up of source and target conductive polymer (PEDOT:PSS) electrodes separated by an electrically insulating but ionically conducting area (overoxidized PEDOT:PSS). As a result, the cations are liberated from the dopant and transferred via the over-oxidized

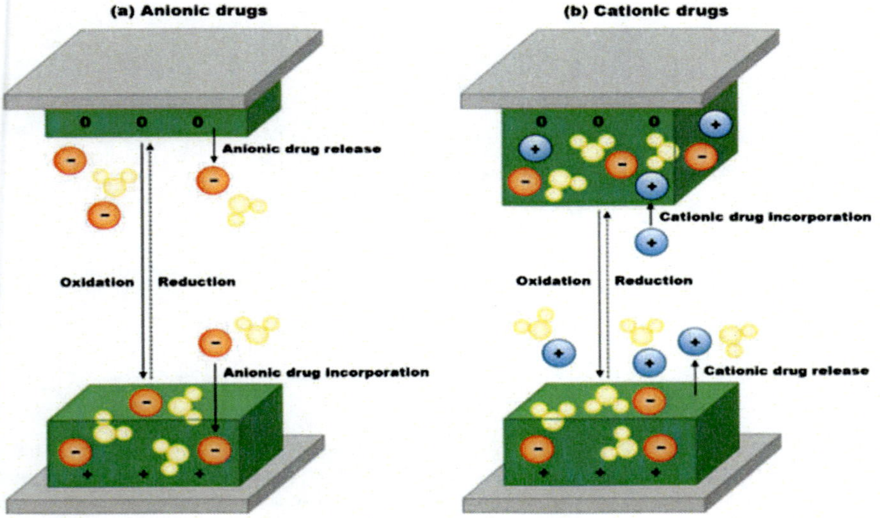

Fig. 5.6 Release of anionic drugs occurs by anion-driven shrinking when a CP is oxidized (**a**) whereas cationic drugs are released during reduction of a CP by cation drive (**b**). Reprinted from ref. [15], copyright 2019, with permission from MDPI

membrane from the source to the target electrolyte (Fig. 5.7a). The electrically insulating area allows reasonably large voltages (1–30 V) to be delivered while avoiding significant electric field exposure to the target (cells or tissues). This kind of delivery is not affected by the concentration of ions at the target and can even be pumped against its concentration gradient. The quantity of ions that diffuse from the target electrode in the OFF state (no applied voltage) is very modest (<10 nmol potassium or calcium after 4 h, ON/OFF ratio 300).

These ion pump devices distribute drugs in a regulated manner, both geographically and temporally. The microfabricated drug outlet shape, as well as the distance between the outlet and the cell or tissue, gives spatial control, while the timing and amplitude of the applied voltage (pulsed or continuous) provide temporal control (Fig. 5.7b, c). Other advantages of ion pumps for drug delivery include high ON/OFF ratios, electronic control of administered dosages, precise dose estimation, and "dry" distribution—without aqueous flow, which might damage neighboring cells or tissues (e.g., cell dissociation, changes in local concentrations of signaling chemicals, and increased pressure in tissues).

5.6.2 *Drug Loading*

Conductive polymers (CPs) are redox-active materials having programmable electrical and physical characteristics. To preserve electrostatic neutrality, the charge of

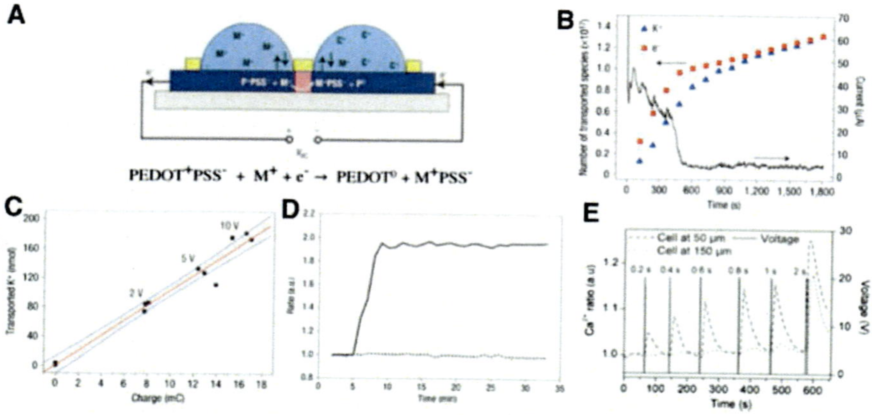

Fig. 5.7 **A** Schematic of the organic ion pump. There are two PEDOT:PSS electrodes (dark blue), source (left) and target (right), and these are separated by an electronically insulating but ionically conducting region (overoxidized PEDOT:PSS, pink). The source electrode is immersed in an electrolyte with a relatively high concentration of the cation M^+ and binds to the PSS in the film. A potential is applied between the electrodes. The left electrode is oxidized; the positively charged PEDOT interacts with negatively charged PSS and a cation is released. The right electrode is reduced, and there is a net flow of cations from left to right. **B** Number of transported potassium ions and electrons with time. **C** Relationship between delivered potassium ions and applied potential. **D** Fluorescence of HCN-2 cells loaded with calcium-sensitive probe. Fluorescence increase indicates change in intracellular calcium concentrations, a consequence of potassium pumping. Cells on top of channel (solid line) affected while those 500 μm away (dotted line) are not. (**A**–**D**). **E** Applying pulsed potential for acetylcholine pumping stimulates SH-SY5Y cells 50 μm away from outlet and generates oscillating intracellular calcium concentrations. Cells 150 μm away are not impacted. Reprinted from ref. [22], copyright 2019, with permission from John Wiley and Sons

the CP backbone may be controlled through redox processes, which include the transport of ions into and out of the polymer. Surface areas of CPs with specified micro- or nanostructures are much higher than those of conventionally manufactured CPs. The ensuing high surface area contact between polymer and liquid medium facilitates ion exchange and can result in greater and faster redox cycling reactions. As platforms for electrically adjustable medication delivery, CP systems are emerging. CPs with specified micro- or nanostructures provide for increased drug delivery while allowing systems to be highly controlled to regulate the extent and pace of drug release.

Because of their expanded surface area, CPs with specified micro- or nanostructures are attractive materials for biological applications due to their greater responsiveness, decreased impedance, and high charge transfer capacity. The effectiveness of an electrically adjustable drug delivery system is dictated by its performance stability, drug loading capacity, and compatibility with surrounding tissues. Electrical stimulation with low amplitudes, generally less than 1 V, can be utilized to trigger drug release from CP systems, making them very appealing. In drug delivery, CPs with specified micro- or nanostructures provide benefits due to their increased surface area, which results in increased drug loading capacity and responsiveness to stimulation.

The capacity to enhance the quantity of medicine that the material can transport is one advantage of employing CPs with specified micro- or nanostructures. Increased drug delivery capacity will improve the value of CP-based drug delivery systems in managing systemic diseases and will extend their operating lifetime. Loading is limited to 1 drug molecule every 3–5 monomer units when a drug is integrated directly as a dopant into the CP. Ion exchange can be predicted to be more efficient if distances between the polymer bulk and the polymer-media interface are shortened due to the increased surface area provided by CPs with defined micro- or nanostructures. However, if electrostatic forces alone determine loading, loading levels cannot surpass 1 drug molecule every 3–5 monomer units. Meanwhile, when pharmaceuticals may be physically entrapped within CPs, using templates to drive polymerization opens up a world of fascinating possibilities. A template containing medicine can be employed, or if the template is removed after polymerization, the empty area can operate as a reservoir containing drug.

5.6.2.1 Drug Loading into CPs with the Template Retained

Drug-containing templates can be utilized to guide CP development. PEDOT-coated PLGA fibers loaded with dexamethasone were created by electrodepositing PEDOT on PLGA fibers. Jeon et al. took a different technique, inserting a model drug, FITC-BSA, into the pores of the AAO membrane template. The pores in the template were confined to a given diameter after being loaded with FITC-BSA by preparing PPy on the top half of the template. The AAO template with consistent pore size functioned as a storage mechanism for huge amounts of medicines. The pores were opened and closed by electroactuation of PPy, regulating drug flow.

Drug delivery has made use of nanocomposite materials incorporating a CP. As codopants, graphene oxide (GO) nanosheets and DexP were introduced into PPy films to generate nanocomposites. Carboxylic groups were generated at the ends of GO during oxidative polymerization, rendering them negatively charged and allowing GO to be taken up as a codopant into the CP with DexP. Because of GO high surface-to-volume ratio, substantial amounts of DexP may be taken up between the layers of GO nanosheets. As a result, DexP may be loaded into PPy not only as a dopant but also as a component of GO, resulting in increased drug loading. Aspirin has also been incorporated into nanocomposites of palygorskite-PPy (Pal-PPy), which combine the electrical characteristics of PPy with the large surface area of Pal. The effective electrochemical surface area of the Pal-PPy nanocomposite was discovered to be roughly five times more than that of the unstructured PPy. Aspirin loading capacity in the Pal-PPy nanocomposite was determined to be 2.5%. Because of the strong ion–dipole interactions between anionic aspirin and cations in Pal, the drug loading was greater than with PPy alone.

Carbon nanotubes are another kind of material that has piqued the curiosity of researchers for medicinal uses such as drug delivery. This is due to their excellent drug loading capacity and biocompatibility. Composites of carbon nanotubes and CPs have better electrical and mechanical characteristics. The drug loading capacity of the interior cavity of different sized carbon nanotubes was measured by Luo et al. Carbon nanotubes were used as nanoreservoirs that could be loaded with DexP after being sonicated in acid to open their ends. However, significant leakage was discovered from the carbon nanotube ends. DexP loading was greatly enhanced, passive release was reduced, and release could be induced electrically by encapsulating the carbon nanotubes with PPy. Carbon nanotubes having a smaller outer diameter but a larger inner diameter had a drug loading level of 78.7 $\pm$ 2.1 mg, compared to 38.8 $\pm$ 3.2 mg in unstructured PPy films. Sirivisoot et al. showed that the anionic medicines DexP and penicillin/streptomycin (P/S) were incorporated into the polymer backbone during the electropolymerization of PPy on multiwalled carbon nanotubes produced on titanium (MWNT-Ti).

Esrafilzadeh et al. developed a unique technique for generating coaxial CP fibers filled with ciprofloxacin. They effectively manufactured ciprofloxacin hydrochloride-containing PPy on PEDOT:PSS/chitosan hybrid fibers. Wet-spinning using chitosan as a coagulant produced PEDOT:PSS/chitosan fibers. Ciprofloxacin (isoelectric point-7.4) was kept in a positive state at the polymerization pH and physically entrapped in the polymer while chloride ions operated as a dopant. The quantity of ciprofloxacin-loaded fiber (diameter 65 $\pm$ 7 mm, weight 254 $\pm$ 13 mg) discovered was 42 $\pm$ 4 mg/cm.

Cetyl pyridinium was utilized by Alizadeh and Shamaeli to direct the formation of PPy films and to increase methotrexate (MTX) loading into nanostructured PPy films. In most cases, MTX has a very poor doping efficiency into CPs. Because of the electrostatic and aromatic interactions between the cationic cetyl pyridinium and anionic MTX, PPy with entrapped cetyl pyridinium increased MTX loading. Ru et al. created a PPy nanofiber network with ATP as a dopant as well as a morphology guiding agent. These nanofiber networks have micro- or nanogaps between the individual nanofibers that can be filled with a second drug. Using this method, DexP, dopamine, hydroxy camptothecin, and paclitaxel were all added as a second drug.

Ge et al. reported the use of a drug-loaded soft template to manufacture CP nanoparticles utilizing an emulsion as a soft template to drive polymerization. They encapsulated fluorescein and daunorubicin in an emulsion made of the cationic surfactant dodecyl trimethylammonium bromide (DTAB) and the cosurfactant decyl alcohol. When pyrrole monomer and ferric chloride were added to the drug-encapsulated micelles, drug-loaded PPy nanoparticles with an average diameter of 60 nm were formed. Fluorescein and daunorubicin entrapment efficiencies were found to be 37.5% and 33.3%, respectively.

5.6.2.2 Drug Loading into CPs After the Template is Removed

Drugs can be put into the vacuum areas of CPs with specified micro- or nanostructures once a hard template is removed. The entrapment of medicines in the voids of porous CP films generated from colloidal templates after selective colloidal removal has been shown. Luo et al. successfully entrapped dexamethasone in the pores of nanoporous PPy films fabricated using a PS nanobead template. They created nanoporous PPy sheets by doping them with fluorescein, a model drug. Following the removal of the PS bead template, they injected a second drug, dexamethasone, into the pores of the nanoporous PPy films. This enabled for the simultaneous loading and release of two medicines, fluorescein and dexamethasone. Sharma et al. published similar results, demonstrating the trapping of the hydrophobic drug risperidone in macroporous inverted opal PPy films generated from a PMMA template. When inverse opal films were compared to non-inverse opal films, the surface area nearly quadrupled and the entrapment effectiveness nearly doubled. Similarly, Pokki et al. created a nanoporous PPy film using colloidally produced PS beads and filled the resultant pores with rhodamine B. Cho and Borgens proposed a similar idea, in which nanoporous PPy/biotin films were produced using a sacrificial PS nanobead template. Biotin-streptavidin interactions then attracted streptavidin-coated gold nanoparticles to the macroporous PPy sheets.

5.6.3 *Drug Release from CPs*

Drug release from CPs is usually governed by the polymer redox state, which affects electrostatic interactions between drug and polymer as well as volume changes. Because of electrostatic interactions between the CP and the drug, anionic medicines are released during reduction. This technique has been used to liberate ferrocyanide, salicylate and naproxen, ATP, and DexP from CP films. Meanwhile, when cationic drugs are oxidized, they are ejected from the polymer. This method has been used to regulate the release of cationic medicines such as dopamine and chlorpromazine. The induced release of risperidone and progesterone has been linked to volume changes rather than electrostatic interactions. The release of pharmaceuticals from CPs with specific micro- or nanostructures takes use of not only electrostatic interactions between drug and polymer, but also volume changes that occur when redox states change. Polymer expansion and contraction are caused by the flow of hydrated ions into and out of the polymer. Surface area is often enhanced in CPs with specified micro- or nanostructures. This increased effective electrochemical surface area accelerates and expands ion transport into and out of the polymer. As a result, reactions to electrostatic switching and volume changes are more accurate and possibly larger in amplitude.

5.6.3.1 Electrostatic-Driven Release

Electrostatic interactions were shown to promote anionic ciprofloxacin release from coaxial PPy and PEDOT:PSS/chitosan fibers, with a twofold increase in release from the decreased state compared to passive release from the as-prepared state. When ciprofloxacin was electrochemically oxidized, there was a 20% reduction in release relative to passive release. Meanwhile, Sirivisoot et al. showed the release of DexP and penicillin/streptomycin (P/S) from nanostructured PPy films on MWNT-Ti.

To improve DexP loading, PPy nanocomposites were codoped with GO nanosheets. When a biphasic electrical pulse was applied to the nanocomposites, more than twice as much DexP was released as in ordinary PPy-DexP films. Because of their great size, the GO nanosheets stayed within the polymer framework. The linear release of DexP from films codoped with DexP and GO was seen over 400 cycles. DexP put into the inner cavity of carbon nanotubes and integrated into PPy films revealed twice as much drug release as drug loaded directly as a dopant into PPy films. The greater drug loading with the GO nanosheets and carbon nanotubes functioning as nanoreservoirs can be attributed to the higher levels of drug release.

In both stimulated and unstimulated samples, the electrostatically regulated release of aspirin from Pal-PPy nanocomposite films was roughly 1.7 times greater than that of ordinary PPy films. The increase in effective electrochemical surface area and electrostatic attraction between anionic aspirin and the cationic groups in Pal resulted in the difference. On stimulation for 45 h, ATP release from PPy nanofibers polymerized with ATP as a dopant and structural guiding agent was found to be 90%, compared to 53% from an unstructured cauliflower-shaped PPy film.

The ejection of anionic DexP from macroporous inverse opal PPy sheets. Although DexP was put as a dopant onto both macroporous and conventional films, stimulation might release more than three times the quantity of medication from the same theoretical amount of PPy. This may be explained by the fact that the inverse opal film was more electroresponsive due to its large surface area, which resulted in better ion exchange between the polymer and the surrounding medium.

When exposed to an electrolyte solution, CP films can work as self-powered galvanic cells when coated with a tiny coating of active metal, such as magnesium. Magnesium served as an anode, while CP film served as a cathode, resulting in a self-powered cell that did not require external stimulation. When the cell was put in an electrolyte medium, such as NaCl, magnesium was oxidized in solution to Mg^{2+} and the CP was reduced, leading in the drug release. This galvanic process was employed to liberate ATP from a PPy nanofiber network, in which ATP served as both a dopant and a morphology guiding agent. The scientists evaluated the release of ATP from a magnesium-coated nanofiber network to a standard PPy film without coating. The nanofiber network released 22% as compared to the traditional PPy film with the release of 13%.

5.6.3.2 Actuation-Driven Release

Redox-driven volume changes in CP systems can be used to drive drug release. The inflow and outflow of hydrated ions determine the pace and amount of actuation. The increased surface area of CPs with specified micro- or nanostructures facilitates ion exchange between polymer and medium. Abidian et al. found that electrical stimulation induced the PEDOT coating to actuate, resulting in fissures that allowed drug release, and that this resulted in bursts of dexamethasone release from PEDOT-coated PLGA fibers.

Jeon et al. exhibited pulsatile release of FITC-BSA from a system consisting of a PPy/NaDBS film electrodeposited over the upper half of the pores of an AAO membrane. Hydrated cations migrate into the CP, causing it to expand and shrink the membrane pores, but on oxidation, cations are ejected and the membrane pores open as the polymer contracts, leading in increased rates of drug release. When the film is switched from the oxidized to reduced states, the pore diameters decline from 190 nm (60 L m 2 flux) to 140 nm (no flux) in less than 10 s.

Sharma et al. discovered that electrostimulation caused fast contraction and expansion of inverse opal films, which resulted in enhanced risperidone release from inverse opal films. On reduction, PPy inverted opal films showed a 10% increase in interplanar spacing. The release was found to be 162.69 ± 3.6 mg on stimulation compared to 119.8 ± 2.5 mg in unstimulated inverse opal films. However, it was only 42.5 ± 737 mg on stimulation compared to 31.3 ± 0.4 mg in unstimulated films in conventional non-inverse opal films. The increased drug loading in porous inverse opal films explains the difference in release levels between inverse opal and conventional films. Similarly, the release of rhodamine B from a PPy film generated from a PS nanobead template placed by colloidal lithography was three times that of non-porous films.

Luo et al. showed the simultaneous release of two medicines (fluorescein and dexamethasone) from nanoporous PPy sheets through electrical stimulation. The anionic fluorescein was released as a result of a dedoping process from the PPy backbone, whereas volume changes in the porous structures during actuation pushed dexamethasone out of the pores. Drug delivery from CPs with specified micro- or nanostructures is more efficient than from conventional CPs, which can be ascribed to enhanced drug loading as well as improved response to electrical stimulation [19].

5.7 Biodegradable and Electrically Conductive Polymers for Biomedical Applications

To overcome the drawbacks of CPs such as poor mechanical properties, poor processability, hydrophobicity, and non-degradability, polymer blends and composites based on conductive polymers such as PPy and PANI and degradable polymers such as polylactide (PLA), polycaprolactone (PCL), poly(lactide-co-glycolide) (PLGA), polycaprolactone fumarate, poly(lactide-co-polycaprolactone) (PLAco-PCL), polyurethane, chitosan, gelatin, collagen, and heparin have been widely studied. These materials were discovered to be very well-suited for applications such as tissue engineering.

When combined with other polymers in blends, PPy is most typically made through chemical polymerization. PPy nanoparticles and poly(d,l-lactide) (PDLLA) were combined to create a new electrically conductive, biocompatible, readily processable, and mostly biodegradable composite material for the growth of human cutaneous fibroblasts with or without electrical stimulation (ES). Cell survival on the PPy/PLLA membranes was 2.2- and 4.0-fold ($p < 0.05$) higher in the presence of a direct electrical field intensity of 100 mV/mm at 2 and 24 h, respectively, than on the same membranes without ES.

Because PPy nanoparticles are not biodegradable, it is preferable to keep the PPy component of the polymer composite as low as feasible. A mostly biodegradable conductor composed of 5 wt% conductive PPy and 95 wt% biodegradable poly(l-lactide) (PPy/PLLA) was made in the same manner as in the previous study. Cell culture data revealed that the PPy/PLLA conductors aided human cutaneous fibroblast adhesion, spreading, and proliferation in the presence and absence of ES. The presence of ES raised the expression of interleukin-6 and interleukin-8 mRNA as well as the secretion of these two kinds of cytokines from human fibroblasts. A cosolution casting approach was used to create a biodegradable conductive membrane composed of conductive PPy nanoparticles (PPy, 2.5 wt%) and biodegradable chitosan (97.5 wt%). Schwann cells were grown on the composite membrane because they provide as a bioactive substrate for axonal migration and release chemicals that regulate axonal development. These findings open up new avenues for improving neuron regeneration in degradable conductive membranes by enhanced neurotrophin release via electrical stimulation. These observations may be explained by rapid electrical stimulation of conducting PPy enhancing protein (fibronectin) adsorption, which altered nerve cell contact and increased neurite extension.

Macroscopic nanomembranes of molecular-scale thickness might be employed for drug delivery vehicles, biomimetic systems, and tissue engineering. Spin-coating was used to create biodegradable free-standing nanomembranes with thicknesses ranging from 20 to 80 nm made of conductive polymer (poly(3-thiophene methyl acetate) and polyester (poly(tetramethylene succinate)). Adhesion and proliferation results from epithelial cell cultures show that these nanomembranes (blends of the two polymers) perform better as cellular matrix than membranes of the two individual polymers, indicating that these nanomembranes could be used as bioactive substrates for tissue regeneration.

Membranes were made from the mainly biodegradable conducting blends or composites. Porous scaffolds are more desirable in tissue engineering and have been produced. Degradable conducting scaffolds have also been created using electrospinning. To create the mostly degradable and conductive scaffolds, in situ chemical polymerization of Py in the presence of a biodegradable scaffold generated by electrospinning has been investigated. This procedure is easy and can keep the porous structure of the scaffolds. This manufacturing technology also overcomes the weak mechanical characteristics and processing challenges of PPy in tissue engineering applications. Degradable and conductive meshes were created by solution polymerizing Py in the presence of the nanofibers and depositing PPy onto random and aligned electrospun poly(lactic-co-glycolic acid) (PLGA) nanofibers (Fig. 5.8). PPy-coated PLGA electrospun meshes grew and differentiated rat pheochromocytoma 12 (PC12) cells and hippocampus neurons better than non-coated PLGA control meshes. PC12 cells activated with 10 mV/cm electrical current on PPy/PLGA scaffolds had 40–50% longer neurites and 40–90% more neurite production than unstimulated cells on the same scaffolds. Furthermore, aligned PPy/PLGA fibers had longer neurites and more neurite-bearing cells than randomly oriented PPy/PLGA fibers. All of these findings suggest that electrical stimulation and topographical guiding have a synergistic impact on PC12 cell differentiation.

PANI has received minimal attention in comparison with PPy, despite being one of the most promising electrically conductive polymers due to its simplicity of synthesis, environmental stability, tunable electrical conductivity, and easy doping/dedoping chemistry. All of these properties indicate that PANI materials might be used as conductive substrates in tissue engineering applications. PANI, for example, has been coelectrospun into nanofibers after being combined with gelatin. When less than 3% PANI was utilized in the blended fibers, SEM revealed that homogenous fibers with no phase segregation were generated. With increasing PANI concentration in the blends from 0 to ~5% w/w, the average fiber length reduced from 803 ± 121 nm to 61 ± 13 nm and the tensile modulus rose from 499 ± 207 MPa to 1384 ± 105 MPa. H9c2 rat cardiac myoblast cells were cultivated on these nanofiber-coated glass cover slips, and all of the blended fibers promoted H9c2 cell adhesion and proliferation to the same extent as the control tissue culture-treated plastic.

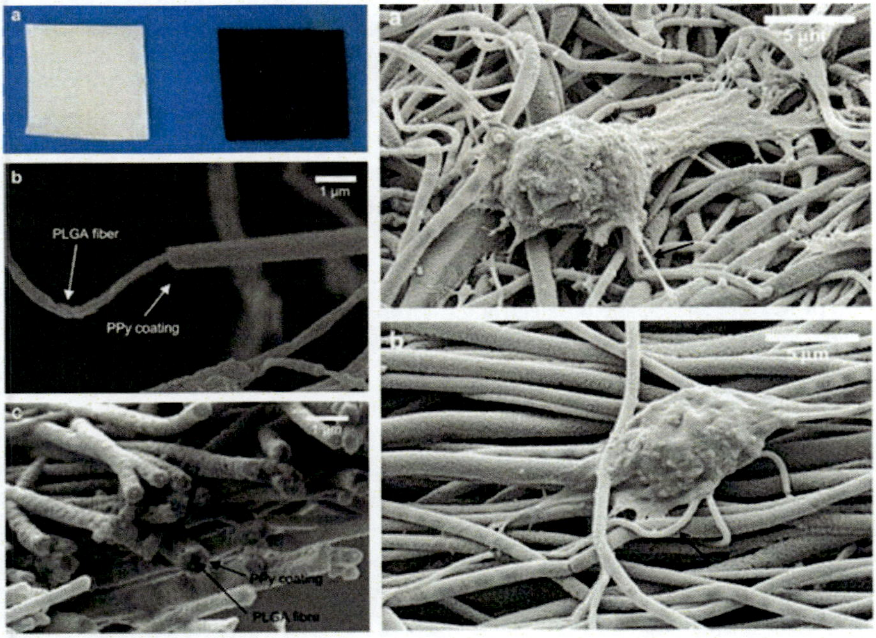

Fig. 5.8 Left column: PPy-coated PLGA meshes. **a** Photographs of uncoated PLGA meshes (white, left) and PPy–PLGA meshes (black, right). **b** SEM micrograph of single strands of PPy–PLGA fibers. **c** SEM image of section of the PPy–PLGA meshes. Right column: SEM images of PC12 cells cultured on **a** PPy-random fiber and **b** PPy-aligned fiber for 2 days; black arrows indicate the neurites. Reprinted from ref. [11], copyright 2009, with the permission from Elsevier

The qualities of the scaffold may be changed for different purposes by using different degradable polymers. Degradable conducting scaffolds with elastic qualities allow for more accurate imitation of the mechanical and electrical properties of the native extracellular matrix (ECM), making them ideal for soft tissue regeneration in areas such as skin, skeletal muscle, and blood vessels. Electrospinning a composite of PANI doped with camphorsulfonic acid (CPSA) and poly(l-lactide-co–εcaprolactone) (PLCL) resulted in electrically conductive nanofibers. The CPSA-PANI/PLCL nanofibers had a smooth fiber structure with no coarse lumps or beds and constant fiber diameters ranging from 100 to 700 nm, with a polyaniline content ranging from 0 to 30% wt.

The high elongation at break of these fibers, ranging from 391.54 ± 9.20 to $207.85 \pm 6.74\%$ depending on the PANI concentration, indicates their elastic capabilities. When the PANI concentration was raised to 30% wt, the electrical conductivity rose dramatically from 0.0015 to 0.0138 S/cm. Cell adhesion assays with human dermal fibroblasts, NIH3T3 fibroblasts, and C2C12 myoblasts revealed that the CPSA-PANI/PLCL nanofibers had substantially greater adherence than pure PLCL nanofibers. Furthermore, activation of diverse direct current flows boosted the proliferation of NIH3T3 fibroblasts. After 4 days of culture, the number of cells positive for sarcomeric myosin on the electrically conductive fibers with 15% PANI was 3.6 times higher than on the PLCL fibers without PANI. This suggested that these electroactive fibers may be used as a guide scaffold for neural tissue engineering.

5.8 Engineering Antifouling Conductive Polymers for Modern Biomedical Applications

In general, the softness of conductive polymers is advantageous for the production of flexible and elastic bioelectronics. Despite the fact that conductive polymer-based organic bioelectronics has shown significant success in a variety of biomedical applications, major obstacles remain, notably for implanted devices and long-term applications. The non-specific adsorption of biomolecules and cells, which produces an inflammatory response and electrode dysfunction, is one of the key problems. It is critical to create antifouling coatings or antifouling biomaterials in order to prevent the damage caused by the process of non-specific adsorption. Conductive polymers having antifouling characteristics are expected to be advantageous in terms of encouraging and driving novel biomedical uses of conductive polymers.

Several antifouling conductive polymers have been reported. Antifouling polymers are broadly categorized into three types: those based on poly(ethylene glycol) (PEG) or oligo(ethylene glycol) (OEG), those based on zwitterions, and those based on other hydrophilic units such as glycans or hydrophilic peptides. Some groups began with a molecular design for synthesizing monomers containing antifouling moieties in order to immobilize these antifouling groups on conductive polymer surfaces. Electropolymerization or oxidation polymerization can be used to directly create antifouling conductive polymers. On conductive polymer surfaces, antifouling moieties create a monolayer-like structure. For a smooth surface, the density of the surface functional group is typically 10^{15} cm^{-2}. The density per unit of antifouling moieties rises as the surface roughness or development of a nanostructure increases. Surface modification has also been employed by other researchers to build antifouling layers on conductive polymer surfaces. Direct covalent bonding or surface-initiated polymerization technologies such as ATRP and RAFT polymerization can be used to introduce the antifouling layer.

5.8.1 In Vivo Electrochemical Biosensing with High Sensitivity

Researchers explored the possible uses of in vivo biosensing and exhibited better performance in terms of sensitivity and reliability from their platform, which has a good antifouling ability and a conductive substrate. Hui et al. grafted PEG onto PANI to generate PEGylated PANI nanofibers, which offered acceptable antifouling capabilities in human blood samples, as shown in Fig. 5.9a. After immobilizing capture probes on PEGylated PANI nanofibers through NHS/EDC coupling, a sensitive DNA electrochemical biosensor for the breast cancer susceptibility gene was created (BRCA1). This platform demonstrated extremely high sensitivity and a wide linear range (from 0.01 pM to 1 nM). Furthermore, BRCA1 was quantifiable in complicated human serum [9]. Wu et al. created a sensitive and stable glucose electrochemical sensor based on a zwitterionic poly(sulfobetaine-3,4-ethylenedioxythiophene) (PSBEDOT) platform, as illustrated in Fig. 5.9b. Through the electropolymerization of PSBEDOT in an aqueous solution containing GO_x, glucose oxidase (GO_x) was directly captured and immobilized on the substrate. In this study, zwitterionic PSBEDOT was employed as a hydrated 3D matrix for encapsulating GO_x, which can increase GO_x stability and longevity. Furthermore, the conductive polymer served as a transducer, transmitting signals generated during glucose oxidation. The zwitterionic PSBEDOT/GO_x electrode outperformed an unfunctionalized PEDOT/GO_x electrode in terms of stability in human blood plasma. After 14 days of storage in human blood plasma, the residual current signal was more than 90%. The importance of antifouling surfaces for continuous electrochemical monitoring of glucose in blood was highlighted in this work [20]. Liu et al. employed a carbon fiber microelectrode (CFE) coated with zwitterionic phosphorylcholine-functionalized PEDOT-PC for in vivo dopamine monitoring in Fig. 5.9c. In comparison with CEF coated with PEDOT or hydroxyl-functionalized PEDOT-OH, CEF coated with PEDOT/PC might resist protein adsorption while maintaining sensitivity and temporal responsiveness for in vivo dopamine monitoring [13].

On other platforms, antifouling PEG or zwitterionic groups were used to improve electrochemical sensing efficacy in a complex biological media or to monitor in vivo electrochemical reactions. The antifouling properties of conductive polymers are

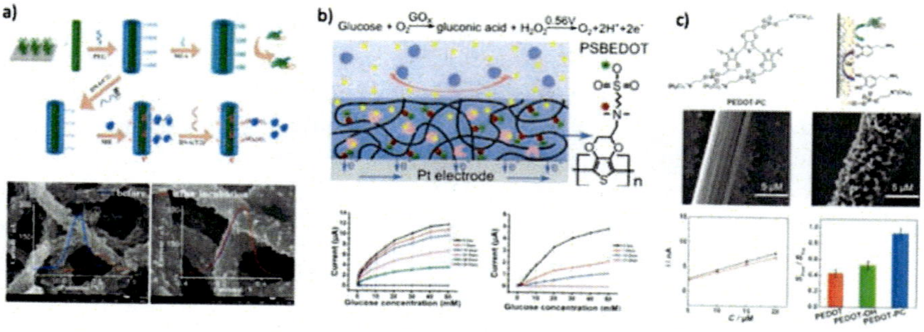

Fig. 5.9 Electrochemical biosensing applications from antifouling conductive polymer platforms: **a** PEGylated PANI nanofibers for sensitive electrochemical DNA sensing in human serum and **b** zwitterionic PSBEDOT-GO_x electrode for glucose monitoring in blood. **c** CFE coated with zwitterionic phosphorylcholine-functionalized PEDOT-PC for in vivo dopamine monitoring. Reprinted from ref. [21], copyright 2019, with permission from American Chemical Society

critical for enhancing the sensitivity and lifespan of in vivo electrochemical sensors. Other parameters, such as electrostatic interaction with analytes or the active surface area of electrodes, may, nevertheless, have a significant impact on sensing performance. It is critical to successfully incorporate these critical qualities when building an improved electrode for in vivo electrochemical sensing or continuous monitoring of human blood.

5.8.2 *Controlled Cell Capture and Release*

The controlled release of ions and pharmaceuticals has been achieved using conductive polymers, which are principally ascribed to the adjustable charged state of the polymer backbones. Different surface potentials were used to alter the redox states of conductive polymers, and ions can migrate in and out of the conductive polymers to preserve electrostatic neutrality. Lin et al. demonstrated an antifouling zwitterionic PEDOT-based platform for regulated cell capture and release (Fig. 5.10a). As illustrated in Fig. 5.10b, a redox-responsive hydroquinone-functionalized EDOT (EDOT-HQ) was produced [12]. The hydroquinone group was employed for oxime ligation in order to immobilize aminooxy-terminated compounds, which may be produced by oxidizing hydroquinone to generate benzoquinone. Under physiological circumstances, the oxime ligation linkage is stable. A reduction potential can then be used to cleave the oxime ligation.

Poly(EDOT-PC-co-EDOT-HQ) films were formed by coelectropolymerizing EDOT-HQ and zwitterionic EDOT-PC. This copolymer film was used to regulate the adhesion and release of NIH3T3 cells. Prior to coupling of an RGD

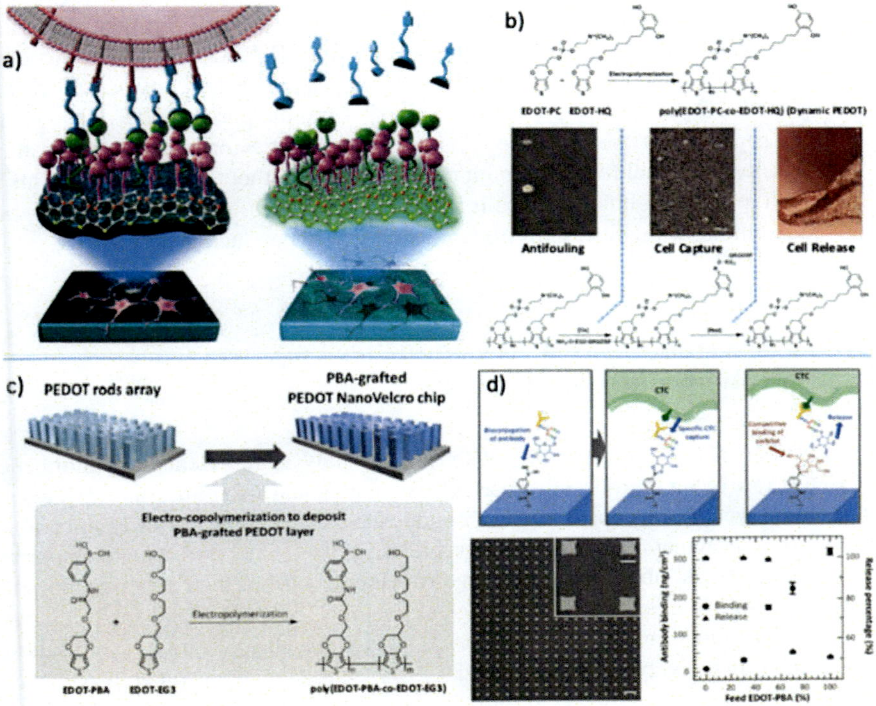

Fig. 5.10 **a** Schematic of controlled release of cells and **b** molecular design of dynamic PEDOT platform, which allows the controlled attach and release of NIH3T3 cells by manipulating the redox state of hydroquinone on polymers. **c** Schematic of fabricating PBA-grafted NanoVelcro chip. **d** Schematic of the mechanism and results of controlled CTC capture and release from this platform. Reprinted from ref. [21], copyright 2019, with permission from American Chemical Society

peptide to poly(EDOT-HQ-co-EDOT-PC), the polymer film demonstrated appropriate antifouling capabilities and resisted NIH3T3 adhesion. Following the application of the oxidation potential on the polymer film to convert hydroquinone to benzoquinone, an RGD peptide with an aminooxy terminus may be attached to the polymer film, resulting in a robust binding of NIH3T3 cells to the polymer. Owing to the breakage of RGD peptide connections caused by the application of a reduction potential, NIH3T3 cells were liberated from poly(EDOT-HQ-co-EDOT-PC) films.

Researchers have previously proven uses of the nanostructured PEDOT technology for improved capture efficiency of circulating tumor cells (CTCs). Shen et al. developed a glycan-stimulated PEDOT-based nanomaterial platform for the separation of CTCs from blood samples by combining antifouling EDOT-EG3 with phenylboronic acid-functionalized EDOT (EDOT-PBA). Figure 5.10c depicts the electropolymerization of a poly(EDOT-PBA-co-EDOT-EG3) layer on a NanoVelcro chip. As illustrated in Fig. 5.10d, in a system for capturing CTCs, the antibody probe can be immobilized by their affinity for surface-grafted PBA, allowing the capture of CTCs. Because of competitive binding, CTCs were liberated from the chip after adding sorbitol with a higher affinity to PBA. The 3D PEDOT NanoVelcro chip provides high capture efficiency and allows for gentle release of CTC cells. This allows for CTC purification for further analysis from the purified CTC [17]. These findings suggest that the presence of antifouling moieties on surfaces, as well as their density, is critical for the smooth release of cells without causing harm. This is attributed to the high binding of cells to culture substrates. Antifouling agents may not be required for the release of tiny molecules from conductive polymers.

5.9 3D Scaffolds Based on Conductive Polymers for Biomedical Applications

Polypyrrole (PPy), polyaniline (PANI), and poly(3,4-ethylenedioxythiophene) (PEDOT) are by far the most often used CPs in biomedical disciplines and are the only ones employed in the construction of conductive 3D structures for bioapplications. Conductive polymer 3D scaffolds have primarily been used in three architectures: porosity scaffolds, fiber meshes, and hydrogels. Because of their superior morphological, physical, electrical, and biological qualities, conductive 3D scaffolds have shown significant promise for a wide variety of biomedical applications such as tissue creation, drug delivery, and biosensing.

5.9.1 Tissue Engineering

Because of their low cost, great repeatability, and adjustable processing morphology and composition that might impact cellular development and morphology, 3D scaffolds are the ideal choice for this purpose. All of these characteristics have been demonstrated to influence cellular development and morphology. Furthermore, the incorporation of chemical groups and bioactive compounds such as medicines, growth factors, and dopants allows them to be tuned to improve cellular adhesion and proliferation. For example, integration of hydroxyapatite (HAp), a bone mineral, demonstrates robust adherence and great proliferation with mouse MC3T3-E1 preosteoblasts.

One of the most important needs for a scaffold used for regeneration is the capacity to keep healthy cells inside its framework for extended periods of time. Cell cultures were carried out for at least 5 days and up to 21 days. The longer the investigation, the more certain the scaffold aptitude for tissue engineering becomes. Three types of cells have been investigated in the literature, all of which show electrically active properties: bone, neural, and cardiac cells. Preosteoblasts, which are mesenchymal stem cells that transform into bone cells, were employed in bone tissue engineering. Rat phaeochromocytoma-derived (PC12) cells are the most often used neural model.

As heart regeneration models, neonatal rat ventricular cardiomyocytes and mouse myoblasts C2C12 cells have also been studied. Non-conductive cells such as epithelial cells and fibroblasts have also been used as models for the cell survival and proliferative capabilities of 3D designs.

5.9.2 Electric Stimulation

Local electrical stimulation (ES) of cells in scaffolds improves cell-to-cell and cell-to-substrate contact, which improves cellular regeneration of electroactive tissues. The combination of CP scaffolds with ES creates a safer environment for the use of targeted and directed regeneration: the applied current flows through the scaffold structure, decreasing electrolysis of the biological medium and lowering the creation of cytotoxic agents.

Zhang et al. found that applying 200 μA (4 h per day for 21 days) to human adipose-derived mesenchymal stem cells (ADM-SCs) grown in PPy/polycaprolactone (PCL) porous scaffolds improved cell migration into deep sections of the scaffold and accelerated osteogenic differentiation. Furthermore, ES on neurons and glial cell cultures has been shown to minimize cluster formation in the PPy-coated structure only when applied early in culture, while boosting proliferation and accelerating neuron maturation. Björninen et al. differentiated adipose stem cells (ASCs) into smooth muscle cells using an ES approach (SMCs). They developed a unique technique for concurrently stimulating 3D scaffolds over lengthy periods of time. The PPy coating was demonstrated to boost the expression and proliferation of SMCs. Although electrical stimulation may not usually have a substantial effect in cell cultures, there have been no instances of cellular culture damage. Furthermore, drug-loaded CPs-based devices can be electrically triggered for regulated release.

5.9.3 Drug Delivery

To enable a longer release, the scaffolds should have a youthful modulus equivalent to the host tissue. Otherwise, only temporary drug liberation is possible. Zanjanizadeh Ezazi et al. created PPy/gel 3D scaffolds and loaded them with a model antibiotic, vancomycin (VCM), for anti-infection objectives in bone regeneration. The resultant scaffolds, with elasticity comparable to cancellous bone, demonstrated a high potential for drug release during a four-month period. Vishnoi et al. created a chitosan/gelatin/PPy scaffold that was loaded with alpha-ketoglutarate (α-KG), an intermediary in the synthesis of ATP in the citric acid cycle, to boost neuro-2a cell growth. Their goal was to investigate the exposure of neuronal cells to varying quantities of medication in the same period using different delivery methods: exogeneous, 2D, and 3D systems with and without PPy. It was found that the 3D system provided longer duration drug release. Furthermore, the liberation of a nerve growth factor

(NGF) packaged in a conductive hydrogel scaffold was evaluated on PC12 cells, which have been shown to develop into neurons in response to NFG uptake. The scaffolds loaded with NGF show more neurite outgrowth than the empty scaffolds, confirming the drug transport capabilities of the material.

There is clear evidence that incorporating CP into a non-conjugated matrix has a favorable effect on drug release. The rougher surfaces formed by the CP coating enhance the surface and hence the drug-polymer interaction. Furthermore, the entrapment inside the matrix and the porosity of the substrate have a substantial influence on the velocity of release, with bigger pore sizes being quicker.

5.9.4 Biosensing

Conductive polymers have mostly been employed as coatings for implanted sensing devices to give electrical conductivity while also improving biocompatibility and avoiding scar formation. The ability of conductive 3D designs to behave as electrodes makes them appealing materials for biosensing devices. Furthermore, 3D substrates provide a substantially wider surface area for cell and biomolecule adhesion due to their porous nature, resulting in higher biosensing sensitivity and concentration range than bidimensional structures. Ravichandran created the first injectable electrochemical sensor for detecting glucose levels in tissue in real time. To create the soft gadget, the authors coupled the hydrogel qualities of collagen with the electrical functionality of PPy. Glucose oxidase was subsequently enclosed within the gel matrix as the sensor bioactive component, enabling for the monitoring of glucose levels up to 10 mM chronoamperometrically. When injected in porcine meat tissue over 5 days, they were able to demonstrate its potential in vitro. The key benefit of the created sensor, according to the scientists, is that the hydrogel may adjust its conformation to the host tissue and offer a well-attached interface in dimensions as small as the puncture incision, therefore decreasing detrimental invasiveness [1].

5.10 Erodible and Electrically Conductive Polymers

Due to the necessity for degradability, PANI or PPy was removed to some extent in the degradable and conducting blends or composites. Small quantities of PANI or PPy injected into the body via these materials are likely to remain in the body. As a result, the synthesis of totally biodegradable and electrically conductive polymers remains a top priority.

Designing erodible conductive polymers is one technique for synthesizing degradable conducting polymers. These polymers degrade gradually rather than shattering the chemical bond, as is the case with most polymers. To develop conducting thin films and colloidal dispersion, β-substituted pyrrole monomers with ionizable and/or hydrolyzable side groups underwent either oxidative electrochemical polymerization or chemical polymerization using ferric chloride as an oxidant. After 80 days of incubation at 37 °C, the weight loss of hydrophilic carboxylic-acid-modified PPy (poly-1) pressed pellets, which are somewhat soluble in water, was 27%, whereas the weight loss of hydrophobic methyl ester modified PPy (poly-2) pellets was 6%. The resistivity of poly-1 thin films was 300 ($\pm$100)Ω, and it rose to 700 ($\pm$200)Ω after two days of incubation in water. This demonstrates that the pellet conductivity is preserved under water circumstances. The cytocompatibility of the poly-1 film was investigated using primary mesenchymal cells taken from human bone marrow, and the results revealed that the poly-1 films aided in the growth, proliferation, and differentiation of primary human cells [6].

5.11 Current Stage and Challenges

Because of its electrical conductivity and biocompatibility, CPs have made great advances in the realm of biomedical engineering. However, significant disadvantages continue to restrict the use of these polymer materials in biomedical engineering. Biodegradability is one of the important challenges with ordinary CPs that has to be addressed, particularly in tissue engineering, where scaffolds are anticipated to disintegrate over time. A scaffold is a temporary artificial support structure that is implanted to stimulate cell differentiation and proliferation until new tissues emerge. As a result, the CPs utilized as scaffolds must disintegrate once they have served their purpose. The CPs have been combined with other degradable polymers to assure the biodegradability of a scaffold. PLA, a typical biodegradable polymer, was combined with conductive PPy nanoparticles, for example, to give both conductivity and degradability.

The hydrophobicity of the CPs is another barrier that must be addressed before the CPs can be used in scaffolds and biosensors. Despite their high mechanical qualities and biocompatibility, the absence of hydrophilicity prevents cells from adhering to their surfaces. Surface characteristics can be altered by functionalization. Guo et al., for example, presented a two-step technique for the functionalization and surface modification of PLA in order to increase hydrophilicity and introduce conductive aniline oligomers onto the PLA surface. Later on, it led to enhanced cell adhesion and controllable cell behaviors including spreading, proliferation, and differentiation. Specifically, photocrosslinking was used to add carboxyl acid and anhydride functions onto a PLA sheet. Following that, aniline tetramers were linked with functionalized PLA sheets. The surface change increased both hydrophilicity and electroactivity.

Actuators based on CPs are electrically controlled at low voltage and biocompatible with liquid electrolytes such as biological fluids, making CPs appealing materials for the application. However, in order to give quicker reactions in the actuators, the mobility of charge carriers in the CPs must be improved. The charge-carrier mobility can be increased by combining CPs with nanofibers and nanotubes due to their porous architectures, but making the mixed materials biocompatible is another difficulty.

Finally, there is a limit to the number of medicinal compounds that can be supplied employing CPs for drug delivery. One method of increasing drug loading capacity is to improve porosity and surface area. By generating nanostructures, films of template-grown CP can improve porosity and surface area.

References

1. Alegret N, Dominguez-Alfaro A, David M (2019) 3D scaffolds based on conductive polymers for biomedical applications. Biomacromolecules 20(1):73–89
2. Alshammary B, Walsh FC, Herrasti P, Leon CPD (2015) Electrodeposited conductive polymers for controlled drug release: polypyrrole. J Solid State Electrochem 20(4):839–859
3. Bendrea AD, Cianga L, Cianga I (2011) Review paper: progress in the field of conducting polymers for tissue engineering applications. J Biomater Appl 26(1):3–84
4. García-Córdova F, Valero L, Ismail YA, Otero TF (2011) Biomimetic polypyrrole based all three-in-one triple layer sensing actuators exchanging cations, J Mater Chem 21:17265–17272
5. Ghasemi-Mobarakeh L, Prabhakaran MP, Morshed M, Nasr-Esfahani MH, Baharvand H, Kiani S et al (2011) Application of conductive polymers, scaffolds and electrical stimulation for nerve tissue engineering. J Tissue Eng Regen Med 5(4):e17–e35
6. Guo B, Glavas L, Albertsson AC (2013) Biodegradable and electrically conducting polymers for biomedical applications. Prog Polym Sci 38(9):1263–1286
7. Guo B, Ma PX (2018) Conducting polymers for tissue engineering. Biomacromol 19(6):1764–1782
8. Hardy JG, Lee JY, Schmidt CE (2013) Biomimetic conducting polymer-based tissue scaffolds. Curr Opin Biotechnol 24(5):847–854
9. Hui N, Sun X, Niu S, Luo X (2017) PEGylated Polyaniline Nanofibers: Antifouling and Conducting Biomaterial for Electrochemical DNA Sensing. Acs Appl Mater Interfaces 9(3):2914–2923
10. Jia M, Rolandi M (2020) Soft and ion-conducting materials in bioelectronics: From conducting polymers to hydrogels. Adv Healthcare Mater 9(5):1901372
11. Lee JY, Bashur CA, Goldstein AS, Schmidt C (2009) Polypyrrole-coated electrospun PLGA nanofibers for neural tissue applications. Biomaterials 30(26):4325–4335
12. Lin H, Zhu B, Wu Y, Sekine J, Nakao A, Luo S et al (2018) Dynamic Poly(3,4-ethylenedioxythiophene)s integrate low impedance with Redoxlogwitchable biofunction. Adv Func Mater 28(12):1703890
13. Liu X, Xiao T, Wu F et al (2017) Ultrathin cell-membrane-mimic Phosphorylcholine polymer film coating enables large improvement for in vivo electrochemical detection. Angew Chem 129(39):11964–11968
14. Otero TF, Martinez JG, Arias-Pardilla J (2012) Biomimetic electrochemistry from conducting polymers. A review artificial muscles, smart membranes, smart drug delivery and computer/neuron interfaces. Electrochimica Acta 84:112–128
15. Park Y, Jung J, Chang M (2019) Research progress on conducting polymer-based biomedical applications. Appl Sci 9(6):1070
16. Reynolds M (1995) Poly(pyrrole adenosine 5′-triphosphate) (PP-ATP) and conducting polymer bilayers for transport of biologically active ions. Synth Met 71(1):2233–2236
17. Shen MY, Chen JF, Luo CH, Lee S, Li CH, Yang YL et al (2017) Glycan stimulation enables purification of prostate cancer circulating tumor cells on PEDOT NanoVelcro chips for RNA biomarker detection. Adv Healthc Mater 1870013

18. Talikowska M, Fu X, Lisak G (2019) Application of conducting polymers to wound care and skin tissue engineering: A review. Biosens Bioelectron 135:50–63
19. Uppalapati D, Boyd BJ, Garg S, Travas-Sejdic J, Svirskis D (2016) Conducting polymers with defined micro- or nanostructures for drug delivery. Biomaterials 149–162
20. Wu H, Lee C-J, Wang H et al (2018) Highly sensitive and stable zwitterionic poly(sulfobetaine-3,4-ethylenedioxythiophene) (PSBEDOT) glucose biosensor. Chem Sci 9(9):2540–2546
21. Wu JG, Chen JH, Liu KT, Luo SC (2019) Engineering antifouling conducting polymers for modern biomedical applications. ACS Appl Mater Interfaces 11(24):21294–21307
22. Zeglio E, Rutz A L, Winkler T E et al (2019) Conjugated polymers for assessing and controlling biological functions. Adv Mater 31(22):e1806712

Chapter 6
Energy Technology Based on Conductive Polymers

In a modern era marked by the unavoidable transition from the use of fossil fuels to greener renewable energy sources, scientists are pursuing new cutting-edge materials for energy storage in order to meet the increasing demand for clean energy. Such materials should be able to store or create large amounts of energy in systems that are ultimately inexpensive, light, and simple to manufacture in order to maximize efficiency. Conducive polymers stand out as attractive new possibilities to replace traditional materials such as metals and metal oxides in the quest to identify solutions that may solve these concerns. This is attributed to its distinct physical and chemical features, which include a wide conductivity range, processability, flexibility, lightweight, cheap cost, and the ability to be made on a large scale.

Furthermore, conductive polymers may be physically and chemically tailored to fit the needs of a wide range of applications. CPs are projected to enhance energy storage devices due to the redox process they experience in order to store charge in the bulk of the material, hence boosting the energy stored. The sluggish passage of ions inside the bulk of the CP electrode, on the other hand, results in comparatively low power (i.e., a low rate of charge/discharge). Nonetheless, CPs offer several benefits that offset this disadvantage, and they have quicker electron transfer kinetics than other materials, such as metal oxides.

6.1 Conductive Polymer-Based Supercapacitor

Supercapacitors are intended to bridge the gap between batteries and capacitors in order to provide quick charging energy storage devices with intermediate specific energy. They are viewed as having a potential market in both hybrid and pure electric cars to increase regenerative braking (through quick charging capabilities) and give greater acceleration (through fast discharge capability).

A supercapacitor is made up of two electrodes, an electrolyte, and a separator. The electrodes are immersed in the electrolyte and play an important part in the

Y. Wang and W. Feng, *Conductive Polymers and Their Composites*,
https://doi.org/10.1007/978-981-19-5363-7_6

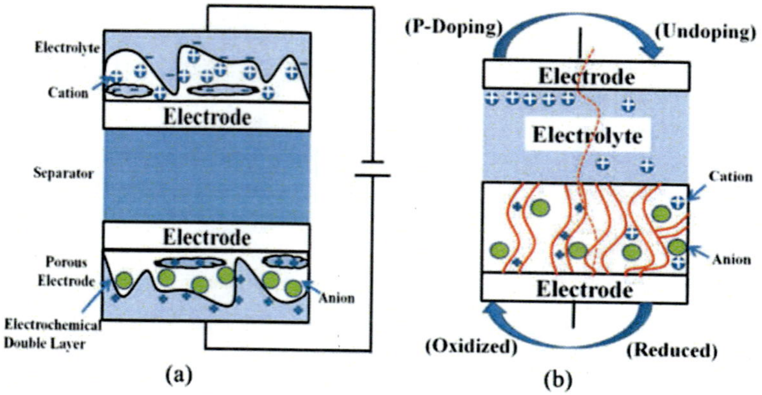

Fig. 6.1 A schematic diagram illustrating the operation principles of: **a** electric double layer capacitor (EDLC); **b** pseudosupercapacitor (PSC). Reprinted from ref. [22], copyright 2017, with permission from Elsevier

supercapacitor properties. Supercapacitors are classified into two varieties based on their energy storage mechanism: electrochemical double layer capacitors (EDLCs) and pseudocapacitors. As illustrated in Fig. 6.1a, the energy storage mechanism of EDLCs is based on the electrostatic interaction of ions on the large specific surface area of active electrode materials and electrolytes. The energy storage process in pseudocapacitors is a rapid and reversible Faradaic redox reaction on the surface and bulk near the surface between electrode materials and electrolyte ions, as shown in Fig. 6.1b.

CPs are regarded as the most promising materials when compared to transition metal oxide/hydroxides owing to their high specific capacitance, high conductivity, ecologically acceptable nature, simplicity of synthesis, high flexibility, and structural variety. CPs provide high specific capacitances, such as 620 Fg^{-1} for PPy, 750 Fg^{-1} for PANI, and 485 Fg^{-1} for PTh, by storing energy from the doping and dedoping processes associated with reversible redox reactions. However, due to the volume changes that occur throughout the doping/dedoping processes, CP-based electrodes often suffer from poor cycle stability. Table 6.1 shows the normal dopant levels for various polymers, as well as their typical specific capacitances and voltage ranges [19, 26].

CPs such as PANI, PPy, and PEDOT are commonly utilized in supercapacitors due to their numerous desired qualities such as high specific capacitance, electroactivity, and doping level. PANI is protonated by a redox process that converts it from a semiconductor (emeraldine base) to a metal (emeraldine salt) state, resulting in a structural change in the unpaired spin per repeat unit with a positive charge.

A flexible supercapacitor based on PANI nanowire arrays with a high capacitance of 588 F cm^{-3} and high rate capability is disclosed with significantly improving the specific area surface. The cyclic stability was improved further by incorporating a crosslinked structure into PANI using a hydrothermal technique. This was attributed

Table 6.1 Theoretical and experimental specific capacitances of conductive polymers

Conductive polymer	Mw ($g\ mol^{-1}$)	Dopant level	Potential range (V)	Theoretical specific capacitance (Fg^{-1})	Measured specific capacitance (Fg^{-1})[a]
PANI	93	0.5	0.7	750	240
PPy	67	0.33	0.8	620	530
PTh	84	0.33	0.8	485	–
PEDOT	142	0.33	1.2	210	92

Mw is molecular weight per unit monomer ($g\ mol^{-1}$), PANI is polyaniline, PPy is polypyrrole, PTh is polythiophene, and PEDOT is poly(3,4-ethylenedioxythiophene).

mostly to the stable net structure in the PANI backbone, which can withstand mechanical strain during cycling.

PPy provided more versatility in electrochemical processing than most CPs. PPy nanorods may be readily coated onto soft cotton textiles to create flexible supercapacitors with high specific capacitance of 325 Fg^{-1} and energy density of 24.7 Wh kg^{-1}. However, for practical applications, the cyclic stability of the supercapacitor needs to be increased. The construction of a hollow capsular structure for PPy can improve stability, with the specific capacitance remaining above 90% after 11,000 charge/discharge cycles. The peculiar tubular structure of PPy with porous capsular walls contributed to its excellent stability. The holes in the hollow fibers and between the capsules provide ample room for volume change during the doping/dedoping process.

6.1.1 Polyaniline-Based Supercapacitor

Polyaniline has been widely researched as a supercapacitor and battery electrode material. Figure 6.2D depicts the structure of polyaniline. Polyaniline possesses a number of favorable qualities for usage in a supercapacitor device, including strong electroactivity, a high doping level, great stability, and a high specific capacitance (400–500 Fg^{-1} in an acidic medium). Furthermore, it has high environmental stability and adjustable electrical conductivity (about 0.1 S cm^{-1} in the doped state with a Li dopant, but can range from 0.1 to 5 S cm^{-1}) and is readily processed. Polyaniline has a significant drawback in that it requires a proton to be effectively charged and discharged, necessitating the use of a protic solvent, an acidic solution, or a protic ionic liquid.

Polyaniline has a capacity range of 44–270 $mAhg^{-1}$. This variance in capacity is caused by a number of factors, including the synthetic process utilized, polymer morphology, the amount and type of binders and additives used, and the electrode thickness. Of all conductive polymers, polyaniline has the most variable-specific

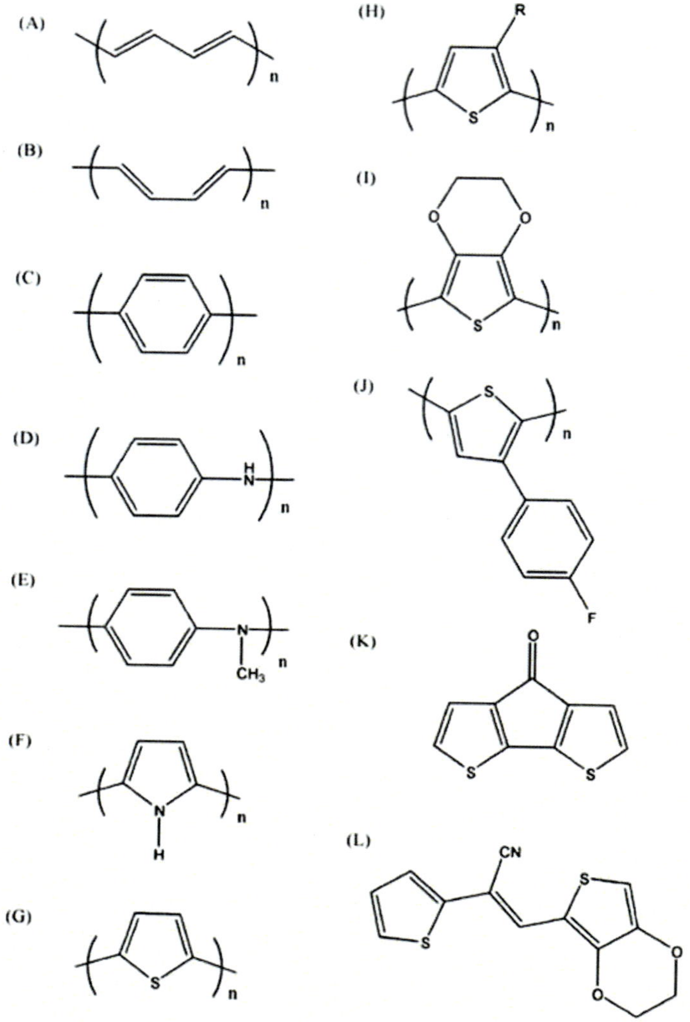

Fig. 6.2 Various conductive polymer structures. **A** Trans-poly(acetylene), **B** cis-poly(acetylene), **C** poly(p-phenylene), **D** polyaniline (PANI), **E** poly(n-methyl aniline) (PNMA), **F** polypyrrole (PPy), **G** polythiophene (PTh), **H** 3-substituted polythiophene, **I** poly(3,4-ethylenedioxythiophene) (PEDOT), **J** poly(3(4-fluorophenyl)thiophene) (PFPT), **K** poly(cyclopenta[2,1-b;3,4-b′-dithiophene-4-one]) (PCDT), and **L** 1-cyano-2-(2-[3,4-ethylenedioxylthienyl])-1-(2-thienyl)vinylene (PThCNVEDT). Reprinted from ref. [27], copyright 2011, with permission from Elsevier

capacitance. The specific capacitance achieved by electrodeposited polyaniline is greater than that of chemically produced polyaniline. The cycle life of polyaniline for a Li-doped positive electrode has been reported to reach over 5000 cycles, during which time the specific capacitance declined from 100 to 70 Fg^{-1}. Another research found that polyaniline doped with $LiPF_6$ had a specific capacitance of 107 Fg^{-1} after 9000 cycles, which fell to 84 Fg^{-1} after 9000 cycles.

Polyaniline can be made more stable by combining it with poly(n-methyl aniline) (Fig. 6.2E). The methyl groups of this polymer block proton exchange sites, making the polymer more resistant to chemical degradation and redox active. When this conductive polymer is combined with a lithium negative electrode, it can obtain a capacity of 52 $mAhg^{-1}$.

Fabrication of ordered and aligned nanostructures, such as PANI nanostructures, has gotten a lot of interest because of its better energy storage applicability. Kulia et al. created ordered arrays of vertically aligned PANI nanorods on a transparent ITO substrate using a nanotemplate based on supramolecular assemblies of block copolymer. The PANI nanorod arrays with nanospacing demonstrated exceptional electrochemical properties, with the maximum capacitance value obtained about 3407 F/g. This ultrahigh specific capacitance finding for PANI nanorods, on the other hand, is inconsistent and calls into question the fundamental understanding of PANI-based supercapacitors.

6.1.2 Polypyrrole-Based Supercapacitor

Because of its unique properties such as high conductivity, quick charge–discharge mechanism, strong thermal stability, low cost, and high energy density, polypyrrole (PPy) is one of the most promising p-type CPs for the Faradaic pseudocapacitor application. The CPs-based research has been heavily focused on PPy for the production of the pseudocapacitor electrode by the chemical or electrochemical methods. PPy, on the other hand, has been heavily concentrated on the creation of flexible electrodes because of its greater flexibility and higher conductivity characteristics when compared to polyaniline and polythiophene.

Polypyrrole has a higher degree of flexibility in electrochemical processing than other conductive polymers; hence, it has been the focus of much investigation as a supercapacitor or battery electrode. Polypyrrole, on the other hand, cannot be *n*-doped like thiophene derivatives and thus can only be utilized as a cathode material. Polypyrrole has a high capacitance per unit volume (400–500 F cm^{-3}) due to its higher density. One downside of dense growth is that it restricts dopant ion access to the interior locations of the polymer. This lowers capacitance per gram, especially for thicker electrode coatings. Polypyrrole is commonly doped with single-charged anions such as Cl^-, ClO_4^-, and SO_3^-, but when doped with multiple-charged anions such as SO_4^{2-}, it becomes unstable and the polymer undergoes physical crosslinking. According to Suematsu et al., the crosslinked materials have high diffusivity and capacitance, owing to the larger porosity of the growth.

6.1.3 *Thiophene-Based Conductive Polymers for Supercapacitor*

Polythiophene (PTh) and its derivatives can be classified as both *p*- and *n*-type CPs. Polythiophene is n-doped at very low potentials (below 2.0 V vs. Ag|Ag$^+$) and has inferior oxygen and water stability as well as conductivity than p-doped polythiophene. Polythiophene has weak conductivity in general; however, *p*-doped polymers are extremely stable in air and humidified environments. The specific capacitance of the PTh-based electrode is typically lower than that of PANI or PPy, but the main benefit is that it can operate in a higher potential window (1.2 V). This large negative potential window aids in the fabrication of an asymmetric type CPs-based supercapacitor device. Furthermore, it has a rapid self-discharge (i.e., it is readily oxidized back to the neutral state) and a poor cycle life in devices.

Polythiophene derivatives with a reduced band gap (i.e., *n*-doped at less negative potentials) were synthesized. The stability to oxygen and water can be increased further by replacing phenyl, ethyl, and alkoxy groups at the 3-position of the thiophene ring. Another strategy is to combine these substituents with electron withdrawing groups. Figure 6.2H depicts the overall structure of these substituted thiophene derivatives such as poly(3,4-ethylenedioxythiophene) (PEDOT) (Fig. 6.2I), poly(3-(4-fluorophenyl)thiophene) (PFPT) (Fig. 6.2J), poly(3-(3,4-difluorophenyl)thiophene) (MPFT), and poly(1-cyano-2-(2-(3,4-ethylenedioxylthienyl)vinylene) (ThCNVEDT). Because of inherent challenges in the *n*-doping process, which results in a considerable reliance of the behavior on the size of the counter ion and the solvent, there are few instances of *n*-doped thiophene derivatives being employed as supercapacitor materials. Tables 6.2 shows the potential range and specific capacitance of several polythiophene derivatives.

Because of its great environmental stability, PEDOT has been investigated as a viable pseudocapacitor material. In the *p*-doped state, the polymer is strongly conductive (300–500S cm^{-1}) and has a greater potential range of 1.4 V, a narrow

Table 6.2 Properties of polythiophene derivatives in p-doped state

Active material	Potential limits (V) versus Ag|AgCl	Voltammetric charge (C g^{-1})	specific capacitance (Fg^{-1})
P-PFPT	0.3 to 0.9 (0.6)	146	244
P-MPFPT	0.3 to 0.9 (0.6)	127	212
PThCNVEDT	0 to 0.8 (0.8)	173	216
PMeT	−0.16 to 1.18 (1.34)	–	220
PEDOT	−0.5 to 0.9 (1.4)	144	103
Carbon	−1.5 to 2.1	82	133

Reprinted from ref. [27], copyright 2011, with permission from Elsevier

PFPT is poly(3-(4-fluorophenyl)thiophene), P-MPFPT is poly(3-(3,4-difluorophenyl)thiophene), and PThCNVEDT is poly(1-cyano-2-(2-(3,4-ethylenedioxylthienyl))-1-(2-thienyl)vinylene)

band gap of 1–3 eV, good thermal and chemical durability, and rapid charge mobility, resulting in quick electrochemical kinetics. The polymer has also been discovered to have strong film-forming capabilities and to be switchable quickly with a minimum of side reactions, resulting in a long cycle life. On the negative side, it has a relatively low specific capacitance of roughly 90 Fg^{-1} due to its huge molecular weight and a doping level of around 0.33. Lota et al. claim that a capacitance of 180 Fg^{-1} is attainable using PEDOT. Although this value was published for a thin film, it is unsuitable for energy storage. When a pellet (10–20 mg) was created, the PEDOT had an estimated capacitance of 80–100 Fg^{-1}. Another drawback of this polymer is that the monomer is insoluble in aqueous solutions. As a result, acetonitrile, a volatile and hazardous organic solvent, is commonly utilized to deposit the polymer.

As shown in Table 6.3, several polythiophene derivatives exhibit lower capacitances in the n-doped state than in the *p*-doped state. It has been discovered that poly(3-(4-trifluoromethylphenyl)thiophene) (PTFMPT) has the same charge for *n*- and *p*-doping but shows "charge trapping," where the electrically isolated areas form when the material is cycled. As a result, less charge is discharged during discharge than during charging. A hybrid supercapacitor with an activated carbon negative electrode has been created using PFPT. The specific energy for 4 cm^2 cells was 48 $Whkg^{-1}$, and the specific power was 9 $kWkg^{-1}$. While the specific energy for 60 cm^2 prototypes was 7.5 $Whkg^{-1}$, the specific power was 250Wkg^{-1}.

Substituting different chemical groups on the polythiophene to increase n-doping leads to larger molecular weight (and hence lower specific capacitance), higher material costs, and the need for a higher injected charge and greater mechanical strength. Poly(3-methyl thiophene) (PMeT) is a well-known thiophene derivative used in supercapacitor electrodes. PMeT is inexpensive and has a reasonably low molecular weight (a decent specific capacitance of 220 Fg^{-1}) and a low resistivity (2 cm^2). In a Type III device, the charge is provided at high potentials and may be cycled thousands of times without substantial capacity loss. Poly(cyclopenta(2,1-b;3,4-b′-dithiophene-4-one)) (PCDT) is another intriguing thiophene-based material that may be employed in either the p- or n-doped state (see Fig. 6.2K). Cyclic voltammetry revealed that the doping level was 0.19 dopants per monomer unit, which was much lower than polythiophene. The p- and n-doping states produce 70 Fg^{-1}, whereas the

Table 6.3 *p*- and *n*-doping of various thiophene derivatives

Polymer	*p*-Doping		*n*-Doping	
	Potential limits (V) vs.SCE	Capacitance (Fg^{-1})	Potential limits (V) vs.SCE	Capacitance (Fg^{-1})
PFPT	−0.2 to 1.0	95	−1.7/−1.0	80
PDTT	−0.2 to 1.0	110	−1.5/−0.2	75
PMeT	−0.2 to 1.15	220	−2.0/−1.0	165

PFPT is poly(3-(4-fluorophenyl)thiophene), PDTT is poly(dithieno(3,4-b:3,4d) thiophene), and PMeT is poly(3-methyl thiophene).

energy and power per mass of active material are approximately 6 $Whkg^{-1}$ and 1 $kWkg^{-1}$ (at 18 s discharge).

6.1.4 Comparison with Other Types of Supercapacitor Materials

Figure 6.3 shows a comparison of the specific capacitances of conductive polymers and their composites with other supercapacitor materials. When compared to carbon-based supercapacitor materials, both conductive polymer and metal oxide materials (usually referred to as pseudocapacitors) enhance the amount of specific capacitance available for a device, but at the expense of cycle life. At the other end of the scale, ruthenium oxide is capable of producing specific capacitances that are significantly superior to those of any other materials. In actuality, ruthenium oxide is simply too costly to be a viable supercapacitor material. When examining the specific capacitance of conductive polymer materials, polyaniline is the most promising material. The specific capacitance, on the other hand, might vary greatly depending on the polymer manufacturing procedure. Furthermore, the use of polyaniline as a supercapacitor material is hampered by the need for enough proton activity in the electrolyte solution to provide appropriate cycling. The capacitance of manganese oxide is provided in a variety of values in Fig. 6.3 because it is reliant on the phase, morphology, and form of the oxide, as depends on the technique of synthesis. Nonetheless, as a supercapacitor material, this material seems to be the most competitive to conductive polymers. Specific capacitances equivalent to polyaniline are achievable with other conductive polymer materials by combining them, for example, with PEDOT and iron oxide. The benefits of such a composite include high specific capacitance and the flexible nature of the film assisting the device production. Other conductive polymer composites (e.g., using carbon compounds) do not always result in higher specific capacitances, but they do have numerous other advantages, such as longer cycle lifetimes and higher charge rates.

6.1.5 Multidimensional Performance Optimization of Conductive Polymer-Based Supercapacitor Electrodes

Strategies for increasing the specific capacitance and rate capability of conductive polymer electrodes are thus largely focused on increasing material utilization efficiency: creating high surface area nanostructures with short ion diffusion length scales that allow for easy electrolyte infiltration throughout the electrode bulk.

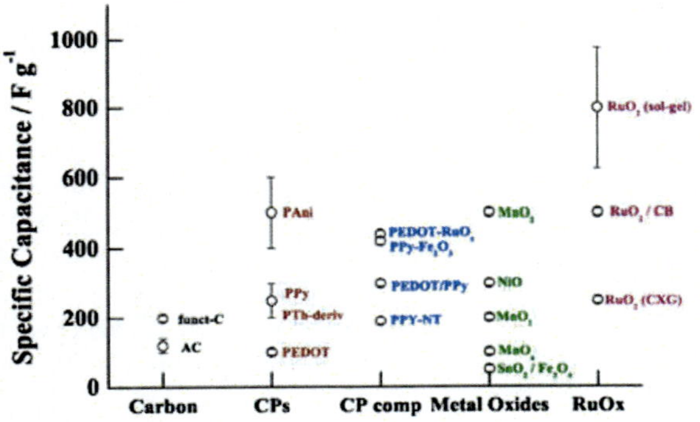

Fig. 6.3 Typical specific capacitances for different supercapacitor materials, where CXG is carbon xerogel and CB is carbon black. Funct-C refers to functionalized carbons, AC refers to activated carbon, CPs refers to conductive polymers, and CP comp refers to conductive polymer composites. Reprinted from ref. [27], copyright 2011, with permission from Elsevier

6.1.5.1 Enhancing Material Utilization Efficiency

The bulk of the pseudocapacitive material must be positioned within nanometers of the electrode–electrolyte contact for efficient material use. It should be noted that using very thin layers of active material might alleviate concerns with ion diffusion kinetics and so theoretically avoid this issue. Zhang et al. created an ultrathin (300 nm) layer of PEDOT:PSS on graphite foil that remained capacitive even at ultrahigh scan speeds of up to 1000 V s^{-1}.

Because of their enormous surface area and aspect ratios, one-dimensional (1D) materials are particularly appealing for this purpose. Vertically aligned PANI nanowire arrays on graphene oxide sheets attained a specific capacitance of 555 Fg^{-1}, but randomly oriented PANI nanowires synthesized by the same process generated only 298 Fg^{-1}. The regular organization of the arrays in the former scenario increased the total quantity of PANI in the supercapacitor that was accessible to the electrolyte and therefore capable of participating in pseudocapacitance processes. Estimating the average diffusion length of ions in the electrode can be important for further optimizing 1D morphologies like this. If this diffusion length is higher than or equal to the diameter of the 1D polymer structure, the system is not restricted by ion diffusion within the polymer.

A high-performance conductive polymer nanostructure must be tailored for electrolyte penetration in addition to low ion diffusion lengths within the active material. Three-dimensional, high surface area morphologies are particularly successful in this application. Most typically, conductive polymers are placed on surfaces with homogeneous macroporous architectures, such as nickel foam, to promote large surface area and easy ion transport.

While easy electrolyte penetration is essential for optimal material usage, excess electrolyte held in porous structures can actually impair specific capacitance by adding inactive mass to the electrode. Modifications were made to the pore size distributions of conductive polymer electrodes. Wang et al. modified nanocellulose fiber substrates with quaternary amine groups prior to PPy polymerization. This cationic surface modification was discovered to diminish the polymer macropore volume while preserving the micro- and mesopore structure. The volumetric capacitances of the resultant pore structure were found to be greater than those of PPy produced on unaltered substrates. The creation of self-doping polymer architectures is another interesting method of enhancing ion accessibility and material use. Christinelli et al. created layer-by-layer electrodes of poly(o-methoxyaniline) (POMA) and poly(3-thiophene acetic acid) (PTTA), with the POMA serving as the pseudocapacitive polymer and the PTTA providing carboxylate anions to balance the POMA positive charges. This technique decreases the requirement for electrolyte ion intercalation, allowing for higher mass loading without losing material utilization. As the number of POMA/PTTA bilayers rose, the specific capacitance of these electrodes increased nearly linearly (up to 140 Fg^{-1} for 112 bilayers).

6.1.5.2 Electrical Conductivity

High electrical conductivity is a critical criterion for supercapacitors since it allows for the quick charge transfer kinetics necessary for high power operation. The equivalent series resistance is the sum of all the electrical and ionic resistances in a device that restrict its power density (ESR). Ion transport in the electrolyte, ion transport inside the electrode, electrical conduction within the electrode, and all other device components are examples of resistance sources (current collector, leads, etc.). Improving specific capacitance is inextricably tied with increasing electrical conductivity in supercapacitor electrodes. The electrical conductivity of a conductive polymer is primarily governed by its doping level, because dopants introduce the free charge carriers that are responsible for polymer conductivity.

High charge carrier mobility, both inside and between polymer chains, is required for high electrical conductivity. Intrachain conductivity is largely restricted by conjugation length and may be increased by reducing the amount of polymer defects that disturb the delocalized π system. Chain alignment and polymer crystallinity promote interchain conductivity. Single crystal PEDOT nanowires grown by vapor phase polymerization, for example, have ultrahigh conductivity of over 8000 S cm^{-1}, which is greater than any achieved for PANI or PPy to date. Aromatic dopants facilitate anisotropic development of polymer films and hence better conductivity compared to films formed with spherical, non-aromatic dopants, which promote interchain π-π stacking and thus strong molecular ordering.

The ESR of a device can also be influenced by the interfacial resistance between the active material and the current collector. Certain anionic dopants can help with this problem by improving the adherence of the polymer active material to the substrate. Tiron (a sulfonate aromatic compound) and sulfanilic acid azochromotrop have been reported to promote uniform and adherent film formation. Their phenolic hydroxyl groups can deprotonate and bind with the metal ions of the current collector, which improves polymer adhesion and facilitates charge transfer.

These binder and current collector adhesion concerns are often solved by synthesizing the electrode active material directly onto the current collector or creating the active material as a free-standing film. The former technique is often implemented via synthesis (generally electrochemical polymerization) directly onto conductive substrates with high surface area such as nickel foam and carbon nanotube-based frameworks.

6.1.5.3 Cycling Stability

Swelling-Induced Degradation

One of the key issues with conducting polymer supercapacitors is long-term cycling stability. Unlike EDLCs, which may maintain stable capacitance values for over 100,000 cycles, many conductive polymer electrodes retain less than half of their initial capacitance values after only 1000 cycles. These long-term stability concerns are mostly caused by mechanical fatigue produced by volumetric changes during cycling. During charging and discharging, ions from the electrolyte intercalate in and out of the polymer to maintain charge neutrality, causing the active material to inflate or shrink. This process can harm the microscopic hierarchical structure of the electrode as well as produce molecular-scale disturbances such as polymer chain disorder or the collapse of ion flow channels, making the polymer unable to store charge effectively.

One of the most commonly used methods for enhancing supercapacitor stability is to include conductive polymers into open and/or flexible networks that can adjust to volumetric changes and thus retain the mechanical integrity of the electrode upon repeated cycling (Fig. 6.4a) [6]. Carbon-based composites are ideal for this application. For example, carbon nanotubes (CNTs) can be formed into open networks, resulting in electrodes with free space for conductive polymer swelling/shrinking. By combining conductive polymers with MXenes, stability may be increased in a similar manner. Cycling stability can also be increased by providing a protective layer to reduce the volumetric changes in the polymer. Cycling stability can be greatly improved by adding a thin carbonaceous shell to PANI or PPy nanowires. After 10,000 cycles, carbon-coated PPy kept 85% of its initial capacitance, but bare PPy tested under the same conditions retained less than 25% of its initial capacitance (Fig. 6.4b) [15]. Cycling stability can be improved via synthesis processes that increase the molecular ordering of a conductive polymer. Huang et al. found that electropolymerized PPy on a stainless steel mesh retained 97% of its specific capacitance after 15,000 cycles and 86% after 100,000 cycles (Fig. 6.4c).

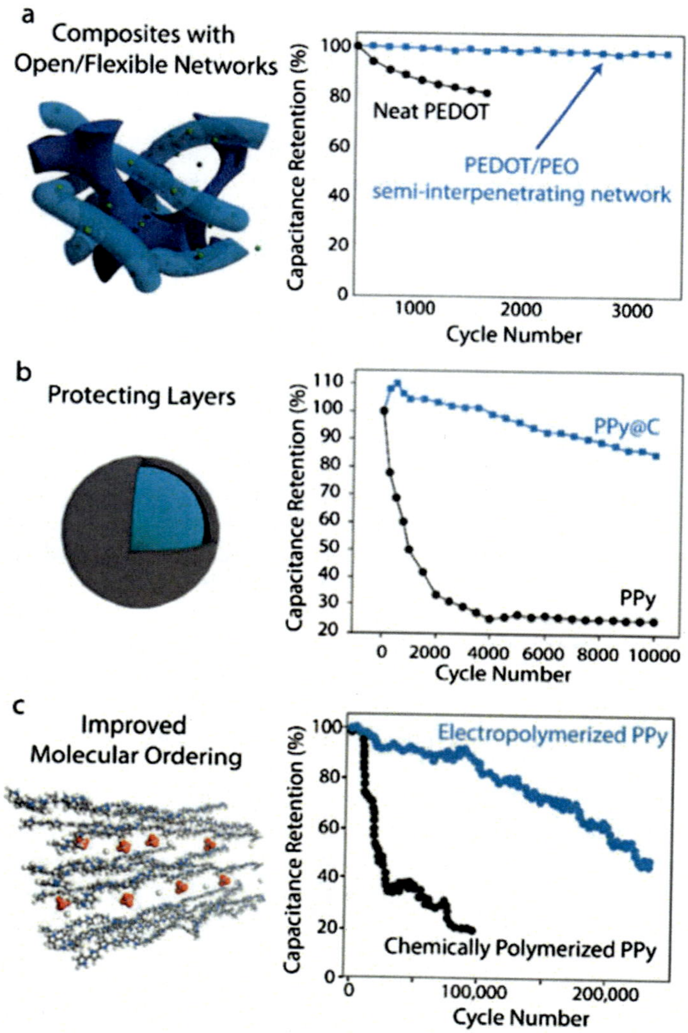

Fig. 6.4 Strategies for enhancing polymer-based supercapacitor cycling stability. **a** Composites with open/flexible networks. Data shows the stability improvements upon interpenetrating PEDOT in a flexible, ionically conducting PEO network. **b** Protecting layers. The stability enhancements imparted by introducing a carbon protecting layer over PPy electrodes are shown. **c** Improved molecular ordering. Data shows the results of improving the molecular ordering of PPy through electropolymerization. Reprinted from ref. [7], copyright 2017, with permission from Royal Society of Chemistry

Overoxidation-Induced Degradation

In addition to the challenges created by volumetric fluctuations, polymer overoxidation caused by operating outside of an adequate potential window might impair long-term device performance. PANI, for example, will break down at high rates to produce hydroquinone and p-aminophenol. Similarly, in polythiophene-based polymers such as PEDOT, sulfur from the thiophene ring can form high-potential bonds with oxygen from the solvent, resulting in the generation of SO_2. In each of these circumstances, irreversible chemical changes disturb the conductive polymer backbone, drastically reducing the material electrical conductivity (and consequently its capacity to store charge).

6.1.6 Nanostructured Conductive Polymers for Supercapacitor

Because of the unique features resulting from their nanoscale size, conductive polymer nanostructures are projected to outperform bulk conductive polymers in technological applications: (i) high electrical conductivity; (ii) large specific surface area; (iii) short path lengths for ion and electron transport; (iv) improved cycle life due to better accommodation of strain caused by electrochemical reaction; (v) mixed conductive mechanism of both electronic and ionic conductivity, which lowers the interfacial impedance between electrodes and electrolyte; (vi) lightweight and large ratio of specific discharge power to weight; and (vii) higher electrode/electrolyte contact area leading to higher charge/discharge rates [23]. In general, a high-performance EC electrode must simultaneously minimize four primary resistances during electrochemical charge/discharge: (1) ion transport in the electrolyte, (2) ion transport in the electrode, (3) electrochemical reactions in the electrode, and (4) electron conduction in the electrode and current collector.

3D nanostructured conductive polymers have been created and used to improve the electrochemical performance of conductive polymer-based supercapacitor devices. Electrochemical polymerization was used to create PPy microcontainers like as bowls and cups utilizing gas bubbles as templates. The novel structures exhibit high oxidation and reduction peaks in the CV test, as well as a capacitance of 200 Fg^{-1} due to the increased surface area.

PANI hydrogels with 3D nanostructures showed considerable promise for high-performance ECs. During charging/discharging, the 3D continuous nanostructured framework promotes rapid ion movement. The associated specific capacitance resulted in a high specific capacitance (480 Fg^{-1} at 0.2 Ag^{-1}). 3D nanostructured conductive PPy hydrogels with structure-derived elasticity were also demonstrated at an organic/aqueous biphasic interface. This technique resulted in PPy hydrogels with adjustable porosity nanostructures, which caused strong mechanical qualities and great performance with a specific capacitance of 380 Fg^{-1} [24].

Conductive polymer nanowire arrays (e.g., PPy and PANI) provide a huge surface area and an optimum ion diffusion channel in ordered nanowire architectures, making them potential materials for high-performance supercapacitors. PPy nanowire arrays, for example, had a capacitance of 566 Fg^{-1} and kept 70% of their original capacitance after hundreds of charge–discharge cycles.

6.1.7 *Flexible Supercapacitors from CP-Based Hydrogels*

Hydrogels are 3D crosslinked networks of hydrophilic polymers that contain a lot of water and have excellent flexibility. They may be made from a variety of hydrophilic monomers, including as acrylamide, acrylic acid, and vinyl acetate, by chain or step growth polymerization, and then crosslinked using a functional crosslinker to produce a 3D porous network. Multiple benefits, such as the intrinsic 3D porous structure, redox activity, and quick ion/charge transfer, may be imparted into the hydrogels as flexible supercapacitor electrodes, by incorporating CPs into flexible hydrogels. In addition to CP-hybrid hydrogels, pure CP hydrogels have been studied as electrode materials for supercapacitors in recent years. To generate flexible CP-based hydrogels with 3D crosslinked conductive structures, a number of techniques, including self-crosslinking, crosslinking using crosslinkers, and oxidant-templating, have been devised.

6.1.7.1 Pristine CP Hydrogel Supercapacitors

Ghosh and Inganäs published the first substantially swollen CP hydrogel electrodes for supercapacitors in 1999, in which crosslinked PEDOT-polystyrene sulfonate (PSS) colloidal particles created a nanoscale continuous conducting network with a large surface area desirable for charge storage. This innovative hydrogel electrode expanded the possibilities for supercapacitors, particularly flexible supercapacitors. Since then, significant progress has been achieved in the preparation of CP hydrogels for supercapacitors, including PANI, PPy, and PEDOT hydrogels. As illustrated in Fig. 6.5a, b, a phytic acid with several hydroxyl groups was utilized as both a crosslinker and a dopant in the creation of PANI hydrogel, which could be inkjet printed or spray-coated to have conductivities as high as 0.11 S cm^{-1}. The resulting PANI hydrogel supercapacitor electrodes had a hierarchical 3D porous structure, a large surface area, and electrical conductivity and produced a high specific capacitance of 480 Fg^{-1} with good rate capability and cycle stability. A symmetric flexible solid-state supercapacitor was created using PANI hydrogel electrodes supported by carbon fiber cloth and a PVA/H_2SO_4 hydrogel as the solid electrolyte.

Shi et al. created a hierarchically 3D nanostructured elastic PPy hydrogel by interfacial polymerization of pyrrole at the isopropanol alcohol/water interface using phytic acid as a crosslinker and dopant and ammonium persulfate (APS) as an oxidant in a similar but distinct work. The resulting PPy hydrogel has a dry conductivity of

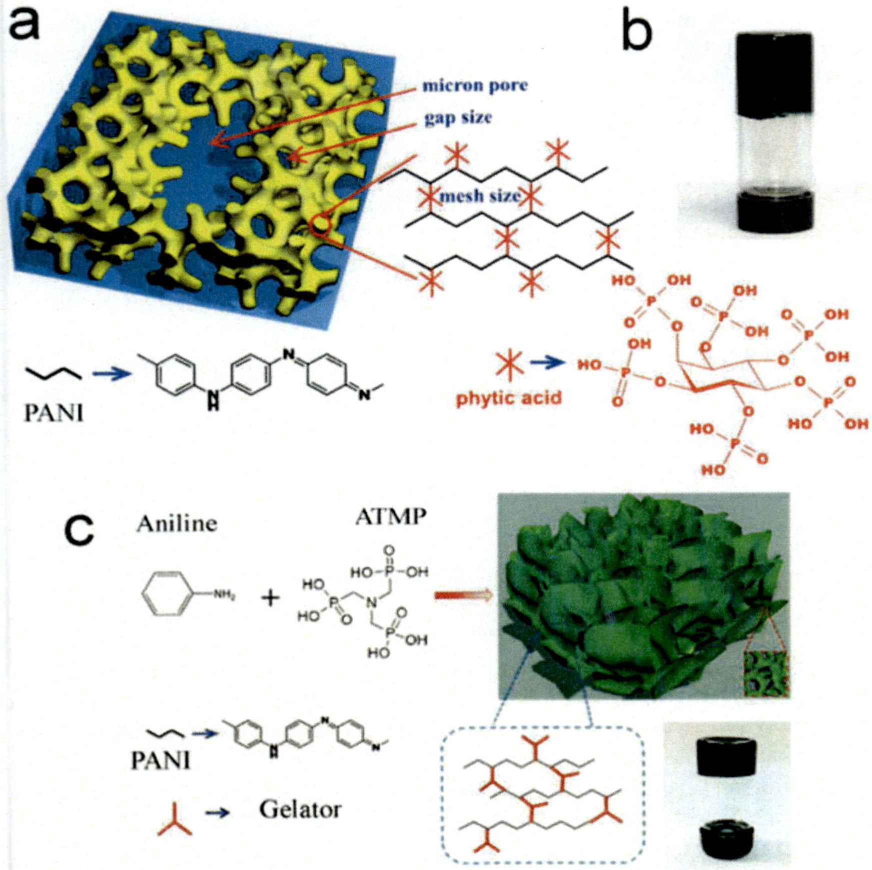

Fig. 6.5 **a** Schematic illustrations of the 3D hierarchical microstructure of the gelated PANI hydrogel where phytic acid plays the role as a dopant and a crosslinker. Three levels of hierarchical porosity from angstrom, nanometer to micron size pores have been highlighted by red arrows. **b** A photograph of the PANI hydrogel inside a glass vial. **c** Schematic illustration of the formation of the 3D hierarchical nanostructured PANI hydrogel and a photograph of the PANI hydrogel inside a glass vial. Reprinted from ref. [11], copyright 2019, with permission from John Wiley and Sons

0.5 S cm^{-1} and a high mechanical strength and flexibility, making it suitable for flexible supercapacitors. The pore size of the 3D porous hierarchical nanostructure may be controlled by altering the molar ratio of pyrrole to phytic acid. The flexible symmetric PPy hydrogel supercapacitors demonstrated a high areal capacitance of 6.4 F cm^{-2} with a strong cycling stability at a mass loading of 20 mg cm^{-2} (less than 7% capacitance degradation after 2000 cycles). Similarly, another polyhydroxy acid (ATMP, Fig. 6.5c) has been used as both a gelator and a dopant in the preparation of PANI hydrogels with high electronic conductivity (0.35 S cm^{-1}) and good electrochemical performance.

Because the addition of non-conducting crosslinkers might diminish the electrical conductivity of PANI hydrogel, self-crosslinked PANI hydrogels were created utilizing aniline hydrochloric salt as the monomer and APS as the oxidant. The 3D porous PANI hydrogel thus created with a coral-like shape of ramous nanofibers is advantageous for improving electrochemical performance. As a consequence, with a current density of 1 Ag^{-1}, the resulting PANI hydrogel electrode had a maximum specific capacitance of 750 Fg^{-1}. Zhou et al. described an in situ polymerization hydrogel of pure PANI nanofibers utilizing $V_2O_5.nH_2O$ as both an oxidizing agent and a hard template. The 3D assembly of the as-synthesized PANI hydrogel with a 3D crosslinked structure and excellent conductivity (0.12 S cm^{-1}) was enhanced by the oxidant template-assisted fast synthesis of ultrathin PANI nanofibers. The resulting PANI hydrogel electrode has a high specific capacitance of 636 and 626 Fg^{-1}, respectively, at current densities of 2.0 and 25 Ag^{-1}. After 10,000 cycles, the capacitance retention achieved 83.3%, showing high cycle stability. This oxidant-templating approach may be used to make PPy or PEDOT hydrogels with 3D porous conducting channels.

6.1.7.2 CP Hybrid Hydrogel Supercapacitors

As electrode materials for flexible supercapacitors, several hybrid CP hydrogels have been created to enhance the mechanical characteristics and flexibility of pure CP-based conductive hydrogels. PVA has been the most often utilized flexible substrate for CPs hybrid hydrogels so far, but PAAM, graphene oxide/graphene, sodium alginate, and G-Zn-tPy have also been employed. These hybrid hydrogels can be highly stretchable, foldable, compressible, and even self-healable, promising for applications in portable/wearable electronics.

Due to the presence of the non-conducting substrate, designing and developing hybrid hydrogels with high conductivity and large surface area remain significantly difficult. Nonetheless, a supramolecular strategy for crosslinking PANI and PVA via dynamic bonding has been developed, in which a functional monomer (i.e., 3-aminophenylboric acid, ABA) bearing both amino and boronic acid groups was used to copolymerize with aniline and crosslink with PVA via the –OH condensation reaction. The resulting hydrogel had a porous network structure made up of linked nanoparticles, a conductivity of 0.1 S cm^{-1}, a high tensile strength (5.3 MPa), stretchability (at elongation at break of 250%), and a high compress stress (30 MPa), all of which were similar to previous hydrogels. Flexible supercapacitors made from the PANI/PVA composite hydrogel demonstrated a high electrochemical capacitance of 306 mF cm^{-2} and 153 Fg^{-1}, as well as a high energy density of 13.6 Wh kg^{-1}. Furthermore, after 1000 mechanical folding cycles and 1000 charge–discharge cycles, respectively, capacitance retentions of 100% and 90% were achieved. These exceptional capabilities are ascribed to the conductive hybrid hydrogel, which is crosslinked between stiff and soft polymer chains with a mesoporous structure. As a result, the supermolecular method offers a viable path for creating conductive, flexible, and durable electrode materials.

Later, the mechanical properties of the PANI/PVA hybrid hydrogel were improved using a post-synthesis freeze–thaw cycle process. The PANI/PVA hybrid hydrogel increased its tensile strength and elongation break from 5.3 to 16.3 MPa and from 250 to 407% after five freeze–thaw cycles, respectively. The resulting flexible supercapacitors had a high energy density as well as good cycle stability. Other approaches for controlled polymerization of CP chains within 3D crosslink networks consisting of soft polymer substrate, such as PVA and PAAM, have been developed in addition to crosslinking the rigid CP chains with soft polymer chains to form a conductive interconnected structure or a homogeneous distribution of CP phase in the soft substrate.

6.1.7.3 "All in One" Hydrogel Supercapacitors

The electrochemical performance of flexible supercapacitors is heavily influenced by electrode materials, electrolytes, and device designs. It is highly desirable to develop "all in one" flexible high-performance supercapacitors that can integrate the flexible electrodes and electrolyte all together for wearable electronics. Wang et al. created an integrated supercapacitor device with two PANI layers (acting as electrodes) immersed in a solid-state electrolyte of a 3D porous PVA-H_2SO_4 hydrogel sheet. The surface and subsurface of a 40-μm thick PVA-H_2SO_4 hydrogel sheet covered with in situ polymerized PANI functioned as both the electrode and electrolytes for flexible supercapacitors with a large areal capacitance of 488 mF cm^{-2}, outstanding cycle stability, and flexibility.

Self-healing matrices for self-healing supercapacitors are usually made from supermolecular networks with many hydrogen bonds. Unfortunately, the majority of the reported self-healing supercapacitors require an extra healable layer(s), resulting in difficult supercapacitor construction and a low energy density. In situ chemical deposition of PANI and single-walled CNT (SWCNT) hybrid onto the two-sided faces of healable PVA-H_2SO_4 hydrogel resulted in the fabrication of a self-healable "all in one" flexible supercapacitor.

6.1.8 Conductive Polymer Composites for Supercapacitor

Pure CP electrodes typically suffer from poor cyclic stability owing to the generation of rupture caused by the large volume variation between the doped/dedoped states during the charge/discharge process. Their electrical conductivity is not high enough, which makes it difficult to reach the standard of high power density required by supercapacitors, especially at high current densities. Great efforts have been made to solve this problem, and one of the most effective approaches is introducing the highly conductive materials such as carbon nanomaterials into the CPs.

Carbon materials are appealing because they are inexpensive and have strong chemical and thermal stability and great electrical conductivity and mechanical characteristics. As a result, combining CPs with carbon materials is an effective way to increase CP stability. After compositing, the rate capability and cycle stability of the supercapacitor are significantly increased without compromising flexibility. The carbon nanomaterial cannot benefit for the improvement of specific capacitance in this case. Many papers on composites of CPs and carbon nanostructures have been published during the last few decades.

Because of their great chemical and electrochemical stability, huge surface area, and high electrical conductivity, CNT and graphene are commonly employed to build composites with CPs for flexible supercapacitors. To create free-standing and flexible PANI nanorod/graphene composites for supercapacitors, PANI nanorods were directly deposited onto high-quality graphene paper or woven fabric. Due to a synergistic effect of graphene and PANI, the PANI/graphene paper-based supercapacitor exhibited higher specific capacitance and better cyclic stability than pure graphene or PANI, with the specific capacitance retaining 96% after 1000 cycles.

However, the particular surface area of the electrode was rather small, limiting the performance. One excellent method for improving the specific region is to introduce a porous construction or interconnected network, e.g., carbon cloth. An interfacial polymerization method, for example, can coat CPs onto a flexible carbon cloth. In addition to flexibility, supercapacitors based on the combination of carbon materials and CPs demonstrated improved specific capacitances and power densities. However, the performance falls well short of what is expected for actual applications.

Furthermore, higher capacitive materials, such as transition metal oxides/hydroxides, are used to create multielement composites with greater energy densities for flexible electrodes. Recently, the use of transition metal oxides/hydroxides such as MnO_2, CoO/Co_3O_4, RuO_2, $NiO/Ni(OH)_2$, Fe_2O_3, and $Co(OH)_2$, as a viable technique for high-performance supercapacitors based on their high specific capacitances, despite their poor electrical conductivity, has been examined. By coating highly ordered 3D $PPy/NiCo_2O_4$ nanowires on carbon textiles, metal oxides were introduced into the composite with a high specific capacitance of 2244 Fg^{-1}. The hydrous RuO_2 nanoparticles that were introduced and adhered to the PEDOT:PSS/graphene surfaces allowed for redox interactions with the surrounding electrolyte, increasing the specific capacitance of the electrode materials. CPs have the ability to significantly increase the electrical conductivity of transition metal oxides/hydroxides. PPy was employed to accelerate metal oxide electron transport, and a supercapacitor containing $NiCo_2O_4$/PPy core–shell nanowires coated on carbon microfibers demonstrated a high pseudocapacitive performance of 2055 Fg^{-1}. PANI was also used in electrochemical deposition to improve the electrical conductivity of cobalt-based metal–organic framework (MOF) crystals and Faradaic processes at the interface (Fig. 6.6). PANI was not only deposited on the surface of MOFs, but it also served as a bridge to unite isolated MOF particles by covering the interparticle free space.

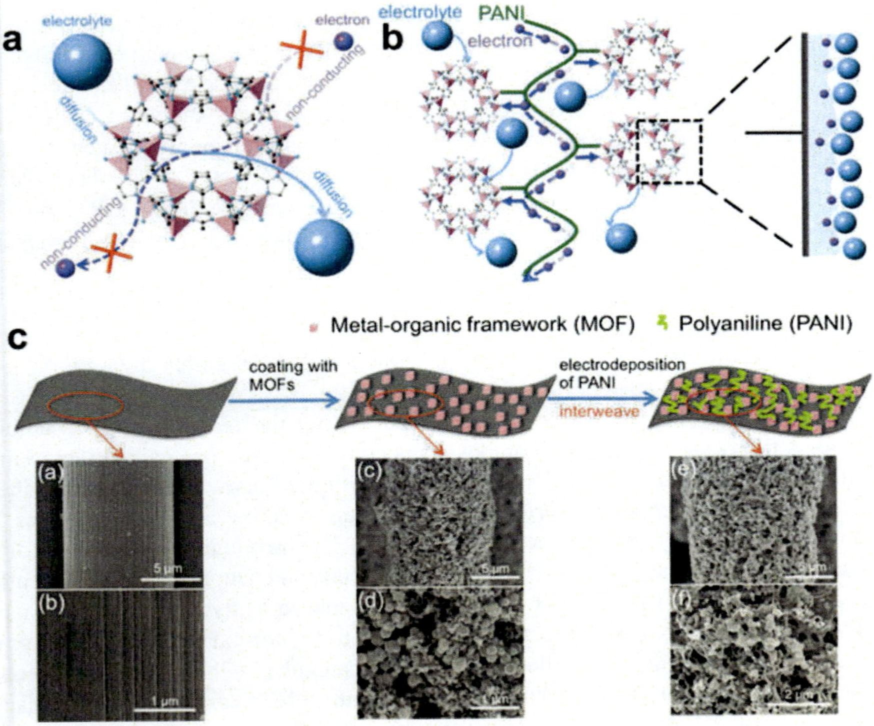

Fig. 6.6 Flexible supercapacitors based on the PANI/MOF/carbon cloth electrode. **a** Schematic illustration showing that electrons cannot migrate or access the skeleton of MOF crystals, while the electrolyte can move in and out of the MOF pores by diffusion effect. **b** Schematic illustration indicating that, after depositing with PANI, both electron and electrolyte can access MOF crystals. **c** Schematic illustrations and SEM images during the preparation of PANI/MOF/carbon cloth electrode. Reprinted from ref. [31], copyright 2015, with permission from American Chemical Society

6.1.8.1 Graphene/Conductive Polymer Composites for the Application of Supercapacitor Electrode Materials

Because of their highly interconnected pore structure, fast ion and mass transfer, high percolated electronic transport, low density, and easy access to adsorption sites, three-dimensional graphene frameworks (3DGFs) such as foams, aerogels, and sponges have emerged as promising electrode materials for EDLCs. However, because of the minimal capacitance contribution from the double layer storage method, pure 3DGF electrodes often have low specific capacitance. Due to their larger theoretical capacitance than carbon materials, their superior redox properties, and easy synthesis, π-functional conductive polymers, particularly polyaniline (PANI) and polypyrrole (PPy), have been employed as electrode materials for pseudocapacitors. Unfortunately, due to swelling and shrinking during the charge/discharge process, their low cycle stability restricts its practical applicability. Incorporating conductive polymers into 3DGFs to create three-dimensional graphene/conductive polymer (3DGCP) composites has the potential to considerably increase overall electrochemical performance by using both components. The synthetic strategies of 3DGCP composites used in supercapacitors can be roughly classified into four categories: (i) directly incorporating polymers during reduction self-assembly of graphene oxide (GO) sheets, (ii) in situ polymerization of monomers in pre-synthesized 3DGFs, (iii) one-step in situ polymerization of monomers and GO assembly, and (iv) template methods.

Directly Incorporating Polymers During Reduction Self-Assembly of GO

Direct incorporation of conductive polymers during the reduction self-assembly process to synthesize 3DGCP composites is convenient and has gained increasing attention because these conductive polymers not only endow 3DGCP composites with novel electrochemical performance and properties, but also act as spacers to further minimize graphene sheet agglomeration. Nanostructured conductive polymers, in particular nanotubes, nanowires, nanofibers, and nanorods, have frequently been engineered to combine with GO, which subsequently self-assembled into 3DGCP composites via a reduction method. Ye et al., for example, dispersed pre-prepared one-dimensional hollow polypyrrole nanotubes (PNTs) in GO aqueous dispersion. Through π-π stacking interactions, the PNTs can interact with many surrounding GO sheets. The PNTs were then effectively wrapped into graphene layers during the 3D self-assembly of reduced graphene oxide (rGO) and functioned as spacers to prevent graphene from restacking and agglomeration via a simple chemical reduction. The as-prepared hierarchically porous 3D rGO/PNT framework had a high specific capacitance of 253 Fg^{-1}, which was superior to pure 3DGFs and PNTs.

In Situ Polymerization of Monomers in Pre-Synthesized 3DGFs

In situ oxidative polymerization or electropolymerization (electrochemical deposition) of monomers in pre-synthesized 3DGFs can also be used to create 3DGCP composites. The approach does not necessitate the difficult functionalization and dispersion of individual GO sheets, which is advantageous for overcoming any process incompatibility, and as a result, the composites display improved mechanical and electrical conductivities. In contrast to in situ oxidative polymerization of monomers, which requires initial synthesis of 3DGCP composites followed by slurry coating on conductive substrates, a facile electrochemical deposition technique may be employed to immediately manufacture 3DGCP electrodes. Furthermore, the electrochemical deposition process was discovered to be successful for creating ordered structures, and as-fabricated hybrid electrodes demonstrated outstanding electrochemical performance.

Yu et al. prepared a 3D graphene/PANI composite electrode through a one-step electrochemical deposition, in which highly ordered polyaniline nanocone arrays were tightly attached to the surface of the 3D graphene electrode using a 3D graphene foam by template-153 directed growth as a free-standing working electrode. The as-prepared electrode had a high specific capacitance of 751 Fg^{-1} in mol/L $HClO_4$ at a current density of 1.0 A^{-1} and retained 93.2% of the initial capacitance of 160 Fg^{-1} after 1000 charging/discharging cycles. Electrochemical deposition has been shown to be an effective and environmentally friendly method of preparing 3DGCP composites, and it is appealing for electrochemical applications due to the high surface area and low electrolyte transport resistance of these electrosynthesized composites, which make them suitable for high-performance electrode materials in electrochemical devices.

The disadvantage of electrodepositing polymers on 3DGFs is that only a thin layer of conductive polymers can be deposited on the surface of graphene, resulting in low areal capacitance. As a result, it is extremely desired to fill the macrosized holes with pseudocapacitive materials that have smaller pores in order to increase the areal capacitance. Zhang et al. created a bicontinuous macroporous graphene foam out of a few-layered graphene sheet and utilized it as a highly conductive substrate to manufacture mesoporous polyaniline using electrodeposition and inkjet methods.

One-Step in Situ Polymerization of Monomers and Assembly of GO

To simplify the preparation method, one-step in situ polymerization of monomers and assembly of GO have also been devised to build 3DGCP composites. 3D rGO/PPy composite hydrogel electrodes, for example, were created using a one-step hydrothermal procedure in the presence of $KMnO_4$. Under heating conditions, pyrrole (Py) monomers were polymerized with $KMnO_4$ as an oxidant to create PPy on the graphene surface. GO sheets were thermally reduced and self-assembled into a 3D porous structure in the meanwhile. It is generally known that aniline may be polymerized by oxidizing it in an acidic environment using a variety of oxidants. On the other hand, GO sheets can be chemically reduced into rGO by certain reductants including aniline. Based on these, Shen et al. fabricated 3D rGO/PANI composites through one-step in situ redox reaction between GO and aniline. At room temperature, the redox reaction might occur immediately.

Hollow micro/nanostructured materials have been identified as a potential form of material in energy-related applications. Directly combining GO with various polymer templates is a quick and easy way to create hollow 3DGCP composites. For example, 3D rGO/PANI hollow spheres could be created by electrostatically wrapping GO on a polyaniline hollow sphere and then electrochemically reducing GO. Wrapping rGO sheets over PANI hollow spheres might provide highly conductive paths by connecting individual PANI hollow spheres, hence improving supercapacitor rate and cycle performance. The specific capacitance of 3D rGO/PANI hybrids could reach 614 Fg^{-1} at a current density of 1 Ag^{-1} and preserved 90% after 500 charging/discharging cycles at the same current density, showing high cycling stability [32].

6.1.8.2 Electrochemical Pseudocapacitors Based on Ternary Nanocomposite of Conductive Polymer/Graphene/Metal Oxide

The use of ternary nanocomposites as a supercapacitor electrode material has become the focus of a significant amount of current scientific research in the field of energy storage devices due to a unique combination of features such as superb electrical conductivity, corrosion resistance in aqueous electrolytes, highly modifiable nanostructures, long cycle life, and the large theoretical specific surface area. Graphene can be used in these nanocomposites not only to provide a substrate for growing nanostructured polymers in a polymer/carbon nanocomposite structure to overcome the insulating nature of conductive polymers in dedoped states, but also to provide a platform for the decoration of metal oxide nanoparticles to avoid agglomeration. Recent research has focused on ternary composites composed of carbon materials, metal oxides, and conductive polymers, such as poly(3,4-ethylene-dioxythiophene) (PEDOT)/graphene/SnO_2, reduced graphene oxide/polyaniline (PANI)/MnO_2, polypyrrole (PPy)/TiO_2/graphene, and SnO_2/reduced graphene oxide/polyaniline. Ternary nanocomposite of conductive polymers stores charge using both Faradaic and non-Faradaic processes, resulting in exceptional electrochemical features such as increased specific capacitance and energy density, higher cell voltage, longer cycle life, and regulated power density. Furthermore, ternary electrodes outperform their individual parts in terms of electrochemical responsiveness due to not only the synergistic impact of the three components, but also the decrease of diffusive route length via increased specific surface area. The usage of ternary nanocomposites as a supercapacitor electrode material has become the focus of a large amount of current scientific study in the field of energy storage devices due to a unique combination of the aforementioned qualities.

Graphene/ZrO_2/PANI, graphene/SnO_2/PPy, rGO/MoO_3/PANI, and graphene/$CoFe_2O_4$/PANI composites, for example, demonstrated specific capacitances of about 1359.99 Fg^{-1}, 553 Fg^{-1}, 616 Fg^{-1}, and 1133.3 Fg^{-1} at 1 mVs^{-1}, respectively. Yu et al. improved the supercapacitor performance of graphene/MnO_2-based nanostructured electrodes by wrapping them in 3D with conductive polymers or carbon nanotubes, resulting in greater specific capacitance and cycling performance with >95% capacitance retention over 3000 cycles. Owing to all of the advantages listed above, it is widely assumed that ternary hybrid electrodes will play an important role on all commercial supercapacitor electrodes in the near future. Khatua et al. created ternary composite electrode material with a high energy density composed of polyaniline (PANI), molybdenum trioxide (MoO_3), and graphene nanoplatelets (GNP) by in situ polymerization of aniline in the presence of MoO_3 and GNP. It exhibited not only a significantly high specific capacitance of ≈593

Fg^{-1} at the current density of 1 Ag^{-1} but also an extraordinary cyclic stability and 92.4% retention of the original specific capacitance even after 1000 cycles.

Preparation of Ternary Nanocomposite of Conductive Polymer/Graphene/Metal Oxide

Chemical Methods

Wang et al. created a ternary electrode material based on graphene, SnO_2, and polypyrrole in a one-pot method, in which GO was produced from powdered flake graphite using a modified Hummers process and used as a graphene precursor. The produced graphene/SnO_2/PPy ternary electrode material not only had a high specific capacitance of 616 Fg^{-1} at 1 mV s^{-1}, but it also had a high specific energy density and power density of 19.4 W hkg^{-1} and 9973.26 Wkg^{-1}, respectively. Wet-chemistry approaches, such as hydrothermal fabrication, solgel procedures, and chemical in situ deposition, are widely used to prepare composites containing graphene and metal oxide, beginning with a dispersed solution of suspended graphene and serving as a two-dimensional precursor for an integrated support network for discrete metal nanoparticles. Moyseowicz et al. created a polypyrrole/Fe_2O_3/rGO ternary composite in two steps, first fabricating the binary Fe_2O_3/rGO composite via hydrothermal treatment of GO in the presence of an iron precursor and then oxidative polymerization of pyrrole on the surface of the as-prepared composite. The existence of iron oxide nanoparticles in both FG and PFG composites was proved by XRD and XPS findings. The produced composite had a specific capacitance of 140 Fg^{-1} and an outstanding capacitance retention of 93% after 5000 cycles at 1 Ag^{-1}, which may be ascribed to the synergistic impact of polypyrrole and FG composite.

In situ polymerization is used to create graphene and polymer nanocomposites. The graphene precursors and monomer are first dissolved in a shared solvent and then ultrasonicated to achieve a uniform dispersion, followed by the addition of an initiator to the mixture to synthesize the polymer. This procedure may be developed or modified to create a variety of categories, depending not only on the type of final result but also on the preparation process used. For example, the addition of a free-radical initiator and a reducing chemical can be done concurrently to promote polymerization while also reducing GO to graphene or preparing ordered layer assembly of the desired composites. Zhu et al. created a self-assembled CuO/PANI/rGO ternary composite using an in situ polymerization process combined with a hydrothermal approach. They used XPS to show that the Cu (II) oxidation state is the lone species and that the ternary composite contains no additional copper states. At a current density of 1.0 Ag^{-1}, the produced ternary composite displayed not only a specific capacitance of 634.4 Fg^{-1}, but also an energy density and power density of 126.8 $Whkg^{-1}$ and 114.2 $kWkg^{-1}$, respectively. Furthermore, the composite demonstrated excellent cycle longevity, with 97.4% specific capacitance retention after 10,000 cycles.

Electrochemical Methods

When compared to chemical polymerization, the electrosynthesis approach offers various benefits. It is an oxidant-free method that results in a conductive polymer that is spontaneously integrated as a continuous uniform coating on the electrode. As a result, there is no need to use binders, which typically operate as an insulator and hydrophobic substance, limiting the electrical and electrochemical characteristics of the electrode.

Electrodeposition is commonly regarded as a simple method for producing nanocomposites by electrochemical processes, such as the electropolymerization of a polymer/graphene composite from an aqueous solution comprising the monomer, doping agent (if required), and graphene oxide. In general, an electrodeposition cell consists of three electrodes: a working electrode onto which the layer is deposited, a counter electrode (often composed of platinum), and a reference electrode. The working electrode is constructed from an electrically conductive substance or substrate, such as ITO glass. The polymer is electrodeposited at a specified voltage and ceases once a certain amount of charge has passed, resulting in the creation of a nanocomposite layer on the surface of the conductive substrate.

Asen et al., for example, created a graphene oxide/polypyrrole/vanadium pentoxide (V_2O_5) ternary nanocomposite using a one-step electrochemical deposition process on a stainless steel substrate using an aqueous solution including vanadyl acetate, GO, and pyrrole. The produced ternary nanocomposite had a specific capacitance of 750 Fg^{-1} at a current density of 5 Ag^{-1}, which was much higher than that of PPy (59.5 Fg^{-1}), V_2O_5/PPy (81.5 Fg^{-1}), and PPy/GO (344.5 Fg^{-1}). Furthermore, after 3000 cycles, the ternary nanocomposite retained about 83% of its initial capacitance value, suggesting exceptional electrochemical stability of the electrode after repeated cycling.

It is possible to make ternary metal oxide/rGO/conductive polymer nanocomposites by both chemical and electrochemical techniques using a two-step strategy, in which a binary metal oxide/rGO nanocomposite is first prepared chemically and then the conductive polymer is electrodeposited on its surface. For example, Jiang et al. produced a free-standing TiO_2/rGO/PPy composite film as a supercapacitor electrode material, in which titania precursor and graphene oxide composite films were initially made by directly mixing and drying approach. The hydriodic acid (HI) reduction process was then used to decrease GO, resulting in the formation of a graphene/TiO_2 precursor composite layer. Heat treatments at various temperatures were then used to create TiO_2/graphene composite films with diverse titania crystalline phases. Finally, the ternary nanocomposite was prepared by electrochemical deposition of PPy on the surface of the as-fabricated TiO_2/graphene composite. The addition of TiO_2 in the nanocomposite greatly improved not only the pseudocapacitance but also the electrochemical stability of the produced electrodes. Furthermore, the PPy coating increased the capacitance of the composite electrode significantly. Different TiO_2 phases had no discernible influence on the capacitance of the

TiO_2 graphene system; however, distinct TiO_2 phases represented different performances after coating with PPy. The anatase/graphene/PPy nanocomposite had a high capacitance, but the rutile/graphene/PPy nanocomposite had a high stability.

Characterization of Ternary Nanocomposite of Conductive Polymer/Graphene/Metal Oxide

Cyclic Voltammogram (CV)

Electrochemical cyclic voltammetry (CV) is a technique for recording cyclic voltammograms of materials and calculating capacitance. It is accomplished by depositing materials onto the surface of an inert electrode and then measuring the electrode in an electrolyte. Han et al. created a sandwich construction ternary hierarchical rGO/MnO_2/PPy nanocomposite (GMP) by coassembling PPy/GO and MnO_2/GO into a layer-by-layer architecture and reducing GO. The synergistic impact of the components, resulting from the unique architecture, led to an amazing capacitance. The specific capacitance of the ternary composite electrode was 404 Fg^{-1}. In addition, cyclic voltammetry (CV) was used to study the capacitive behavior of the produced composite electrode. At low scan speeds, the fairly rectangular and symmetric CV curves revealed not only quick reversible Faradic reactions, but also exceptional capacitive performance of the produced GMP electrodes.

Electrochemical Impedance Spectroscopy

Electrochemical supercapacitor capacitance may also be determined using electrochemical impedance spectroscopy (EIS), which is regarded as an excellent instrument. Pan et al. used a simple and straightforward technique to create a unique ternary hierarchical MnO_2/PANI/graphene nanocomposite with an increased reversible capacitance of 695 Fg^{-1} after 1000 cycles at a current density of 4 Ag^{-1}. They plotted the Nyquist diagrams to study the electrochemical impedance characteristics of the ternary nanocomposite, PANI, graphene, and the binary PANI/graphene nanocomposite at an applied AC potential of 0.4 V across a frequency range of 100 kHz–0.01 Hz.

The Nyquist plots revealed that the internal resistance of the ternary nanocomposite was lower than that of the other materials described above, indicating that the ternary nanocomposites increased ion transport between the solution and the electrode active center. Furthermore, in the high frequency ranges, the semicircle radius of the ternary MnO_2/PANI/graphene nanocomposite was less than that of PANI and graphene, revealing the shortest charge transfer resistance (R_{ct}) and improved electrochemical performance of the produced ternary nanocomposite.

Galvanostatic Charging/Discharging Measurements

Galvanostatic charge/discharge experiments are regarded as effective methods for assessing the electrochemical characteristics of electrode materials under regulated current conditions. Chee et al. used potentiostatic electrochemical polymerization to create a ternary PPy/graphene oxide/ZnO nanocomposite that was then coated on nickel foam. At a current density of 1 Ag^{-1}, constant current charge/discharge experiments yielded a specific capacitance of up to 94.6 Fg^{-1}. Because current density has a strong influence on specific capacitance, the specific capacitance *versus* current density plot showed a decrease in specific capacitance as current density increased, owing to an insufficient response time for electrolyte ions to reach the electrode surface at a significant current rate. The cycling stability and capacitive retention of the ternary PPy/GO/ZnO nanocomposite electrode material were also studied, with the specific capacitance increasing steadily up to 100 cycles due to the slow infusion of electrolyte ions within the matrix of the ternary nanocomposite and then decreasing. After 1000 continuous charge/discharge cycles at a current density of 1 Ag^{-1}, it retained 73.6% of its initial quantity. Furthermore, the pseudocapacitive performance of PPy and ZnO, along with the EDLC of graphene oxide, resulted in energy and power density of 10.65 $Whkg^{-1}$ and 258.26 Wkg^{-1} at current density of 1 Ag^{-1}, respectively. According to the Ragone plot, the insertion of ZnO into the PPy/GO increased the energy density of the system, most likely owing to the addition of a pseudocapacitance, whereas the increase in power density represented the subsequent usage of graphene oxide [5].

6.1.8.3 Conductive Polymer Composites-Based Unconventional Solid-State Supercapacitors

Conventional supercapacitors are made up of two electrodes separated by a separator in a liquid electrolyte that can be aqueous, organic, or ionic liquid. In order to minimize leakage and ensure safety and performance, they must be hermetically sealed in sturdy casings. To eliminate leakage issues while maintaining appropriate performance, solid-state electrolytes, particularly polymer gel electrolytes, are presented. In addition to the requisite performance, the advent of unusual multifunctional electronic devices such as smart clothes, foldable and transparent screens, and E-skins necessitates energy storage devices that are structurally compatible with their designs.

Lightweight thin unconventional supercapacitors that are flexible, stretchy, and compressible, or even with novel functionalities (such as transparency, electrochromism, and self-healing), have piqued the interest of researchers. For integration with wearable electronics, flexible and lightweight solid-state supercapacitors with sandwiched, interdigitated, and fiber-type designs are in great demand. Aside from its great energy storage capabilities in many configurations, the continual development of new materials has given solid-state supercapacitors with smart functionalities such as electrochromism, self-healing, and transparency. These intelligent functionalities will encourage the use of solid-state supercapacitors in intelligent wearable electronics.

The gel electrolyte is often sandwiched between two flexible electrode materials to form the flexible supercapacitor. Poly(ethylene terephthalate), poly(dimethylsiloxane), textile and cloth, and gel electrolyte are all very flexible

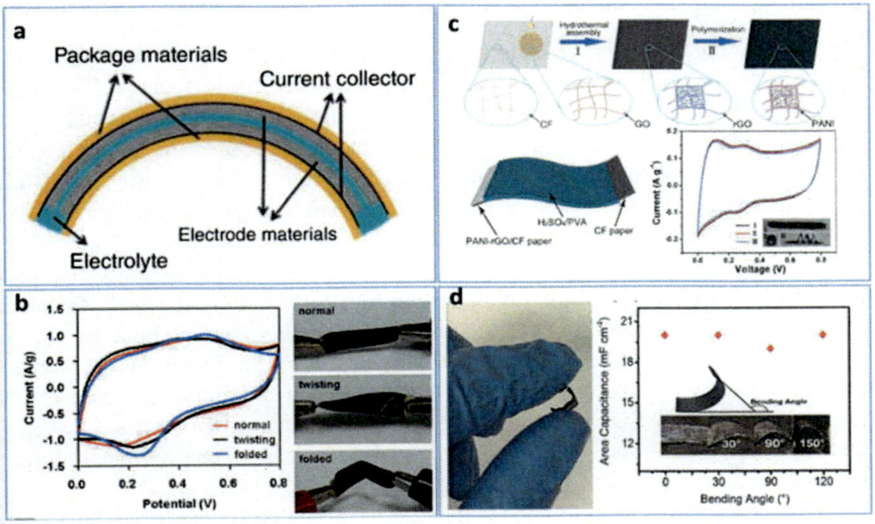

Fig. 6.7 **a** Schematic diagram of flexible solid-state supercapacitors with a sandwiched structure. **b** CNT/PANI-based paper-like flexible supercapacitor. **c** Cellulose paper/rGO/PANI-based flexible supercapacitor. **d** MXene/PPy film-based flexible supercapacitor. Reprinted from ref. [37], copyright 2017, with permission from Royal Society of Chemistry

substrates, and hence, the recent advancement of flexible electrode materials based on CPs in flexible supercapacitors was primarily emphasized.

Flexible Solid-State Supercapacitors with Different Configurations

Due to their basic form and relatively simple fabrication technique utilizing film electrodes, sandwich type flexible solid-state supercapacitors (fSSCs) have achieved remarkable advances. Microsupercapacitors with in-plane interdigitated electrodes and fiber-shaped supercapacitors have been developed in response to the recent advent of on-chip and textile electronics.

Flexible Solid-state Supercapacitors with a Sandwiched Structure

Flexible electrodes, solid-state electrolytes, and flexible packaging are commonly used to make fSSCs. Flexible electrodes should have high capacitance, low self-discharge, extended cycle stability, and mechanical flexibility. The solid-state electrolyte must be extremely ionically conductive, thermally and mechanically stable, and capable of serving as a separator without the need for an extra one. It uses the standard sandwich structure design with a layered configuration of current collector–film electrode–solid electrolyte–film electrode–current collector (Fig. 6.7a). When the electrodes are strong and have a high conductivity, the current collector can be removed [18].

Film-like CP composites are frequently employed in the fabrication of fSSCs, and the CNT/CP composite is particularly appealing. Meng et al., for example, described a highly flexible solid-state paper-like supercapacitor made from a CNT/PANI composite film formed by chemically depositing PANI onto a thin CNT film via a filtering process utilizing CNT dispersion. Two gel electrolyte-impregnated CNT/PANI composite sheets were pressed face-to-face under pressure to construct an integrated solid-state device [21]. The device had a specific capacitance of 31.4 Fg^{-1}, and no capacitance fading was detected when twisted or folded (Fig. 6.7b).

Because they are ultrathin, flexible, and lightweight, free-standing graphene sheets have showed significant promise as electrodes for flexible supercapacitors. Graphene sheets, on the other hand, tend to agglomerate and restack throughout the filming process. Hu et al. used in situ chemical polymerization and subsequent chemical reduction to create a graphene/PANI composite film. The built solid-state supercapacitor with PVA/H_2SO_4 gel electrolyte has an areal capacitance of 0.425 F cm^{-2} and essentially unaltered energy storage capabilities when bent, demonstrating its exceptional flexibility.

Graphene may also be combined with other substrates with vast surface areas, such as paper and textiles, for the deposition of CPs. Liu et al., for example, showed a fSSC built with cellulose paper/graphene/PANI nanocomposite electrodes. The fSSC was created by sandwiching two such composite sheets together with a PVA/H_2SO_4 polymer electrolyte [14]. The specific capacitance of this composite electrode was 224 Fg^{-1}, and the solid-state device performed consistently in the folding or bending phases (Fig. 6.7c).

CPs may be used with 2D materials like TMDs and MXenes to generate flexible composite electrodes that inherit the features of each individual component while also providing synergistic benefits. Ge et al. used a simple filtering approach to create free-standing MoS_2/PEDOT:PSS. When bent to 90°, the fSSC with PVA/H_3PO_4 gel electrolyte could keep 87.8% of capacitance, and 80% of capacitance could be kept after 1000 bends. The areal capacitance of the solid-state supercapacitor with MXene/PPy films as electrodes and PVA/H_2SO_4 as electrolyte was 35.6 mF cm^{-2}, and the performance was maintained in different bending states (Fig. 6.7d).

Microsupercapacitors

The fast advancement of smaller personal wearable gadgets has accelerated the search for ultrathin, flexible, and lightweight energy storage devices that can be directly combined with other electronic components. It has been proposed to use microsupercapacitors (MSCs) with in-plane interdigitated microelectrodes. MSCs have all of the advantages that make them the most viable alternative for energy storage components in small wearable electronics.

The manufacturing of interdigitated microelectrode arrays may be divided into two ways based on active material deposition methods: bottom-up and top-down. Current collectors and active materials are deposited sequentially onto a flexible substrate with a predetermined pattern in the bottom-up technique. It is often accomplished by patterning current collectors and then depositing active materials using different processes such as spray coating, electrodeposition, electrophoretic deposition, and mechanical pressing. Gao et al. used a simple templating and imprinting process to create a graphene/PPy composite-based flexible solid-state MSC on paper. At 0.4 mA cm^{-2}, this paper-based MSC has an areal capacitance of 22 mF cm^{-2}. When the MSC was bent to 180°, there was no discernible change in capacitance.

In the top-down technique, a thin film of active materials is etched to eliminate undesired components, resulting in interdigitated electrodes that are immediately built into solid-state MSCs after coating with polymer gel electrolyte. Various etching techniques, including dry etching (e.g., plasma, laser, scribing) and wet etching (microfluidic or mechanical), have been used. Liu et al., for example, used laser etching to create a flexible solid-state MSC with graphene/PEDOT:PSS sheets. The flexible MSC with a PVA/H_3PO_4 gel electrolyte coating demonstrated a high areal capacitance of 84.7 mF cm^{-2} as well as good flexibility with no change during repeated bending.

Fiber-shaped Supercapacitors

In comparison with sandwiched supercapacitors or in-plane interdigitated MSCs, fiber-shaped supercapacitors provide unparalleled benefits for direct usage as wearable devices or integration with textile structures for smart garments. Plastic wires, metal wires, carbon fibers, CNT or graphene fibers, conductive polymer fibers, and composite fibers are among the fiber electrodes that have been produced.

For fiber-shaped supercapacitors, many device topologies have been devised, most notably parallel, twisted, and coaxial configurations (Fig. 6.8a) [8, 36]. Similar to planar supercapacitors, two fiber electrodes are positioned parallel to the separator or polymer gel electrolyte in the parallel construction. Huang et al. created a fiber supercapacitor using parallel architectures of CP composite-coated metal yarns. One fiber electrode has a specific capacitance of 31 mF cm^{-1} or 411 mF cm^{-2} [12]. Due to the high strength of fiber supercapacitors, the constructed fiber supercapacitor demonstrated excellent flexibility and could be knitted into a wrist band with woolen threads using a standard knitting machine (Fig. 6.8b).

Twisting two fiber electrodes separated by a separator or gel electrolyte results in a twisted structure. Twisted structures feature more surfaces than parallel structures, which improves electrochemical performance. Cai et al. described a twisted fiber supercapacitor with PANI-coated CNT fibers created by electrochemically depositing PANI onto CNT fibers [2]. This fiber supercapacitor also showed strong mechanical characteristics. After 50 cycles of bending, the capacitance had dropped by less than 3%. The fiber supercapacitor may be woven into textile due to its outstanding flexibility and stability (Fig. 6.8c). Layer by layer, the core electrode, separator or polymer gel electrolyte, and outside electrode are built for the coaxial construction. When compared to the twisted construction, the coaxial structure has more contact surface and less internal resistance, resulting in better performance. Shen et al. described a coaxial fiber supercapacitor using MnO_2/PPy-coated carbon fibers as the inner electrode, PVA/H_3PO_4 as the electrolyte, and dip-coated carbon nanotubes as the outside electrode. The capacitance of the fiber supercapacitor was 25.8 mF cm^{-1}, with exceptional bending flexibility and tensile strength (Fig. 6.8d).

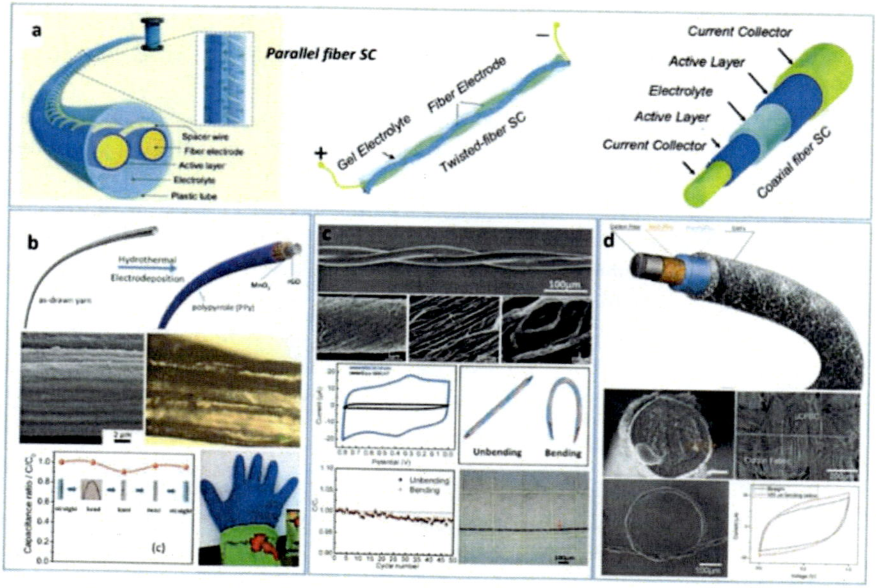

Fig. 6.8 **a** Schematic illustration of three typical device configurations of fiber-shaped supercapacitors. **b** A parallel fiber supercapacitor with rGO/MnO_2/PPy composite-coated stainless steel yarn electrodes. Supercapacitor with CNT/PANI composite fibers. **c** A twisted fiber. **d** A coaxial fiber supercapacitor with MnO_2/PPy-coated carbon fibers as the inner electrode and dip-coated CNTs as the outer electrode. Reprinted from ref. [37], copyright 2017, with permission from Royal Society of Chemistry

Multifunctional Solid-State Supercapacitors

It is of considerable importance, with the development of functional materials with high performance, to combine adaptable materials into devices to accomplish multiple purposes. These multifunctional supercapacitors may be able to withstand more severe mechanical deformations (stretching and compressing) than bending, or they may be intelligent and responsive to external stimuli such as voltage (electrochromism) and physical damage (self-healing), or they may have special physical properties (transparency). Wearable devices can benefit from high degrees of customization thanks to their multifunction.

Stretchable Supercapacitors

To date, much effort has been expended in constructing CP composite-based stretchable supercapacitors with various topologies (e.g., 1D fiber, 2D planar, and 3D) to provide a stable performance output at multidimensional stretching states. Peng's group created a stretchable coaxial fiber supercapacitor employing a PANI-coated CNT sheet for 1D fiber supercapacitors. The resulting fiber supercapacitor has a stretchability of 400% and an areal specific capacitance of 3.08 mF cm^{-2}, allowing it to be woven into elastic fabrics.

Buckled electrodes produced on pre-stretched substrates or electrolytes have been frequently used to generate stretchy supercapacitors in 2D planar supercapacitors. Furthermore, materials with certain architectures (embedded in elastomers, 2D reentrant auxetic, 3D porous) or on intrinsically stretchy substrates (textiles) can be directly employed in planar stretchable supercapacitors. A buckled electrode is typically generated using a mechanical buckling method, in which a thin electrode sheet is layered over a pre-stretched substrate or electrolyte and a buckled structure forms spontaneously after releasing the pre-strain owing to mechanical competition in-between.

The strain imparted to the electrodes may be handled by adjusting the wave amplitude and wavelength of the buckled structure, preventing damage to the active material layer. Zhang et al. created a very flexible supercapacitor using a mechanical buckling technique paired with a gel electrolyte. It had a high stretchability of 120% and a specific capacitance of 106 Fg^{-1}.

Stretchable supercapacitors with inherent stretchability have also been produced. Stretchable knitted cloth can be utilized as the substrate for supercapacitors with inherent stretchability. CNT-coated cotton textiles were employed as stretchy electrodes after being decorated with MnO_2 nanoparticles and PPy. The created solid-state supercapacitor with poly(ethylene oxide) (PEO)-based electrolyte could be stretched by 21% while retaining 98.5% of its capacitance.

Origami, kirigami, and patterning techniques may be used to create 3D stretchy supercapacitors. Peng's team created a pyramid-shaped CNT film by arranging the deposition of the catalyst for CVD development of CNT. The pyramid-shaped CNT film could be deformed in three dimensions and also support complicated deformations costretching in two directions. The pattern structure might also influence this distortion. PANI electrodes were formed by electrodeposition on such a CNT sheet, and the built solid-state supercapacitor with PVA/H_3PO_4 gel electrolyte could withstand both in-plane and out-of-plane deformations.

Compressible Supercapacitors

Compressible supercapacitors are similar to stretchable supercapacitors in that they can withstand compression. CNTs, graphene, and composite foams of these materials are commonly used in compressible supercapacitors. The inclusion of CPs with pseudocapacitance can effectively increase the energy density of such foams.

He et al. described a compressible solid-state supercapacitor made of PEDOT-coated CNT sponge. CVD was used to create the CNT sponge, and then, electrochemical polymerization was used to deposit PEDOT, which dramatically increased the specific capacitance. The supercapacitor, which used composite sponge electrodes and a PVA/H_2SO_4 gel electrolyte, had a specific capacitance of 51 Fg^{-1} and could preserve 97% of its capacitance under 50% compression. Sponge great compression tolerance is attributed to the linked fibers. By using van der Waals interaction, carbon nanotubes or graphene may connect to the sponge skeleton. The sponge/CNT/PANI composite with great compressibility was created by "dipping and drying" CNTs onto the sponge skeleton and then chemically polymerizing PANI onto the CNTs. A compressible all-solid-state supercapacitor was created using a PVA/H_2SO_4 gel electrolyte. When squeezed to 60% strain, the performance of this supercapacitor remained practically unaltered.

CP composite-based compressible supercapacitors, like CP composite-based stretchable supercapacitors, are primarily made of carbon/CP composites. CVD synthesis or coating with commercial polymer sponges can be used to create 3D carbon supports. Because of the poor capacitance of such carbon compounds, energy storage capacity is restricted.

Electrochromic Supercapacitors

The use of electrochromism in a supercapacitor makes it intelligent and allows for the signaling of the stored energy level. As a result, electrochromic supercapacitors may combine energy storage and color changes in a single device. Colors associated with the oxidation/reduction processes can be changed by CPs. Because of its high energy storage capacity and visible color changes at different doping states, PANI is commonly employed in electrochromic supercapacitors. Peng's team created an electrochromic and stretchy supercapacitor using PANI-coated aligned-CNT sheet electrodes and a PVA/H_3PO_4 gel electrolyte. The composite electrode had a high specific capacitance of 308.4 Fg^{-1}, and the capacitance remained stable after stretching up to 100% for 200 cycles. At the corresponding voltages of 1, 0.5, and 0 V, the positive electrode displayed a rapid and reversible color shift between blue, green, and yellow. The high conductivity of the carbon/CP composites allows for the simple manufacturing of electrochromic supercapacitors without current collectors. Metal oxides/CP composites with greater specific capacitance require enhanced conductivity to avoid the use of transparent current collectors.

Self-Healable Supercapacitors

As a power source for wearable/portable electronics, solid-state supercapacitors have showed considerable potential. However, electronics that have been in operation for a long time may sustain physical damage or even breakage as a result of constant mechanical deformations such as bending, twisting, and stretching. All of these failures would lead to a decrease in performance and even breakdown. As a result, developing supercapacitors with self-healing capabilities that can repair damage and restore structural integrity and electrochemical performance by adding healable materials into the system is desirable.

Peng et al. created a self-healing fiber supercapacitor with a healable conducting wire by wrapping an aligned-CNT sheet around healable polymer fibers. The reconnecting of aligned-CNTs enabled the self-healing process. During the cutting process, an aligned-CNT strand was created along the cross-section of broken ends and acted as a "bridge" to join two broken ends of the CNTs upon contact. The fiber supercapacitor was created by twisting two healable wires with a PVA/H_2SO_4 gel electrolyte. With a PANI coating on the CNT sheet, the specific capacitance may be increased from 11.1 Fg^{-1} to 140.0 Fg^{-1}. The self-healing of such CP composite-based supercapacitors is mostly caused by the substrates or packaging shells, which contribute nothing to the total capacitance but increase the volume. Self-healing supercapacitors require CP composites with intrinsic self-healing capabilities, which may be obtained by integrating dynamic hydrogen bonds into the composites.

Transparent Supercapacitors

Each component of a transparent supercapacitor must be transparent. To obtain the best transparent supercapacitors, capacitance and transparency must be balanced and optimized. CNT sheets produced from a CVD-synthesized spinnable CNT array had a thickness of 25 nm and exceptional transparency. The transmittance value for the composite film was greater than 70% over a wavelength range of 300–900 nm with a layer of electrochemically produced PANI at a weight percentage of 70%. The specific capacitance of the composite film was 233 Fg^{-1}. With these electrodes and a PVA/H_3PO_4 gel electrolyte, a solid-state transparent supercapacitor was created.

Because of their increased specific capacitance and higher cycle stability, CP composites have been frequently employed as active materials for supercapacitor electrodes. These composites have aided in the creation of unconventional supercapacitors. Table 6.4 summarizes the device designs, composite electrodes, and electrochemical performances of such supercapacitors.

6.1.9 *Present Efforts and Future Developments*

The CPs-based redox materials are ushering in a new era of flexible pseudocapacitor applications. The future of CPs-based flexible energy storage devices promises even more revolutionary low-cost, lightweight, and extremely flexible systems. These CPs-based supercapacitors are predicted to take a significant market share from conventional Li-ion batteries during the next ten years in order to meet the enormous need for flexible electronic gadgets. High Faradaic capacitance, high flexibility, high conductivity, and simplicity of manufacture for thin and wearable electronic devices distinguish CPs-based supercapacitors. Because of these distinct features, CPs-based flexible supercapacitors may be directly attached or incorporated into stretchy or rolled-up garments that can be hooked into an electrical storage device. Although recent advances in CPs-based flexible supercapacitors look exceedingly promising, various difficulties remain that limit their real-world applicability.

Several potential resolutions have been proposed, with the following goals in mind: (1) improved mechanical stability using highly flexible current collectors;

Table 6.4 Configurations, CP composite electrodes, and the electrochemical performances of unconventional solid-state supercapacitors

Devices	Configurations	Electrode composites	Electrolyte	Specific capacitance (device)
Flexible SCs[b]	Sandwiched	CNT/PANI film	PVA/H_2SO_4	31.4 Fg^{-1} (whole device)
		CNT/PANI	PVA/H_2SO_4	184.6 mF cm^{-2} at 1 mA cm^{-2}
		Cotton/CNT/PPy	PVA/H_3PO_4	50.09 mF cm^{-2} at 25 mV s^{-1}
		rGO/PANI	PVA/H_2SO_4	[a]425 mF cm^{-2} at 0.5 Ag^{-1}
		Paper/graphene/PANI	PVA/H_2SO_4	123 mF cm^{-2} at 0.5 mA cm^{-2}
		Textile/rGO/PPy	PVA/H_2SO_4	230 mF cm^{-2} at 1 mV s^{-1}
		Paper/rGO/PANI	PVA/H_2SO_4	[a]56 Fg^{-1} at 0.1Ag^{-1}
		MoS_2/PEDOT:PSS	PVA/H_3PO_4	[a]39 F cm^{-3} at 50 mV s^{-1}
		MXene/PPy	PVA/H_2SO_4	35.6 mF cm^{-2} at 0.3 mA cm^{-2}
	Interdigitated	Graphene/PPy	PVA/H_2SO_4	22 mF cm^{-2} at 0.4 mA cm^{-2}
		rGO/PEDOT:PSS	PVA/H_3PO_4	84.7 F cm^{-3} at 5 mV s^{-1}
	Fiber-shaped	rGO/PPy	PVA/H_2SO_4	107.2 mF cm^{-2} at 0.24 mA cm^{-2}
		rGO/PEDOT:PSS	PVA/H_3PO_4	304.5 mF cm^{-2} at 0.08 mA cm^{-2}
		CNT/PANI	PVA/H_3PO_4	263 mF cm^{-2} at 2 Ag^{-1}
Stretchable SCs	Fiber-shaped	CNT/PANI	PVA/H_3PO_4	3.08 mF cm^{-2} at 1 Ag^{-1}
		CNT/graphene/PANI	PVA/H_3PO_4	273.7 mF cm^{-2} at 1 Ag^{-1}

(continued)

Table 6.4 (continued)

Devices	Configurations	Electrode composites	Electrolyte	Specific capacitance (device)
	Sandwiched	CNT/PANI	PVA/H_3PO_4	61.4 mF cm^{-2} at 1 mA cm^{-2}
		Graphene foam/PPy	PVA/H_3PO_4	258 F cm^{-2} at 5 mV s^{-1}
		CNT/PANI	PVA/H_3PO_4	106 Fg^{-1} at 1 Ag^{-1}
Compressible SCs	Sandwiched	Sponge/CNT/PANI	PVA/H_2SO_4	[a]54 Fg^{-1} at 0.4 Ag^{-1}
		CNT/sponge/PEDOT	PVA/H_2SO_4	51 Fg^{-1} at 0.5 Ag^{-1}
Electrochromic SCs	Sandwiched	CNT/PANI	PVA/H_3PO_4	[a]77.1 Fg^{-1} at 1 Ag^{-1}
		WO_3/PEDOT:PSS	Polyacrylamide/LiCl	[a]117.75 Fg^{-1} at 1 Ag^{-1}
Self-healable SCs	Fiber-shaped	CNT/PANI	PVA/H_2SO_4	1.34 mF cm^{-1} at 0.33 Ag^{-1}
		Fe_3O_4/PPy	PVA/H_3PO_4	61.4 F cm^{-2} at 10 mV s^{-1}
Transparent SCs	Sandwiched	CNT/PANI	PVA/H_3PO_4	[a]58.25 Fg^{-1} at 1 Ag^{-1}
		RuO_2/PEDOT:PSS	PVA/H_2SO_4	0.84 mF cm^{-2} at 10 mV s^{-1}
		$Ni_3V_2O_8$/PANI	PVA/KOH	58.5 F cm^{-2} at 5 mV s^{-1}

[a]Calculated based on the results given in reference when specific results were not given. [b]SCs: supercapacitors

(2) improved cycle stability of CPs-based flexible supercapacitors using composites with carbon nanomaterials or self-healing ionic polymers; (3) adaptability of nanoscale active materials with high power characteristics and solid/gel-type or ionic or organic electrolyte systems ensuring safety and improving power density; (4) improved charge/ionic transport, lowering the interfacility; and (5) hybrid asymmetric flexible supercapacitors based on *p*-type and *n*-type CPs. However, in the current circumstance, *n*-type CPs derivatives are shown to be much worse in quality than their *p*-type counterparts.

It is well acknowledged that the interaction of the active electrode materials with the current collector greatly influences the efficiency of CPs-based supercapacitors. A variety of materials were used as current collectors, with CC, CFs, porous carbon paper, and porous GOs composites seeming particularly promising for future flexible ES applications. One of the most crucial concerns for commercial applications is the stability of the CPs-based supercapacitor. Creating composites of C-based nanomaterials (CNTs, graphene, and GOs) is a typical technique for improving the stability of electroactive materials. Furthermore, the creation of copolymers composited with self-healing ionic polymers and other compatible CPs may be a potential method of producing mechanically stable polymer composites for flexible supercapacitor applications. In both scenarios, self-healing ionic polymers or carbon nanostructures might shield the CPs structures from mechanical rupture during the electrochemical charge/discharge process, hence improving supercapacitor cycle stability.

According to the electrode-design considerations, in addition to the suitable selection of active electrode materials, the appropriate selection of electrolytes is also necessary to increase ES performance. It is well acknowledged that increasing the working potential window is the most effective technique to boost the energy and power density of supercapacitors. The safe or withstanding working potential window in any electrochemical process is restricted due to the electrochemical stability of the electrolytes and the formation of gaseous products at the cathode and anode (positive or negative electrodes) during voltammetric measurements. According to advanced supercapacitor electrolyte research, organic (acetonitrile, propylene carbonate, linear sulfones, etc.) and ionic liquid (imidazolium, pyrrolidinium, etc.)-based electrolytes could improve the working potential window nearly 1.5–2 times that of acid-based aqueous electrolytes, despite having poorer ionic conductivity (by a factor of 1/2) than the aqueous type.

Meanwhile, substantial research has been concentrated on the creation of nanostructured active electrode materials, which improve ES performance by increasing the electrode/electrolyte contact regions and decreasing cationic diffusion length. Several ways have been tried to synthesize CPs in nanostructures in order to increase their electrochemical capacitance performance. The specific method of charge storage and delivery for a supercapacitor, however, is quite complex, because the form, connectivity, and size distribution of CPs, as well as the nature of the electrolyte and substrate-polymer interface, must all be taken into account. Another critical consideration is the equivalent series resistance (ESR) of the electrode, which is inversely related to the maximum power density of a supercapacitor. As a result, decreasing both the internal resistance of the active materials (CPs) and the contact (interfacial) resistance between the CPs and the current collector would be critical for improving charge/ionic transfer and thus achieving high power density. Recently, several approaches to overcome the aforementioned issue have been attempted, including developing core/shell-type CPs/C or CPs/MO_x composites on the current collector, growing C-based nanostructures or MO_x nanostructures directly onto the current collector, and subsequent coating of CPs on carbon or MOx nanostructures. The whole procedure is depicted schematically in Fig. 6.9. In contrast to supercapacitors made of powder-type active materials, the direct electrode-design approach can improve high power performance by lowering the ESR with low contact resistance between the CPs collector and the components of CPs/C or CPs/MO_x composites, as well as the absence of insulating binders.

CPs are often classed as *p*-type or *n*-type based on their doping technique. Because of their electrical properties, these two types of CPs operate in distinct

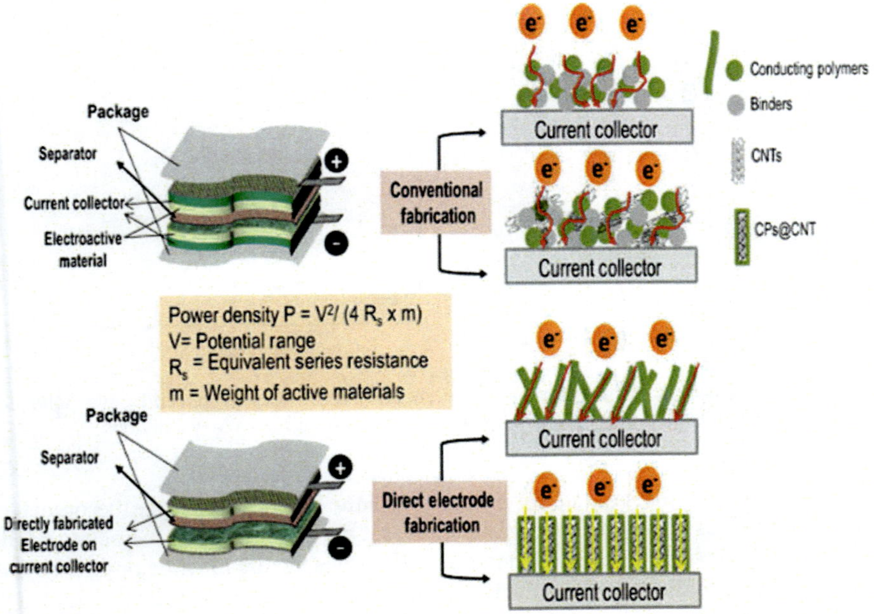

Fig. 6.9 Schematic diagram of the CPs-based flexible electrode fabrication process. CPs, conductive polymers. Reprinted from ref. [25], copyright 2014, with permission from John Wiley and Sons

potential ranges. Because of these properties, it is feasible to create an asymmetric hybrid supercapacitor, as shown schematically in Fig. 6.10. The maximum withstanding voltage limits (cell working potential window) of the supercapacitor would be improved, and the specific power density would be increased by combining the positive potential window of *p*-type polymers with a comparably negative potential window of *n*-type derivatives. In fact, various hurdles remain, despite the fact that, based on recent developments in CPs-based flexible supercapacitor research, significant improvement may be expected in the near future.

6.2 Conductive Polymer-Based Solar Cells

Massive progress has been made in the development of conductive polymer-based solar cells. In university labs, the power conversion efficiency (η)—the proportion of solar light input transformed into useful energy output—has reached 12%, making polymer solar cells a credible rival to the far more established amorphous silicon solar cells (13%).

A polymer solar cell is a sophisticated device that generally consists of a polymer:fullerene blend in a bulk heterojunction (BHJ) structure for converting light

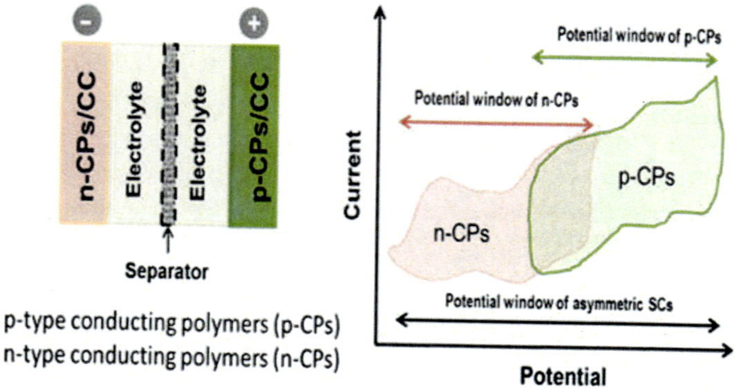

Fig. 6.10 Schematic diagram of asymmetric supercapacitors based on conductive polymers. Reprinted from ref. [25], copyright 2014, with permission from John Wiley and Sons

into power. Though the material properties of the active layer (i.e., the optoelectronic properties of polymers and fullerenes) set the upper limit on the output of the photovoltaic device (i.e., the efficiency), many other factors must be optimized and work together to maximize the device characteristics (i.e., short-circuit current, J_{sc}; open circuit voltage, V_{oc}; and fill factor, FF) within the upper limit. The inherent complexity of BHJ solar cells necessitates a wide range of knowledge in synthesis, morphology, and device physics (including physical chemistry).

Because there is a limit to the amount of functionalization that can be applied to the synthesis of fullerenes, the design and synthesis of materials have mostly centered on polymers. Nonetheless, there are several elements to consider in the synthetic design of these polymers. From the standpoint of molecular engineering, selecting the appropriate conjugated repeating unit determines the energy levels and band gap; selecting the appropriate side chains largely determines the solubility in the processing solvent; and both the conjugated backbone and the side chains have a significant impact on aggregation and intermolecular interactions in the active layer. From the standpoint of polymer engineering, the molecular weight, dispersity, solubility in the processing solvents, and viscosity of the resultant solution (i.e., the "active ink") must be properly balanced.

The "active ink" is a polymer:fullerene mix in a processing solvent that then goes through one of many solution-processing processes (e.g., spin coating, slot-die, doctor-blading, etc.) to create a thin film (i.e., the active layer) with the correct shape to optimize device output. Given the importance of the morphology, a range of approaches, including weight ratio of polymer versus fullerene, processing additives, solvent annealing, and thermal annealing, have been used to regulate the kinetically trapped morphology of the active layer. However, the optimal morphology for a given BHJ blend for achieving the maximum device characteristics is largely material dependent. Therefore, fundamentally understanding and characterizing the morphology, particularly correlating morphological features with device characteristics (J_{sc}, V_{oc}, and FF), are becoming increasingly important. Finally, the material properties and morphological features of the active layer determine the fundamental optical and electronic processes in such a thin film, such as exciton generation and separation, charge generation, transport, collection, and recombination, all of which provide direct and quantitative insights into device performance. As a result, examining and comprehending these device-related physical processes are a critical component of gaining a full understanding of the link between structures (of materials) and attributes (of devices).

6.2.1 Molecular Engineering (Backbone, Substituents, and Side Chains)

In terms of molecular engineering, the conductive polymers for solution-processed BHJ solar cells usually include three structural components: a conjugated backbone, side chains, and, in certain cases, substituents (Fig. 6.11a) [43]. The substituents are described individually because they frequently have intriguing (sometimes considerable) effects on the BHJ system that cannot be explained just by their electrical influence (e.g., electron donating/withdrawing) on the conjugated backbone. Among the three components, the conjugated backbone is without a doubt the most important in determining the inherent optoelectronic characteristics of conductive polymers, such as energy levels, band gap, mobilities, and so on. However, as shown in Fig. 6.11c and d, side chains (and substituents) have a considerable impact on the optoelectronic characteristics of these conductive polymers, particularly in the solid form, where the morphological aspects (aggregation, packing/stacking, and so on) are strongly affected by the choice of side chains (and substituents) [16, 28, 35].

6.2.1.1 Backbone

Because the "weak donor–strong acceptor" technique may (nearly) independently fine tune the energy levels and band gaps of conductive polymers, it has been widely used in the construction of conjugated backbones. On the donor side, fused aromatics are dominating the scene due to advantages such as reduced steric hindrance between neighboring conjugated units, adjustable HOMO levels, and greater solid-state stacking/packing (thus higher charge carrier mobility), among others. However, overly fused aromatics have created problems such as dramatically reduced solubility in processing solvents, as well as synthetic challenges in creating highly fused aromatics with appropriately decorated side chains.

Balancing "investment versus return," it is not surprising that the majority of such fused aromatics have been a relatively simple fusion of three or four simple aromatics, such as benzodithiophene (BnDT), which is perhaps the most popular fused aromatic in the design of conductive polymers for solar cells. A few major changes to the core framework of BnDT have lately occurred. First, linking the alkyl side chains of BnDT

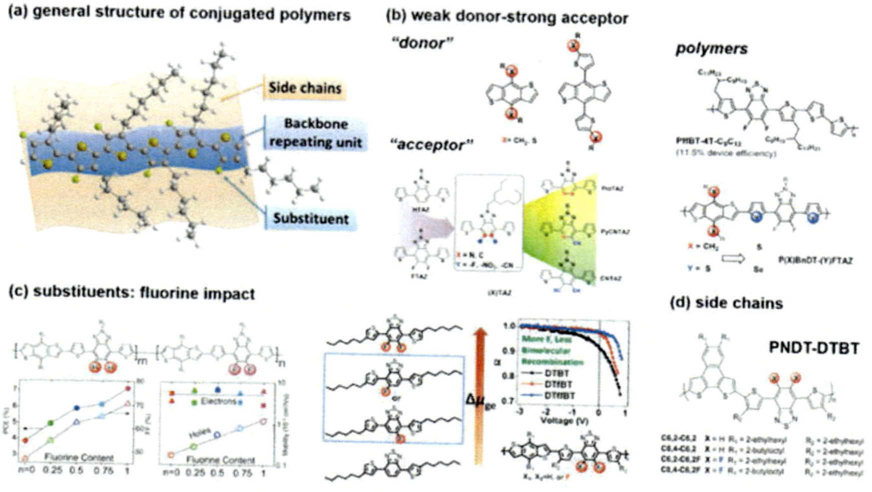

Fig. 6.11 **a** The three constituents to conductive polymers; **b** recent developments in donor moieties and acceptor moieties, and representative polymers. **c** Two detailed studies exploring the fluorine impact. **d** One example disentangling the influence of side chains and fluorine substitutions. Reprinted from ref. [33], copyright 2016, with permission from John Wiley and Sons

to the core benzene through a sulfur atom can significantly raise the V_{oc} of such polymer-based BHJ devices to ≈0.9 V, an effect that has not been well investigated. Second, it appears that adding an extra (alkylated) thiophenyl unit on either side of the core benzene of BnDT increases the J_{sc} of related BHJ devices, potentially due to the two-dimensional transport channel (from the additional conjugation on the side chains) that might aid hole transport. The most recent breakthrough is to combine these two tactics, i.e., to add alkylthio side chains to these two thiophenyl units on the core benzene of BnDT. A few polymers based on this improved form of BnDT demonstrated device efficiencies of up to 9.52%, with V_{oc} of 0.807 V, J_{sc} of 17.55 mA cm^{-2}, and FF of 0.672.

On the acceptor side, a few fused aromatics, such as FTT, DPP, isoindigo, DTffBT, and FTAZ, have gained popularity and demonstrated outstanding device performance with their polymers (Fig. 6.11b). The discovery of a few new electron-deficient triazole units (Fig. 6.11b) may be the most recent achievement in the design and synthesis of novel acceptors [17, 30]. Li et al. developed a broad technique for synthesizing three novel electron-deficient triazole units and three new polymers by combining them with BnDT. At a thickness of ≈300 nm, one novel polymer, PyCNTAZ, was able to provide an overall device efficiency of 8.4%, with both V_{oc} and J_{sc} greater than those of the original PBnDTFTAZ-based devices.

Another major advancement is the incorporation of non-common aromatic heteroatoms (e.g., Ge, Si, Se) into the conjugated backbone. Perhaps the most effective technique is to substitute sulfur (S) with selenium (e.g., in thiophene) (Se). For example, substituting the thiophene units in the conjugated backbone with selenophene can significantly decrease the band gap of conductive polymers (mostly by reducing the LUMO level), potentially resulting in a larger J_{sc} in the device output. However, due to lower V_{oc} and FF in the former, the overall efficiency of the selenophene polymer-based device was lower than that of the thiophene-based counterpart. Some used a similar method when designing acceptors, such as substituting the sulfur in the benzothiadiazole unit with selenium/tellurium. Jiang et al. investigated the influence of selenium substitution on energy levels, band gaps, and associated device performance by varying the placement and amount of selenium atoms in 2D conductive polymers.

Though the electron-drawing nature of the acceptor predominantly dictates the LUMO energy level of such "donor–acceptor" conductive polymers, the acceptor also has some influence on the HOMO energy level of the "donor–acceptor" conductive polymer (similarly, the donor moiety also has some impact on the LUMO energy level of the polymer). As a result, one can only estimate the energy levels and band gap of a "donor–acceptor" conductive polymer with known donor and acceptor moieties semi-empirically while the exact values of these energy levels must be determined experimentally (e.g., via a combination of cyclic voltammetry and UV–Vis absorption profile).

6.2.1.2 Substituents

Many substituents have a comparable electronic impact on various conjugated units, and their influence can occasionally go beyond the electrical impact. For example, electron-rich substituents (such as alkoxy) can raise the energy level of the conjugated backbone, whereas electron-drawing substituents (such as cyano) can reduce it. As a result, substituents provide extra control/tuning of conductive polymer energy levels and band gap. However, some substituents, such as fluorine, can not only have the predicted electron-drawing effect on the conjugated backbone, but also—probably more crucially—have a dramatic impact on the morphology and charge/exciton associated physical processes in the BHJ film.

The sole difference between PBnDTDTfBT ("1F") and PBnDTDTffBT ("2F") is the amount of fluorine substitutions on the benzothiadiazole. For a number of reasons, adding fluorine atoms to the conjugated backbone can significantly improve the three essential characteristics of the device output (V_{oc}, J_{sc}, and FF). While the suppressed HOMO level by the increased fluorine substitution primarily enhanced the V_{oc}, a synergistic impact of good morphological traits and beneficial exciton/charge behaviors improves J_{sc} and FF with more fluorine substituents. Li et al. used a series of five PBnDT-(X)TAZ copolymers to gradually increase the fluorine concentration.

6.2.1.3 Side Chains

The position of these side chains on the conjugated backbone has a strong impact on the band gap of the conductive polymers, because the steric hindrance induced twist in the adjacent conjugated units. More importantly, the shape and size of the alkyl side chains have a profound impact on modulating the intermolecular interaction within the BHJ blend in the solid state, which has a direct impact on the device characteristics (V_{oc}, $J_{sc,}$ and FF).

It is well acknowledged that in order to attain the maximum potential device efficiency for any given polymer, morphology must be adjusted. Because these side chains can have a significant impact on the morphological parameters of the BHJ blend in the solid state (via carefully controlled processing conditions), engineering side chains has become an increasingly important approach to maximize device efficiency for a given conductive polymer.

Structural optimization of the conductive polymer to achieve maximum device efficiency is a very complex process that is frequently guided by trial and error. A small structural perturbation to the existing conductive polymer can indeed improve certain device characteristics (V_{oc}, J_{sc}, or FF), but other desirable properties may be compromised in the meantime.

6.2.2 Polymer Engineering

While the band gap and energy levels of a conductive polymer may be completely exhibited with only a few repeating units, a large, yet acceptable, molecular weight is required to achieve the best device efficiency. Li et al., for example, used properly purified monomers and catalysts, as well as tailored reaction conditions, to accurately regulate the molecular weight of PBnDTFTAZ using the standard Carothers equation. The polymer with the highest device effectiveness (7.3%) was achieved with a molecular weight of 40 kg mol^{-1}.

To completely exhibit the real device behavior of a specific conductive polymer for solar cells, a sufficiently high molecular weight is (nearly) required. The "threshold" molecular weight is determined by the conductive polymer. Nonetheless, an empirical "threshold" of 20 kg mol^{-1} was proposed for conductive polymers (for solar cells), which would equal to ≈20 repeating units.

6.2.3 Conductive Polymers as Hole Transporting Materials for Solar Cells

Conductive polymers can be used as effective hole transporting materials (HTMs) in dye-sensitized solar cells (DSSCs) or directly in perovskite solar cells (PSCs) and silicon (Si)/organic heterojunction solar cells (HSCs). The primary role of these HTMs is to efficiently collect and transport holes created by excited light harvesters, resulting in efficient separation of electrons and holes for increased photo-to-electricity conversion efficiency (PCE). Furthermore, HTMs may operate as a barrier to inhibit the intrusion of moisture and/or oxygen, protecting the photoactive layer from ambient-induced deterioration and thus increasing the long-term stability of the device.

6.2.3.1 Polymers as HTMs in PSCs

While many conductive polymers were verified as potential HTMs in DSSCs, their performances were generally poor, and their PCEs were much lower than those of DSSCs based on liquid electrolytes. Until the introduction of PSCs, this remained an unresolved problem. The performance of conductive polymer HTMs is strongly reliant on the type of photosensitizer used; hence, careful sensitizer selection is required for high PCE devices based on conductive polymer HTMs. Fortunately, lead perovskites appear to be ideal partners for these polymeric HTMs. The device functioned with the pricey, hygroscopic, and heat-sensitive HTM of spiro-OMeTAD, and the maximum validated PCE of PSCs was 25.2%. In comparison, PCEs of more than 20% may be produced reliably by substituting spiro-OMeTAD with low-cost, air-stable conductive polymer HTMs. Currently, conductive polymers are appealing HTM options in PSCs, with materials based on polytriarylamine, P3HT, and PEDOT:PSS receiving the most attention.

Polytriarylamine-Based HTMs

Because of their high affinity to holes caused by triarylamine units, polytriarylamines are an important type of hole conducting materials in organic photovoltaics. Because of its high hole mobility ($10^{-2} - 10^{-3}$ $cm^2 \cdot V^{-1} \cdot s^{-1}$), low thermal degradation rate, and strong interaction with perovskites, poly-[bis(4phenyl)(2,4,6-trimethylphenyl)amine] (PTAA) has been recognized as one of the most effective polymeric hole conductors. Under typical AM 1.5G illumination, PSCs based on a three-dimensional nanocomposite of mesoporous $TiO_2/CH_3NH_3PbI_3$ and PTAA HTM were found to have a maximum PCE of 12.0% in 2013. In comparison, the PCE of spiro-OMeTAD-based devices was just 8.4%. The enhanced performance of PTAA was ascribed to its reduced series resistance and wider work function, which resulted in higher fill factor (FF) and V_{oc}.

P3HT-Based HTMs

P3HT is an intriguing HTM candidate in PSCs, similar to circumstances in DSSCs. Previously, the performance of P3HT-based HTM was inadequate, mostly owing to its comparatively low hole mobility (3×10^{-4} $cm^2 \cdot V^{-1} \cdot s^{-1}$) and quick non-radiative recombination at the perovskite/P3HT interface. Later, Snaith et al. enhanced the PCE to 6.7% by adding a perovskite-sensitized TiO_2 layer functionalized with a fullerene self-assembled monolayer (C_{60}SAM) to the P3HT HTM. Furthermore, PCEs of

10.8%, 13.6%, and 13.7% were obtained for PSCs with P3HT-based HTMs using highly concentrated $CH_3NH_3PbI_2Cl$ perovskite precursors, atomic-layer-deposited compact TiO_2 film, and high-efficiency sequentially vapor-grown $CH_3NH_3PbI_3Cl$ film, respectively.

Despite these advancements, limited hole mobility of the P3HT remained a concern, and PSCs with P3HT HTMs performed usually worse than those based on Li-doped spiro-OMeTAD. Part of the problem can be overcome by doping P3HT with LiTFSI and/or t-BP. When LiTFSI and t-BP were introduced to the P3HT HTM, the PCE of the device was twice (from 6.5% to 13.5%). Hybridizing P3HT with carbon nanomaterials is another feasible option for increasing the charge mobility of the HTM layer. For example, PCEs of 6.45%, 8.3%, and 14.58% were reported for PSCs using P3HT-based HTMs with additions of multiwalled carbon nanotubes, bamboo-structured carbon nanotubes, and graphdiyne, respectively. Snaith's group used a single-walled carbon nanotubes/P3HT composite HTM in combination with hydrophobic PMMA to simultaneously enhance hole percolation and device stability (Fig. 6.12a). The resulting device exhibited a maximum PCE of 15.3% and excellent stability with no significant change in color after 96 h of exposure to air at 80 °C (Fig. 6.12b), proving the great potential of polymeric HTMs in the future development of PSCs. Seo et al. introduced a double-layered halide architecture (DHA) in PSCs (Fig. 6.12c) and reported a certified PCE of 22.7% when P3HT was used as the HTM. The addition of the wide-band gap halide (WBH) layer not only substantially decreased recombination at the perovskite/P3HT interface but also enhanced P3HT self-assembly, resulting in the significant improvement in PCE. Furthermore, the DHA-based PSCs demonstrated good long-term stability, retaining almost 80% of their original PCE after 1008 h at 85% relative humidity without encapsulation and 95% of their initial PCE after 1370 h under 1 sun irradiation after encapsulation (Fig. 6.12d) [10, 13].

PEDOT:PSS-Based HTMs

Because of its superior conductivity, high transparency, and high work function, poly(3,4-ethylenedioxythiophene) and poly(4-styrenesulfonate) (PEDOT:PSS) have been frequently used as the HTM in inverted PSCs. Chen et al. employed PEDOT:PSS as the HTM for the first time in 2013 to manufacture inverted planar heterojunction PSCs, achieving a PCE of 3.9% under simulated AM 1.5G illumination when PCBM was employed as the electron transport layer. By sandwiching solution-processed $CH_3NH_3PbI_{3-x}Cl_x$ perovskite absorbers between PEDOT:PSS and PCBM in inverted planar PSCs, the PCE of such a device was swiftly increased to 10% on glass substrates and over 6% on flexible polymer substrates.

Unfortunately, the hydrophilic and acidic characteristics of PEDOT:PSS continue to disrupt the interface between ITO and PEDOT:PSS. To remedy the problem, a MoO_3 layer was added between ITO and PEDOT:PSS. It not only improved the hole transport capabilities from perovskite to ITO, but also totally separated ITO and PEDOT:PSS. As a consequence, a maximum PCE of 14.87% was reached, and

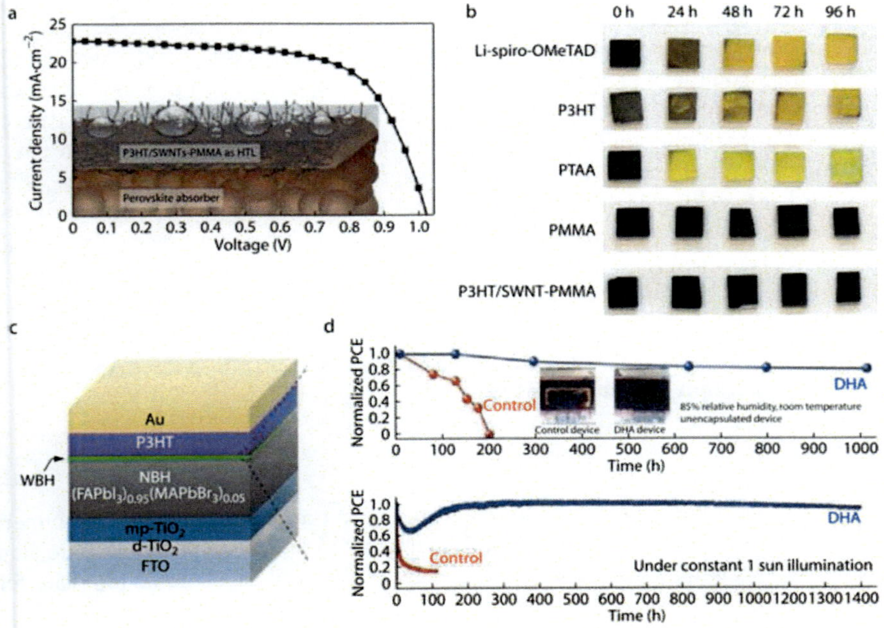

Fig. 6.12 **a** Schematic illustration of the perovskite solar cell with P3HT/SWNTs-PMMA as the HTM; **b** photograph illustrating the visible degradation of the perovskite layer. The color shifts from almost black to yellow for all organic HTLs except for the films covered with PMMA only or a composite of carbon nanotubes and PMMA; **c** the structure of an n-i-p perovskite solar cell based on a DHA using P3HT as the hole transport material; **d** stability tests of control and DHA devices at 85% relative humidity without encapsulation and under 1 sun illumination upon encapsulation. Reprinted from ref. [29], copyright 2019, with permission from Springer Nature

the device kept 93% of its initial PCE after being stored in the dark under ambient settings for 10 days. A PCE of 18.8% was attained by further improving the thickness of the MoO_3 layer. A thin layer of perylene, in addition to MoO_3, may be employed to separate ITO and PEDOT:PSS, resulting in stable PSCs with a maximum PCE of 17.06% and 69.9% retention of the original PCE after 200 h under ambient conditions even without encapsulation.

6.2.4 *Solution-Processable Conductive Polymers as Anode Interfacial Layer Materials for Organic Solar Cells*

Because of their good film-forming properties and low-temperature processing benefits, solution-processed conductive polymers have been widely employed as anode interfacial layer (AIL) materials in organic solar cells (OSCs). Conductive polymers are more suitable for serving as AILs in OSCs. It is worth noting that conductive polymers may be easily manufactured into smooth, homogeneous, and durable films from solutions, endowing the materials with high solution processability and paving the way for possible commercialization and large-scale manufacturing of OSCs. Furthermore, unlike transition metal oxides (TMO), the chemical structure of conductive polymers may be easily manipulated by organic synthesis, allowing for further tweaking of the material chemical, optical, and electrical characteristics. As a result, investigating solution-processable conductive polymers as efficient AIL materials is critical for advancing the development of OSCs.

6.2.4.1 Solution-Processable Conductive Polymers as AIL Materials in Single-Junction OSCs

PEDOT:PSS

PEDOT:PSS is widely regarded as one of the most promising conductive polymer-based AIL materials in OSCs. The PSS is integrated into PEDOT by protonic acid doping, in which the PSS serves as a counter ion as well as a soluble template for the PEDOT.

In general, standard solution processing may be used to create PEDOT:PSS films with remarkable optical transparency in the visible range and strong electrical conductivity. PEDOT:PSS exhibits electrical characteristics similar to inorganic materials such as semiconductors and metals since it is an innately conductive polymer. Furthermore, the appropriate W_F of PEDOT:PSS ($\approx$5.1 eV) corresponds well with several common polymer donors to make excellent ohmic contact at the anode/active layer interface. However, PEDOT:PSS has several obvious drawbacks, such as its high acidity and hygroscopicity, which might create corrosion difficulties on the ITO electrode. Despite these flaws, substantial research was conducted to enhance the AIL performance of PEDOT:PSS. The following are the key ways for enhancing PEDOT:PSS:

Tuning the W_F of PEDOT:PSS

Because the HOMO level of polymer donors in fullerene-free OSCs has been continually reduced to achieve high V_{oc}, the W_F of PEDOT:PSS must be increased further to meet the HOMO level of the newly created donors. The use of organic solvents was discovered to enhance the W_F of PEDOT:PSS films. Despite the fact that polar organic solvents are commonly employed to increase the electrical conductivity of PEDOT:PSS, the influence of polar organic solvent on the W_F of PEDOT:PSS has received little attention. According to Yeo et al., the W_F of PEDOT:PSS may be boosted using an unique polar-solvent vapor annealing (PSVA) technique. The W_F of the PSVA-treated PEDOT:PSS was improved from 5.0 to 5.35 eV by simply immersing a spin-coated PEDOT:PSS film in a glass jar saturated with dimethyl sulfoxide (DMSO) vapor, with an enhanced conductivity of up to 1057 S cm^{-1}.

The PSVA treatment might cause phase separation of the PEDOT:PSS film, concentrating the PSS at the film surface. The surface-enriched PSS reduced the density of the filled states around the Fermi level, increasing the W_F of PEDOT:PSS. A PCE of 3.5% was attained in the P3HT:$PC_{61}BM$-based OSCs using the PSVA-treated PEDOT:PSS anode interfacial layer. Inorganic acid doping, in addition to organic solvent treatment, can improve the W_F of PEDOT:PSS; however, acid doping is primarily employed to improve the conductivity of PEDOT:PSS. Wu et al. discovered a significantly increased W_F of 5.5 eV by adding H_2PtCl_6 to the PEDOT:PSS. The X-ray photoelectron spectroscopy (XPS) study revealed that PEDOT might be oxidized by H_2PtCl_6, which could explain the increased W_F. The H_2PtCl_6-doped PEDOT:PSS is a viable contender for application as flexible electrodes in OSCs due to its high W_F, conductivity, and transparency.

Another method for increasing the W_F of PEDOT:PSS is ultraviolet-ozone (UV-ozone) treatment. AFM pictures indicated that UV irradiation may create a morphological change in the PEDOT:PSS film, resulting in a modest reduction in surface roughness. Changes in surface morphology can affect the surface dipole structure and, as a result, the Femi energy in materials. However, the UV-ozone treatment may cause certain issues with PEDOT:PSS films. Helander et al. proposed that the enhanced W_F might be attributed to the metastable surface dipole from the adsorbed oxygen species and that the W_F of the UV-ozone treated PEDOT:PSS was very unstable and reduced dramatically after prolonged UV-ozone treatment time. Furthermore, while UV-ozone treatment increases the apparent W_F of PEDOT:PSS, it decreases the intrinsic chemical potential (i.e., Fermi level), which may result in a bigger net energy barrier at the anode/active layer interface. Furthermore, UV radiation has the potential to break down the chemical bonds in the PEDOT:PSS film, resulting in increased resistance and film etching. These issues limit the use of UV-ozone therapy on PEDOT:PSS.

Chemical changes are employed to modify the electron structure of PEDOT-based conductive polymers, which permanently alters the electronic energy level, in order to get a reliable boost in the W_F of PEDOT:PSS. By substituting the dopant PSS with a water-soluble non-aromatic resin and grafting sulfonated-acetone-formaldehyde lignin (GSL), Hong et al. created a novel anode interfacial layer material, PEDOT:GSL. The W_F of PEDOT:GSL may be increased from 4.94 to 5.10 eV by increasing the mass ratio of PEDOT to GSL. When the PEDOT:GSL mass ratio was 1:4, the PTB7:PCBM-based OSCs had a V_{oc} of 0.77 V and a PCE of 8.47%. Furthermore, the non-aromatic resin-doped PEDOT interfacial materials are employed to lower the energy barrier at the anode/active layer interface in perovskite solar cells and organic light-emitting diodes, proving their broad use in organic electronic devices.

Introducing Optical Effects to Enhance Light-Trapping in OSC Devices

To optimize the optical distribution within a device, many light-trapping methods have been investigated. Incorporating metal nanoparticles (NPs) into the PEDOT:PSS film has been a popular method for improving the light-trapping properties of OSCs. Light scattering, the localized surface plasmons produced "antenna" effect, and the lateral propagation of surface plasmon polaritons are three mechanisms all involved

Table 6.5 Device characteristics of some representative OSCs using PEDOT:PSS AILs modified with Au or Ag nanoparticles

Nanoparticles	Active layer	V_{oc} [V]	J_{sc} [mA cm^{-2}]	FF	PCE [%]
Ag	PTB7:PC$_{71}$BM	0.75	16.3	0.70	8.6
Ag and Au mixture	PTB7:PC$_{71}$BM	0.71	17.7	0.69	8.67
Ag	PIDTT-DFBT:PC$_{71}$BM	0.96	14.3	0.63	9.02
	PCDTBT: PC$_{71}$BM	0.90	12.2	0.62	7.06
Ag	PTB7-Th:PC$_{71}$BM	0.81	17.9	0.64	9.45
Au	P3HT:PC$_{61}$BM	0.63	8.94	0.62	3.50
Au	PTB7-F20:PC$_{71}$BM	0.68	17.2	0.67	7.92
Au@SiO$_2$ core/shell	P3HT:PC$_{61}$BM	0.62	10.6	0.57	3.80
Au	P3HT:PC$_{61}$BM	0.59	11.4	0.43	2.88
Au	PIDT-PhanQ:PC$_{71}$BM	0.87	12.9	0.67	7.50
Au@Ag core/shell	PCDTBT:PC$_{71}$BM	0.87	11.3	0.64	6.31
	PTB7:PC$_{71}$BM	0.75	17.5	0.70	9.19

in the plasmonic light-trapping action of metal NPs. Direct blending of metal NPs into a photoactive layer, on the other hand, will promote non-radiative decay processes and boost carrier recombination. To address this issue, metal nanoparticles (NPs) can be incorporated in the interfacial layers, possibly reducing charge quenching at the nanoparticle/active layer interface while increasing light harvesting in the active layer. Because of their high localized surface plasmon resonance (LSPR) capabilities, silver (Ag) and gold (Au) NPs have been widely researched among several metal NPs. The LSPR area of Ag NPs is mostly found between 400 and 450 nm, whereas the LSPR impact of Au NPs spans a wider wavelength range between 450 and 650 nm. Furthermore, the LSPR peaks of metal NPs may be easily modified by careful control of size and shape to improve the light harvesting of the organic absorbers used. Table 6.5 summarizes the device parameters of certain typical OSCs employing PEDOT:PSS AILs enhanced with Au or Ag nanoparticles.

Au NPs are also frequently employed as a plasmonic material in several papers due to their higher LSPR capabilities across a wider visible wavelength range than Ag. However, the dispersion efficiency of Au NPs is modest when compared to its self-absorption. In contrast, the scattering efficiency of Ag NPs is great, but the LSPR zone is narrower than that of Au. To address this issue, Lu et al. added a combination of Au and Ag NPs into the PEDOT:PSS interlayer to achieve a greater absorption enhancement over a wide wavelength range of 350 to 750 nm, resulting in a 20% increase in PCE.

Aside from the size effect of metal nanoparticles on the LSPR, the form of metal nanoparticles has a significant impact on the LSPR. Li et al. investigated the use of shape-dependent broad-band plasmonic absorption from metal nanoparticles for

OSC performance improvement. The greatest absorption peaks of Ag nanoparticles and nanoprisms were at 400 and 600 nm, respectively, induced by the LPR of two distinct forms of NPs. More intriguingly, by combining two different shaped Ag NPs, a larger absorption spectrum was achieved, extending the spectral absorption range from 400 to 600 nm. The combination of variously shaped Ag nanoparticles results in broad-band absorption by endowing the active layer with varying LPR and scattering area. As a result, when compared to the control OSC device, the J_{sc} value enhancement in the OSC device doped with Ag nanoparticles, Ag nanoprisms, and mixed Ag NPs is 9.0%, 10.1%, and 17.91%, respectively.

Polyaniline

Polyanilines have been intensively explored as a form of traditional conductive polymer in recent decades due to their high conductivities, environmental stabilities, and simplicity of synthesis for a variety of energy-related applications such as OSCs, supercapacitors, and OLEDs. Furthermore, PANI, a p-type semiconductor, has been chosen as the most promising alternative to PEDOT:PSS due to its high solution processability, optical transparency, and readily programmable work function.

Han et al. used electrodeposited PANI films as AILs in an OSC device with a basic Al/P3HT:PCBM/ITO architecture and compared the AIL performance of PANI to that of PEDOT:PSS. The PCE of the device improved from 0.75% (bare ITO; meaning, AIL-free) to 1.78% when the PANI films were used as the AIL, demonstrating the effective AIL performance of the PANI films. However, when PEDOT:PSS was used as the AIL in the OSC device, the PCE of the device increased to 2.31%, indicating that PANI AIL performance was obviously inferior to that of PEDOT:PSS.

Furthermore, chemical modification of PANI in conjunction with molecular doping has been demonstrated to be an effective technique for further increasing PANI hole transport capability by creating intramolecular covalent bonding or supramolecular contact. Kang et al. created a novel type of PANI-based AIL material by combining sulfonated polyaniline (SPANs) with molecular doping. For example, *p*-toluene sulfonic acid (TSA^-) was used as a dopant in SPANs to create TSA-doped SPANs (SPAN-TSA^-). The PCE of the device using SPAN-TSA as the AIL exhibited a 3.2-fold improvement compared to that of the pristine SPAN in the conventionally structural OSC with a P3HT:PCBM active layer, indicating that molecular doping played a significant role in the performance improvement of PANI-based AIL materials. SPAN-TSA^- was also used as a contact interface layer, which was electrostatically assembled at the interface between the AIL and the active layer. A contact interface layer of this type might create a brief dipole at the active layer, facilitating hole transport. Furthermore, some studies discovered that using 9-anthracene methyl acid and hydrochloric acid as dopants increased the performance of PANI-based AIL materials.

Due to its low-temperature solution processability, high transparency in the 450–650 nm wavelength range, and high electrical conductivity, a water-soluble and self-doped conducting polyaniline graft copolymer based on poly(styrenesulfonic acid)-graft-polyaniline (PSSA-g-PANI) has been developed as a hole transport material to replace PEDOT:PSS. PSSA-g-PANI seemed to be more transparent and electrically conductive than PEDOT:PSS. The higher transparency of PSSA-g-PANI may have resulted in increased light absorption of P3HT, resulting in a higher J_{sc}, while the higher conductivity of PSSA-g-PANI may have resulted in more effectively transporting holes, resulting in a lower leakage current. As a result, the device with PSSA-g-PANI demonstrated ≈4% PCE, which is 20% greater than the device with PEDOT:PSS. In addition to being used as the AIL in the OSC device, PSSA-g-PANI was also used as the AIL in organic–inorganic hybrid perovskite solar cells.

Furthermore, AIL materials similar to PSSA-g-PANI have been reported, such as PANIgr-PSSA, in which PSSA was grafted on the PANI surface via radical polymerization, and PANI:PSS, in which PANI was blended with PSS, implying that combining PANI with PSS was an effective strategy for improving PANI-based AIL materials. In addition to the PANI:PSS composite, PANI mixed with poly(vinylidene difluoride) (PVDF) to produce the PANI:PVDF composite was utilized as the AIL in the OSC device, allowing the ITO surface to be smoothed and the series resistance R_s, lowered to 50%. Graphene (G) and graphene oxide (GO) were also considered as effective additives for further improving hole transport performance in the development of PANI-based AIL materials, because the incorporation of G or GO not only increases electrical conductivity but also facilitates efficient hole transport while effectively blocking electron transport.

Gribkova et al. synthesized the PANI/PAMPS complex by polymerizing aniline with an aqueous solution of poly(2-acrylamido-2-methyl-1-propane sulfonic and (PAMPS) and then adding 1 wt% graphene into the PANI-PAMPS complex to form the nanocomposite material PANI-PAMPS/G to explore new AIL materials comparable to PEDOT:PSS. The results show that PANI-based AIL materials are attractive candidates to replace PEDOT:PSS because they have various benefits, such as cheap cost, aqueous solution processability, high transparency, and improved conductivity. Among the benefits, the cheap cost and aqueous solution-processable preparation are unquestionably advantageous to the future commercialization of OSCs. PANI, in particular, has a high potential to improve its electrical conductivity through optimal molecular doping and protonation levels, as electrical conductivity is a significant factor in AIL performance enhancement. However, in comparison with PEDOT:PSS, future research on PANI-based AIL materials should concentrate on achieving a homogenous morphology and a high work function.

6.2.5 Conductive Polymers as Electrodes for OSC Devices

Polyanilines, like PEDOT:PSS, have a high conductivity that permits them to be employed as transparent electrodes in the fabrication of flexible OSC devices. Lee et al. described the creation of a flexible electrode doped with camphorsulfonic acid (CSA) and coated on a plastic substrate polyethylene terephthalate (PET) to generate ITO-free electrode materials for flexible organic solar cells, specifically PANI:CSA/PET. The power conversion efficiency (PCE) of the PCDTBT:$PC_{71}BM$ bulk heterojunction solar cell reached 3.51% when the PANI:CSA/PET electrode was used, which was similar to the 3.61% achieved by the ITO/PET electrode-based device. The PANI:CSA/PET electrode, on the other hand, has a high sheet resistance, rendering it unsuitable for large-area devices. Lim et al. reported the addition of organic additives into PANI to increase the crystallinity of PANI in order to boost its electrical conductivity even more. The best additive was a self-assembly supramolecule called 3,6-dioctyldecyloxy-1,4-benzenedicarboxylic acid (SAS18), which could engage with the PANI chains and prevent them from recoiling. Thus, the PCE of the flexible OSC device with the PANI:CSA/SAS18 electrode was 1.73%, which was greater than the 0.40% attained by the pure PANI:CSA electrode. PANI-based nanocomposite film electrodes that are flexible and stretchy have gotten a lot of interest because PANI is not only capable of achieving the best conductivity by secondary doping, but it is also compatible with other nanomaterials.

6.2.6 Conducting Polymer-Based Anode Buffer Layers in Solar Cells

Conducting polymers have also been utilized as anode buffer layers in SCs, although they have low W_F (e.g., PEDOT:PSS and CleviosTM PH have a W_F of 4.86 eV), limiting the V_{oc} of SCs. W_F modification of anode buffer layers based on conducting polymer improved the SC properties. The surface W_F of the hole extraction layer was enhanced by a tiny amount of PFI to 5.39 eV. The PeSC with the PEDOT:PSS:PFI anode buffer layer had a PCE of 11.7%, whereas the device with the traditional PEDOT:PSS buffer layer had a PCE of 8.1%. The rise in PCE was attributable to increases in V_{oc}, fill factor (FF), and short-circuit current density (J_{sc}). The higher V_{oc} validated the effect of the polymeric anode buffer layer on enhanced W_F. Furthermore, increases in V_{oc}, FF, and J_{sc} were connected to increases in built-in potential (V_{bi}). Due to the higher V_{bi}, the PeSCs that employed the PEDOT:PSS:PFI anode buffer layer produced a greater photocurrent differential between under illumination and in the dark. Enhanced V_{bi} signifies increased forward diffusion of photogenerated carriers and internal electrical field in the PEDOT:PSS:PFI PeSC. The capacitance–voltage (C–V) characteristics likewise verified the higher V_{bi}. A DMSO treatment was shown to significantly increase the electrical conductivity of PEDOT:PSS:PFI films. After the PEDOT:PSS:PFI films were formed, they were fluxed using DMSO and subsequently annealed. The DMSO-treated PEDOT:PSS:PFI film improved charge collection in PSCs due to the increased W_F by PFI. As a consequence, the J_{sc} of the PSCs was enhanced to 22.10 mA cm^{-2}, and the PCE was improved to 9.82% (the J_{sc} was 14.31 mA cm^{-2}, and the PCE was 7.35% for the PSC employing the pristine PEDOT:PSS). PSS-g-PANI, a water-soluble conducting polymer, was also modified with PFI to improve surface W_F. For PeSCs with the $MAPbI_3$ photoactive layer, the PSS-g-PANI:PFI film was employed as a hole extraction anode buffer layer. The PSS-g-PANI:PFI-based PeSC had a PCE of 12.4%, whereas the conventional PEDOT:PSS-based device had a PCE of 7.8% and the PSS-g-PANI-based device had a PCE of 9.7%.

Material engineering of conducting polymer-based anode buffer layers such as PEDOT:PSS:MoO_3, PEDOT:PSS:GO, PEDOT:PSS:PFI, and PSS-g-PANI:PFI has improved the photovoltaic efficiency and stability of SCs independent of the photoactive materials utilized. Table 6.6 summarizes the characteristics of SCs based on the kind of conducting polymer-based anode buffer layers.

6.2.7 *Conductive Polymer: Fullerene-Based Solar Cells*

In these organic solar cells, photoinduced electron transfer from donor-type semiconductive polymers to acceptor-type polymers or molecules, such as C_{60}, is used. The photovoltaic conversion efficiency is poor for bilayers of semiconductive polymers containing fullerenes. Using phase-separated composite materials resulted in significant improvements. Controlling the nanomorphology of phase separation in an interpenetrating network ("bulk heterojunction") considerably boosted the power conversion efficiency of MDMO-PPV/C_{60} solar cells.

Typical phase separation dimensions must be within the exciton diffusion length (on the order of 10 nm). On the other hand, bicontinuous, undisturbed paths must be guaranteed for charge carrier delivery to the electrodes. The mesoscopic organization and crystallinity of organic and polymeric materials must be improved to enhance charge carrier transit. As a result, for the active layer of polymer photovoltaic devices, a nanoscale interpenetrating network with crystalline order of both ingredients appears to be a desired architecture. To collect more light from the solar spectrum, the electronic band gaps of the components in the photoactive layer need eventually to be tweaked.

6.2.7.1 PPV:PCBM Bulk Heterojunction Solar Cells

2.5% effective solar cells may be made from soluble phenylene-vinylene derivatives, such as poly[2-methoxy-5-(3′,7′-dimethyloctyloxy)-1,4-phenylene vinylene] (MDMO-PPV) coupled with soluble fullerene derivatives, such as 1-(3-methoxycarbonyl)propyl-1-phenyl-[6,6]methanofullerene (PCBM). Changing the solvent from toluene to chlorobenzene enhances efficiency by roughly a factor of three, which was attributed to differences in nanomorphology.

Because PCBM has essentially negligible absorption in the visible-near-infrared range, MDMO-PPV is expected to be the primary light absorber in these solar cells. As a result, increasing the volume concentration of MDMO-PPV for greater solar

Table 6.6 Summary of characteristics of SCs with different conducting polymer-based anode buffer layers

Anode/conducting polymer-based anode buffer layers	Device architecture	V_{oc}[a] [V]	J_{sc}[b] [mA cm^{-2}]	FF[c] [%]	PCE[d] [%]	Device degradation rate[e]
ITO/PEDOT:PSS	$CH_3NH_3PbI_3/C_{60}$/Bphen/Ag	0.97	16.11	63	9.81	≈10 d/20%
ITO/MoO_3/PEDOT:PSS		0.96	20.06	67	12.78	10 d/93%
ITO/PEDOT:PSS	TQ1:PC_{70}BM/LiF/Al	0.90	9.2	66	5.5	10 d/0%
ITO/PEDOT:PSS: MoO_3		0.89	10.3	70	6.4	10 d/80%
ITO/PEDOT:PSS	$CH_3NH_3PbI_{3-x}Cl_x$/PCBM/Bphen/Ag	0.92	20.91	72.61	13.90	
ITO/PEDOT:PSS: MoO_3		0.97	21.59	75.42	15.79	
ITO/PEDOT:PSS	$CH_3NH_3PbI_3$/PCBM/LiF/Ag	0.80	14.35	71.61	8.23	≈300 h/0%
ITO/GO/PEDOT:PSS		0.84	15.72	73.71	9.74	≈500 h/20%
ITO/PEDOT:PSS	PTB7:PC_{70}BM/Al	0.75	14.75	63.6	7.04	
ITO/PEDOT:PSS:GO		0.76	16.42	65.8	8.21	
ITO/PEDOT:PSS	CH3NH_3PbI_3/PCBM/Al	0.835	14.1	68.5	8.1	
ITO/PEDOT:PSS:PEI		0.982	16.7	70.5	11.7	
ITO/PEDOT:PSS	PCDTBT:PC_{70}BM	0.738	10.2	56.4	4.2	
ITO/PEDOT:PSS:PEI		0.880	11.9	59.7	6.3	
ITO/PEDOT:PSS	CH3NH_3PbI_3/PCBM/Al	0.903 ± 0.06	13.15 ± 64	62.8 ± 5.34	7.08 ± 1.88	
ITO/PEDOT:PSS:PFI		1.022 ± 0.023	15.29 ± 1.26	70.9 ± 3.41	10.90 ± 0.77	
ITO/PEDOT:PSS	PTB7-Th:PC_{71}BM	0.80	14.31	64.20	7.35	
ITO/PEDOT:PSS:PEI		0.80	22.10	55.51	9.82	
ITO/PEDOT:PSS	CH3NH_3PbI_3/PCBM/Al	0.923	11.8	73	7.8	186 h/50%

(continued)

Table 6.6 (continued)

Anode/conducting polymer-based anode buffer layers	Device architecture	V_{oc}[a] [V]	J_{sc}[b] [mA cm^{-2}]	FF[c] [%]	PCE[d] [%]	Device degradation rate[e]
ITO/PSS-g-PANI		1.04	14.1	67.3	9.7	267 h/50%
ITO/PSS-g-PANI:PFI		1.07	14.9	77.6	12.4	570 h/50%
Graphene/PEDOT:PSS:PFI	P3HT:PC_{60}BM/Ca/Al	0.546	8.67	59.7	2.83	
	PCDTBT:PC_{60}BM/Ca/Al	0.896	8.83	54.7	4.33	

Reprinted from ref. [1], copyright 2016, with permission from John Wiley and Sons
[a]V_{oc}: open circuit voltage; [b]J_{sc}: short-circuit current; [c]FF: fill factor; [d]PCE: power conversion efficiency. [e]Device degradation rate: device operation time until the PCE maintained certain percentage of its initial PCE

light absorption is preferable. MDMO-PPV/PCBM blends were thoroughly explored to evaluate the relationship between morphology and performance in bulk heterojunction solar cells. The following experimental parameters have been identified as influencing the nanoscale morphology in these polymer-fullerene blends: (i) the solvent used, (ii) processing temperature, (iii) the relative composition ratio of polymer and fullerene, (iv) solution concentration, (v) thermal annealing, and (vi) the primary chemical structure of the materials determining tertiary organization structures.

The main chemical structures of polymer and fullerene affect their solubility in organic solvents as well as their miscibility. Furthermore, the solvent effects the drying time during film formation, whereas heat annealing allows for recrystallization. Diffusion of one or both components in the blend changes the phase separation. Replacing the C_{60} moiety of [60]PCBM with a C_{70} fullerene derivative allows for slightly more HOMO–LUMO transitions and boosts light absorption/harvesting.

6.2.7.2 Poly(3-alkylthiophene):PCBM Bulk Heterojunction Solar Cells

P3ATs (poly(3-alkylthiophenes)) are conductive polymers with excellent solubility, processability, and environmental stability. P3HT:poly(3-hexylthiophene), P3OT:poly(3-octylthiophene), and P3DDT:poly(3-dodecylthiophene) are the regioregular poly(3-alkylthiophenes) (RRP3AT) employed. Al Ibrahim et al. studied the effect of alkyl side chain length on the electrochemical and optical characteristics of regioregular P3HT, P3OT, and P3DDT. The energy levels for P3OT, P3HT, and P3DDT were almost identical, having an optical band gap energy of roughly 1.9 eV. Their electrochemical band gaps were somewhat enlarged as their side chain length grew. Because of chromophore dilution, polythiophenes with larger side chain cause a systematic drop in absorbance coefficient (i.e., conjugated segments in ratio to non-conjugated segments decrease upon increasing the side chain length).

Bulk heterojunction solar cells with external quantum efficiencies of over 75% and power conversion efficiencies of up to 5% have been produced using regioregular poly(3-hexylthiophene) (RR-P3HT) as donor and PCBM as acceptor. It is claimed that the great efficiency of these devices is due to microcrystalline lamellar stacking in the solid-state packing. RR-P3HT has a hole mobility of up to $\sim$0.1 $cm^2/V{\cdot}s$, resulting in decreased recombination. Furthermore, because of this stacking, interchain interactions create a red shift in the optical absorption of RR-P3HT. The electron mobility of the second component, PCBM, is 2×10^{-3} $cm^2/V{\cdot}s$.

A thermal annealing process enhanced the efficiency of P3HT-based solar cells. Padinger et al. found that annealing the devices while simultaneously applying an external voltage improves the properties of such plastic solar cells. Annealing not only produces recrystallization but also decreases the free volume and defect density at surfaces. The mobility of holes is improved by P3HT recrystallization. Thermal annealing using chloroform vapor or a simple thermal treatment increases total power conversion efficiency. The absorbance maxima of PCBM:P3HT films cast from chloroform and chlorobenzene solutions are 600 and 630 nm, respectively. These variations pointed to a greater degree of P3HT side chain ordering when chlorobenzene was used. Thermal annealing and/or changing the solvent from chloroform to chlorobenzene can be used to promote structural order.

Using the flash photolysis time-resolved microwave conductivity approach, Savenije et al. investigated the link between morphology and charge carrier mobility (FP-TRMC). Annealing at 80 °C resulted in the production of crystalline P3HT fibrils, which significantly increased hole mobility. Typically, the mobility of electrons and holes in organic solar cells is imbalanced. The mobility of holes in MDMO-PPV thin films is orders of magnitude lower than the mobility of electrons in PCBM thin films. Simply annealing the devices at temperatures over 110 °C for 4 min improved hole mobility in the P3HT phase by more than three orders of magnitude [9].

6.2.8 Conductive Polymers for Flexible Solar Cells

With significant advancements in the performance of energy harvesting and storage devices, scientists are working to improve their mechanical malleability, which allows them to be suitable for scalable roll-to-roll printing technology and compatible with flexible or non-planar substrates, allowing them to expand their applications, such as portable and wearable electronic technologies. Flexible energy harvesting and storage systems have advanced dramatically, owing mostly to the remarkable invention of flexible electronic materials, notably CPs. CPs with high physical qualities such as electrical conductivity, tensile strength, flexibility, and lightweight, as well as outstanding chemical properties such as quick kinetics of oxidation/reduction reactions, are promising for the development of flexible energy harvesting and storage systems. These flexible devices are often manufactured on flexible substrates such as poly(ethylene terephthalate), poly(dimethylsiloxane), polyurethane, textile, and cloth using layer-by-layer coating methods before being enclosed by barrier materials. Table 6.7 summarizes the physical parameters of common CPs, polymer substrates, and barrier materials.

Mechanical characteristics of the CPs, including as elastic modulus, Poisson's ratio, toughness, and thermal expansion coefficient, are vital for the implementation of flexible devices. Despite the fact that several polymers have been investigated, it is uncommon to establish quantitative criteria or ranges within which the devices can be made flexible. Tracing the major performance characteristic of the device during repeated bending with varied angles can be used to measure its flexibility. A flexible solar cell, for example, was claimed to retain over 90% photoelectric conversion efficiency after 1000 cycles at a bending angle of 180° without sealing, whereas a supercapacitor demonstrated 94% storage efficiency after 1000 cycles at a bending angle of 180°. Because the dimensions of the devices being tested vary, the

Table 6.7 Physical properties of typical CPs, polymer substrates, and barriers

Material[a]	Type	Conductivity	Mechanical properties	Contribution
PA	Conducting/semiconducting	1.70×10^{-11}–3.8×10^{-2} S m^{-1}	Bendable (in a nanocomposite film)	Active material and electrode material
PPy	Conducting	0.02–1 S m^{-1}	No difference upon application of 13% bending strain (in a planar supercapacitor)	Active material and electrode material
PEDOT	Conducting	550–1000 S m^{-1}	Slightly decreased after bending for 1000 cycles (in a fiber-shaped PSC)	Active material and electrode material
PPV	Conducting	10^{-5}–1 S m^{-1}	Bendable (in an organic planar solar cell)	Active material and electrode material
PANI	Conducting/semiconducting	6.3×10^{-9}–4.6×10^{-5} S m^{-1}	Well maintained after being stretched for 100 cycles to a strain of 100% (in a fiber-shaped supercapacitor)	Active material and electrode material
PET	Insulating	–	Elongation at breaking >50%; Young's modulus of 2.8–3.1Gpa	Substrate and flexible barrier
PDMS	Insulating	–	Elongation at breaking >200%; Young's modulus of 0.01–0.07Gpa	Substrate and flexible barrier

Reprinted from ref. [39], copyright 2018, with permission from John Wiley and Sons
[a] Calculated based on the results given in reference when specific results were not given
SCs: supercapacitors

flexibility gained at the same bending angle is not comparable among devices with various architectures and sizes.

A variety of scalable printing methods, such as inkjet, doctor blade, and roll-to-roll processes, may easily produce flexible PSCs on the flexible polymer substrate. As a result, they are the most appealing option for lightweight, low-cost, solution-processable, and adaptable photoelectric conversion devices suitable for industrial mass production. The PCEs of flexible PSCs have been enhanced by as much as 9.4%, indicating a promising future for their use in flexible energy harvesting integrated systems.

6.2.8.1 Conductive Polymer as Light Harvesting Material

The advancement of flexible PSCs is frequently followed by the discovery and design of new high-performance materials. Donor and acceptor materials are two important elements in the photoactive layer that determine the PCEs of PSCs. In most cases, p-type CPs are used as the donor materials, which are then combined with a commercial acceptor material, commonly [6,6]-phenylC61-butyric acid methyl ester (PCBM), to produce the bulk heterojunction for photoelectric conversion.

In general, the electrical characteristics of the donor and acceptor components must be optimized in order to absorb light, create excitons, and separate and transfer charges with the least amount of energy loss. Because the PCBM is the most often used acceptor with the highest efficiency, most researches concentrate on the creation of donor materials. Figure 6.13 summarizes the typically utilized *p*-type CPs and PCBM in the photoactive layer of flexible PSCs.

The first donor polymer used for bulk heterojunction was synthesized from poly(2-methoxy-5-(3′,7′-dimethyloctyloxy)1,4-phenylene vinylene) with a PCE of less than 1%. However, this PCE falls well short of what is expected for practical applications; hence, many studies concentrated on improving PSC PCEs by optimizing the structure of donor materials. Because of its high hole mobility, broad spectrum coverage, and low cost, soluble polythiophenes, particularly poly(3-hexylthiophene) (P3HT), have risen to prominence as the primary donor polymer for PSCs. Many flexible PSCs have been built using the P3HT:PCBM with a maximum PCE of 4.3%.

Because the performance of PSCs is essentially governed by three factors: V_{oc}, J_{sc}, and FF, enormous efforts have been made to build novel CPs with appropriate band gaps and energy levels in order to attain high performance. P3HT, the most often utilized donor, has an electron-rich group with a low V_{oc} of 0.6 V. To boost the V_{oc}, the HOMO energy level of the donor polymer should be reduced by creating less electron-rich regions. For example, as compared to non-fluorinated polymers, CPs with fluorine substituents have a lower HOMO energy level of up to 0.29 eV. As a result, for the difluorinated polymers, the V_{oc} of the resulting PSCs rose by 0.07 V.

Another defining factor for a high-performance PSC is J_{sc}. Narrowing the band gap for larger solar spectrum coverage is a good method for achieving high J_{sc}. Many ways were used, including developing an appropriate donor–acceptor structure, changing the conjugation duration, and managing the polymer chain planarity. The easiest way is to create a good donor–acceptor structure. A PSC using thieno[3,4-b]thiophene/benzodithiophene (PTB7) as a donor polymer, for example, demonstrated a PCE of 8.34% and was maintained by 96% even after 400 bending cycles. The third parameter for achieving high PCE in a PSC is the FF, which is the ratio between the maximum obtainable power and the product of J_{sc} and V_{oc}, which is determined by several factors such as charge carrier mobility and balance, interface recombination, internal resistances, and donor–acceptor film morphology.

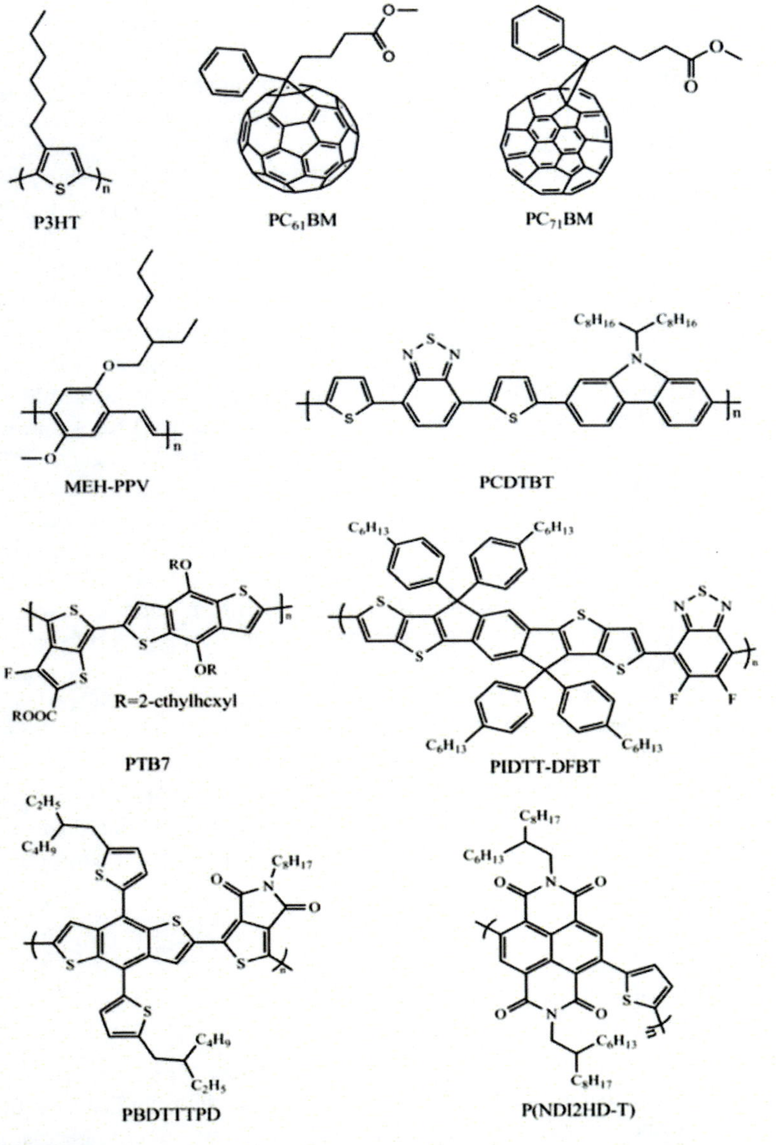

Fig. 6.13 Chemical structures of commonly used donor and accepter CP materials and PCBM in flexible PSCs. Reprinted from ref. [39], copyright 2018, with permission from John Wiley and Sons

6.2.8.2 Conductive Polymer as Charge Transport Material

Because of its strong hole extraction and transport capabilities, CPs may often be used as both a hole transporting layer and a transparent conductive electrode to replace the rigid indium tin oxide electrode in flexible PSCs. Among them, the most researched hole transporting and electrode material is PEDOT:PSS, which is already marketed in a variety of regularly utilized varieties such as 4083, PH500, and PH1000. PEDOT:PSS was employed as both a hole transporting layer and a conductive electrode in several situations to manufacture flexible PSCs using complete roll-to-roll processing on the flexible substrate.

Although the aforementioned PSCs demonstrated considerable adaptability, they are unable to match the criteria for many specialized applications, such as textile technology. PSCs in textiles were necessary to be flexible and twistable in order to conform to the various deformations encountered in actual use. As a result, by constantly dip-coating the PEDOT:PSS layer and wrapping the aligned-CNT sheet or CNT fiber on titania wire coated with the photoactive layer, a novel family of fiber-shaped PSCs was generated. These fiber-shaped PSCs displayed good flexibility, with the PCE remaining almost 90% after 1000 cycles of bending (Fig. 6.14a, b) [38, 42]. They can be made stretchable further by developing a spring-like structure that can withstand external stress (Fig. 6.14c) [41]. Despite significant attempts to weave these fiber-shaped PSCs into textiles, high-efficiency PSC textiles have yet to be achieved due to PSC surface degradation, short circuits produced by weaving procedures, restricted effective areas, and poor PCEs. Direct fabrication of the PSC on the textile-like electrode to generate the PSC textile (Fig. 6.14d) was a potentially successful technique for addressing these issues [40]. As predicted, the resulting PSC fabrics were flexible, with PCEs changing by less than 3% after 200 cycles of bending, paving the way for applications in portable and wearable energy harvesting devices. Table 6.8 summarizes the comparison of photovoltaic properties in the aforementioned PSCs.

6.3 Thermoelectric Generator

Aside from PSCs, thermoelectric generators are another type of efficient energy harvesting device that generates electricity from heat, with applications in solar thermal usage and waste heat recovery. Conventional thermoelectric generators, on the other hand, are made of inorganic materials that are stiff and need a sophisticated production process. CPs might be a promising alternative for accelerating the development of thermoelectric generators with great flexibility. Meanwhile, they are light and low cost and can be easily scaled up using roll-to-roll printing processes.

Because of the consistent nanoscale shape and readily changeable doping level by adjusting the voltage, the CPs utilized in thermoelectric generators were often made using electrochemical polymerization. By adjusting the doping amount, post-treatments can improve the electrical conductivity of CPs even more. When a

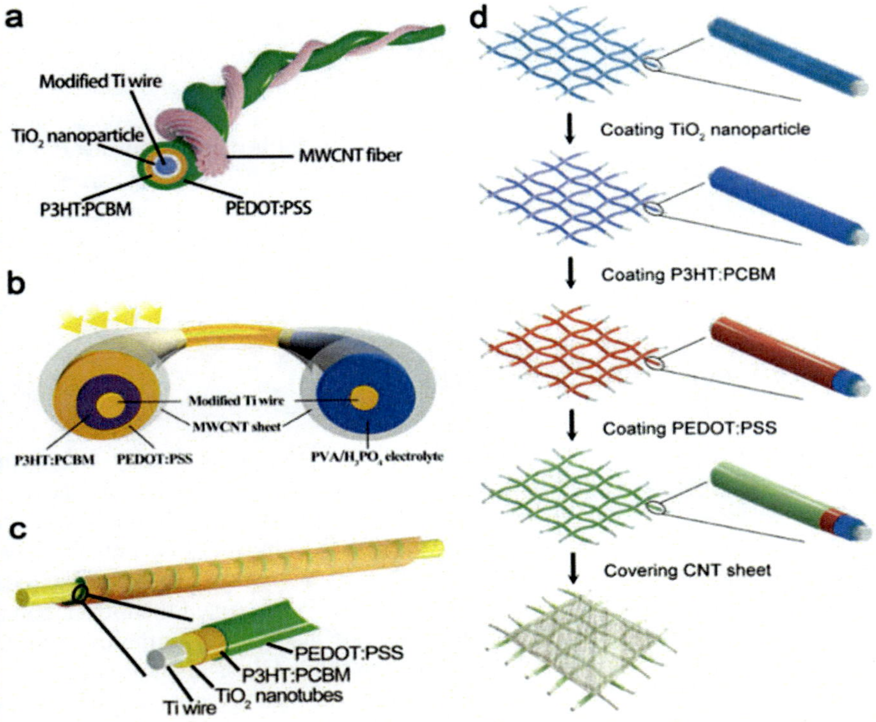

Fig. 6.14 Schematic of the structure of flexible PSC fibers and textiles. **a** A fiber-shaped PSC with a twisted structure. **b** A fiber-shaped PSC with a coaxial structure. **c** A stretchable fiber-shaped PSC. **d** Fabrication of the PSC textile. Reprinted from ref. [39], copyright 2018, with permission from John Wiley and Sons

Table 6.8 Comparison of photovoltaic parameters in PSCs

Light-harvesting material	V_{oc} [mV]	J_{sc} [mA cm^{-2}]	FF [%]	η [%]
P3HT/PCBM	580	11.9	61	4.2
BDT/DTBT	760	9.6	51.2	3.8
PCDTBT/PCBM	3320	2.61	57	4.93
PIDTT-DFBT/PCBM	960	15.75	60	9.4
PTB7/PCBM	750	14.66	75.7	8.34
PBDTTTPD/P(NDI2HD-T)	1060	11.22	56	6.64

Reprinted from ref. [39], copyright 2018, with permission from John Wiley and Sons

PEDOT:PSS film is exposed to ethylene glycol or formic acid, the PSS on the surface can be selectively eliminated, resulting in a reduction in doping level. Furthermore, by doping with DMSO/*p*-toluenesulfonic acid monohydrate and dedoping with hydrazine/DMSO, the PSS concentration and oxidation level on the surface of the PEDOT:PSS film may be precisely regulated. The former boosted PEDOT electrical conductivity, while the latter selectively removed PSS insulating barrier while controlling PEDOT oxidation level. Furthermore, due to the increase in conductivity, PEDOT:PSS combined with CNT demonstrated an increase in thermoelectric power factors that were orders of magnitude larger than those of normal CPs.

6.4 Lithium-Ion Batteries (LIBs)

LIBs are being studied extensively due to their high energy density, long-term stability, and minimal charge loss. Flexibility realized may substantially increase their applications in various growing industries like as electronic skins and sensors. CPs are now being investigated as active materials of electrodes for flexible LIBs. Their resulting capacities, however, were much lower than those of standard electrode materials. Meanwhile, the poor cyclic stability of CPs prevented them from being used in LIBs, because the structural units became inactive due to the irreversible overoxidation reaction at high voltage during the charge/discharge process.

Combining CPs with inorganic materials to create flexible composite electrodes for LIBs is an effective strategy for addressing this pressing issue. Because CPs may interact synergistically with inorganic compounds, considerable gains in rate capability, electrode lifespan, and mechanical and thermal stability have been realized. By increasing the electrical conductivity of the electrode, the LIB with the PPy/$LiFePO_4$ composite electrode had a larger discharge capacity beyond 50 cycles (80 mA h g^{-1}) than the LIB with pure $LiFePO_4$ (60 mA h g^{-1}). Although the present performance of flexible LIBs based on CPs is limited and may be considerably improved, the advent of new CPs with high performance will accelerate the development of flexible LIBs.

6.5 Fuel Cells

6.5.1 *Utilization of Conductive Polymers in Fabricating Polymer Electrolyte Membranes (PEMs) for Direct Methanol Fuel Cells*

The proper operation of a direct methanol fuel cell (DMFC) is critical for its development as a potential alternative to rechargeable batteries for use as a power source in portable electronic devices and transportation vehicles, as well as for future integration into stationary power generation units. In the fabrication of PEMs, the use of an aromatic ring incorporating conventional conductive polymers (CPs) such as polyaniline (PANI), polypyrrole (PPy), and polythiophene (PTh) and its derivatives has led in significant increases in the performance of DMFCs.

6.5.1.1 PANI

Because the repetitive imine groups are weak Lewis base sites favorable to H^+ hopping, CPs with a conjugated backbone, such as PANI, are advantageous for H^+ transport. Furthermore, substantial conjugation-induced structural coplanarity lowers steric barrier in H^+ transport. The perm selective behavior of PEMs was discovered to be site dependent on the PANI layer inside the PEM matrix. Another significant aspect of using CPs in PEMs is their electronic conductivity. Furthermore, the electrochemical stability of the CPs makes them more appealing for use in the corrosive conditions of fuel cells.

The acid-doped *meta*-PANI membrane has a high ion-exchange capacity of 5.35.3 meq.g^{-1} compared to 0.9 meq.g^{-1} attained for Nafion. The oxidation state of PANI has a significant influence on the physicochemical and transport characteristics of Nafion/PANI composite membranes. The Nafion/emeraldine salt oxidation form of PANI (ES) membranes was discovered to have improved selectivity and superior cell performances when compared to other existing oxidation states of PANI. The ES version of PANI also has a much higher H^+ conductivity. This was demonstrated to be the outcome of an effective functional group interaction between Nafion ionic channels and ES. As a result, the microphase-separated structures of Nafion, which are made up of hydrophobic backbone groups and hydrophilic clusters, were altered. As a consequence, CH_3OH crossover decreased substantially more than H^+ conductivity. The open circuit voltage (OCV) levels were also discovered to fluctuate based on PANI oxidation states. The OCV values of Nafion were 705 mV, 688 mV for Nafion-leucoemeraldine base form of PANI (LEB), 742 mV for Nafion-ES, and 722 mV for Nafion-pernigraniline base form of PANI (LEB) (PB). The selectivity ratio (i.e., the ratio of proton conductivity to methanol permeability) is the primary parameter used to determine the efficiency of a DMFC membrane. The published selectivity ratio values for Nafion, Nafion-LEB, Nafion-ES, and Nafion-PB, respectively, are 1.57×10^4, 0.69×10^4, 3.42×10^4, and 2.15×10^4. As a result, despite having the lowest water and methanol absorption, Nafion-ES had the highest OCV, peak power density, and selectivity among the membranes investigated. After electrochemical treatment with PANI ES, Nafion membranes showed a tenfold reduction in CH_3OH permeability at the price of only a small drop in H^+ conductivity.

6.5.1.2 Polypyrrole

Polypyrrole is another major aromatic-based CP that has been widely used in the production of PEMs. Fabricating PPy in situ inside the membrane holes resulted in modified Nafion membranes. SEM images revealed that PPy was present both on the membrane surface and within the membrane pores. This change effectively reduced the permeability of CH_3OH. Further investigation into the influence of three different electrolytes on output current densities revealed that the electrolyte containing a CH_3OH/H_2O mixture produced the highest current density, followed by CH_3OH, and the presence of H_2SO_4 produced the lowest current density. In CH_3OH-containing electrolytes, a new oxidative peak developed, which was attributed to the electrocatalytic activity of PPy toward CH_3OH oxidation. Under the identical experimental circumstances, a composite PPy/Nafion 115 membrane performed much better than Nafion 115 alone. This improvement in performance can be attributed to the presence of PPy within the pores of the Nafion membrane, which results in the creation of electrostatic crosslinking and a decrease in pore volume. This electrostatic interaction between the doped positive PPy chains and the Nafion sulfonate groups, on the other hand, is expected to result in a reduction in the concentration of mobile H^+ ions and thus increased membrane and cell resistances. As a result, in a DMFC running at high current densities, this disadvantage must outweigh the benefits received from a lower CH_3OH crossover. Park et al., on the other hand, found that the presence of PPy particles on the surface of the Nafion membrane, rather than in the interior gaps, was beneficial in terms of minimizing CH_3OH crossover while preserving H^+ conductivity. They discovered that the interaction between the polar phases of Nafion and the secondary ammonium groups of PPy improved the mechanical and thermal characteristics of the used Nafion/PPy composite membrane. Poly(1-methyl pyrrole), a PPy derivative, was coated over Nafion 117 membrane to generate a composite membrane with variable CH_3OH penetration when the operating time was changed.

6.5.1.3 Other Conductive Polymers

Polythiophene and its derivatives, particularly poly(3,4-ethylenedioxythiophene) (PEDOT), are essential components of CP-based PEM materials. The modification of Nafion 117 membranes with in situ polymerized 3,4-ethylenedioxythiophene using H_2O_2 as an oxidant resulted in a membrane with an Arrhenius-type temperature dependency of CH_3OH permeability and H^+ conductivity. The activation energies of CH_3OH penetration were found to be larger in the modified membranes than in the unmodified membranes, whereas the activation energies of H^+ migration were found to be identical in both. When compared to Nafion 117, the CH_3OH permeability of the modified membranes was reduced from 72 to 30% when tested at 60 °C. The resulting selectivity ratio was 1.67×10^4. The following critical points can be drawn from the use of CPs in the development of PEMs for use in DMFC technology: (a) incorporation of CPs in the PEM significantly reduces the critical problem of CH_3OH crossover, with a slight compromise on H^+ conductivity; (b) the porous nature of CPs proves useful in diffusing gases out of the MEA; (c) reducing the use of costly Nafion by replacement with CPs leads to cost reduction; and (d) CPs exhibit thermal stability in the temperature range within which DMFC operates, thus rendering stability to the overall system [4].

6.5.2 *Aromatic Conductive Polymers-Based Catalyst Supporting Matrices for Microbial Fuel Cells*

Aromatic conductive polymers (ACPs) have high electronic conductivity in the range of 10^{-6} Scm^{-1} to 10^{3} Scm^{-1}, which is approximately two orders of magnitude greater than that of conventionally used carbon catalyst supports, combined with their high surface area and stability under fuel cell operating conditions. As a result, they play an important role in the development of catalytic supports in PEMFCs. Furthermore, significant advancements in the solubility features of ACPs have aided in their production capability. Other significant benefits of using ACPs as a support constituent include (a) adequate catalyst binding on the inherently present heteroatoms in common ACPs such as polyaniline (PANI), polypyrrole (PPy), and polythiophene (PTh) and its derivatives, avoiding the need for additional functionalization steps, and (b) fine dispersion and uniform distribution of expensive catalysts, exhibiting the maximum active surface area for catalytic reactions leading to their reduced consumption and maximum utilization.

6.5.2.1 Cathode Catalyst Supports

PANI and PPy ACPs have emerged as viable materials in the manufacture of cathode catalyst layers. Cathode modification and bacterial inoculum optimization have been carried out in order to increase the power output of MFCs. PANI/C-supported Fe-phthalocyanine (FePc) in an air–cathode MFC produced a maximum power density of 630.5 mWm^{-2}, which was higher than the values of 336.6 mWm^{-2} and 575.6mWm^{-2} obtained with FePc/C and Pt cathode, respectively. The power density of the PANI/vanadium pentoxide (V_2O_5) nanocomposite cathode catalyst was 79.26 mWm^{-2}, which was greater than the power density of V_2O_5 (65.31 mWm^{-2}) and PANI (42.4 mWm^{-2}) measured individually.

The ORR activity of a PPy/CB composite in an air–cathode MFC revealed that this electrocatalyst moved the ORR peak potential of the CB electrode to the positive by 260 mV. A maximum power density of 401.8 mW m^{-2} was reached by utilizing a PPy/C loading of 1 mg cm^{-2} (0.2 mg cm^{-2} PPy loading) on the cathode, which was greater than 90.9 mW m^{-2} obtained with a CB cathode and 336.6 mW m^{-2} obtained with a non-pyrolyzed FePc/C cathode. The effect of PPy loading on power production revealed that a loading of 0.2 mg cm^{-2} generated the greatest results.

Furthermore, the power gained per cost of a PPy/C cathode was determined to be 15 times larger than that of a Pt/C cathode with 0.5 mg cm^{-2} Pt loading. Anthraquinone-2-sulfonate-doped PPy-coated stainless steel mesh as a cathode for membraneless MFC produced power for more than 250 h and a maximum power density of 575 mW m^{-2}. A manganese (Mn)/PPy/carbon nanotube (CNT) composite was used to create an efficient and stable cathode catalyst for air–cathode MFCs, with maximal power densities of 169 mW m^{-2} and 213 mW m^{-2} at 1 mg cm^{-2} and 2 mg cm^{-2} loading, respectively.

6.5.2.2 Anode Catalyst Supports

In comparison with cathodes, ACPs are more commonly used in the manufacture of MFC anodes, which is another major driver of cell performance and frequently acts as the limiting factor for a high power output.

Polyaniline

ACP PANI, when employed in MFC, not only protects bacteria but also contributes directly to electrocatalysis, resulting in high current densities. In effect, the presence of ACPs on the surfaces of MFC anodes delays the occurrence of anode deactivation induced by interferences from bacterial metabolic products and electrocatalytic oxidation byproducts. PANI/Pt composites, in fact, have been proven to boost current generation by assisting in the direct oxidation of microbial metabolites. Carbon cloth (CC) anode upgraded with H_2SO_4-doped PANI outperformed the original anode. The PANI adjustment also enhanced roughness and decreased over-potential. Power and current densities in hydroxylated and aminated PANI nanowire networks reached 2800 mW m^{-2} and 2.9×10^4 mA m^{-2}, respectively.

The maximum power density of the nanogranular-structured PANI/glassy carbon anode was 820 mW m^{-2}, which was approximately fivefold higher than the value of 160 mW m^{-2} obtained for the bare glassy carbon anode and 2.6-fold higher than the value of 310 mW m^{-2} obtained with the microgranular-structured PANI/glassy carbon anode. In consequence, nanostructures can cause considerable bacterial growth on the PANI surface, leading in a reduction in anode over-potential, while its high electrical conductivity can result in a reduction in impedance. The most recent development in this study field is the fabrication of MFC anodes using ACPs in combination with CNTs and graphene. CNTs identified inside CNT/PANI multilayer films made by a layer-by-layer assembly approach were shown to be responsible for increasing PANI electroactivity. The CNT/PANI composite produced a maximum power density of 42 mW m^{-2} at 450 mV cell voltage using Escherichia coli as the microbial catalyst. A higher power density of 1574 mW m^{-2} may be obtained by applying a voltage of +1.18 V. A macroporous and monolithic MFC anode based on PANI hybridized 3D graphene outperformed a planar carbon electrode due to its capacity to build a three-dimensional interface with bacterial biofilm and promote electron transport via multiplexed and highly conductive routes.

Polypyrrole

ACP PPy has also attracted interest in the manufacturing of anodes for MFCs. The use of a PPy/CB composite material coated on a CP anode with Proteus vulgaris as the biocatalyst resulted in a synergistic effect and a power density of 452 mW m^{-2} after modification. A PPy-coated CNT composite anode with Escherichia coli as the biocatalyst demonstrated a direct relationship between cell power output and composite loading. In the presence of Shewanella decolorationis S12, an anode manufactured from anthraquinone-2,6-disulfonic (AQDS) acid-doped PPy achieved a maximum power density of 1303 mW m^{-2}, which was 13 times more than that obtained for the unmodified carbon felt anode. At the seventh day of operation, the PPy NFs-coated PMFC anode generated 3.4 mWm^{-2}, whereas the granular PPy-coated anode produced 3.1 mW m^{-2}. This improved performance of NFs was ascribed to a significant improvement in electrochemical characteristics, including stronger anodic redox current and reduced interface electron transfer resistance due to improved contact with the cell membrane of the microorganisms, in comparison with the granular PPy. It was also discovered that a PPy NFs-coated PMFC anode with a PPy NF loading of 3 mg cm^{-2} provided a 450% improvement in power density when compared to previous research utilizing the same photosynthetic culture.

Polythiophene-Based Catalyst Supports

PTH derivatives have found application in fuel cells that employ glucose as a substrate. Kuwahara et al. created an enzyme electrode with an ACP film made of a copolymer of 3-methylthiophene (3MT) and thiophene-3-acetic acid (T3A), which has a high conductivity of 10^{-3} S cm^{-1}. The carboxyl groups of the copolymer structure functioned as binding sites for covalently immobilizing glucose oxidase (GO_x) and bilirubin oxidase (BO_x) on the copolymer film surface. In effect, the conducting copolymer modification increased the active surface area of the electrode and consequently the amount of the attached enzyme. The anode in this investigation was the GO_x/copolymer electrode, whereas the cathode was the BO_x/copolymer electrode [3].

References

1. Ahn S, Jeong SH, Han TH et al (2017) Conducting polymers as anode buffer materials in organic and perovskite optoelectronics. Adv Opt Mater 5(3):1600512
2. Cai ZB, Li L, Ren J et al (2013) Flexible, weavable and efficient microsupercapacitor wires based on polyaniline composite fibers incorporated with aligned carbon nanotubes. J Mater Chem A 1(2):258–261
3. Dutta K, Kundu PP (2014) A review on aromatic conducting polymers-based catalyst supporting matrices for application in microbial fuel cells. Poly Rev 54(3):401–435
4. Dutta KS, Kumar P, Das S et al (2014) Utilization of conducting polymers in fabricating polymer electrolyte membranes for application in direct methanol fuel cells. Polym Rev 54(1):1–32
5. Ehsani A, Heidari AA, Shiri HM (2019) Electrochemical pseudocapacitors based on ternary nanocomposite of conductive polymer/graphene/metal oxide: an introduction and review to it in recent studies. Chem Rec 19(5):908–926
6. Fong KD, Wang TS, Kim HK et al (2017) Semi-interpenetrating polymer networks for enhanced supercapacitor electrodes. ACS energy lett 2(9):2014–20
7. Fong KD, Wang TS, Smoukov SK (2017) Multidimensional performance optimization of conducting polymer-based supercapacitor electrodes. Sustainable Energy Fuels 1(9):1857–1874

8. Fu YP, Cai X, Wu HW et al (2012) Fiber supercapacitors utilizing pen ink for flexible/wearable energy storage. Adv mater 24(42):5713–5718
9. Günes S, Neugebauer H, Sariciftci NS (2007) Conjugated polymer-based organic solar cells. Chem Rev 107(4):1324–1338
10. Habisreutinger SN, Leijtens T, Eperon GE et al (2014) Carbon nanotube/polymer composites as a highly stable hole collection layer in perovskite solar cells. Nano Lett 14(10):5561–5568
11. Han YQ, Dai LM (2019) Conducting polymers for flexible supercapacitors. Macromol Chem Phys 220(3):1800355
12. Huang Y, Hu H, Huang Y et al (2015) From industrially weavable and knittable highly conductive yarns to large wearable energy storage textiles. ACS Nano 9(5):4766–4775
13. Jung EH, Jeon NJ, Park EY et al (2019) Efficient, stable and scalable perovskite solar cells using poly (3-hexylthiophene). Nature 567(7749):511–515
14. Liu LL, Niu ZQ, Zhang L et al (2014) Nanostructured graphene composite papers for highly flexible and foldable supercapacitors. Adv Mater 26(28):4855–4862
15. Liu TY, Finn L, Yu MH et al (2014) Polyaniline and polypyrrole pseudocapacitor electrodes with excellent cycling stability. Nano Lett 14(5):2522–2527
16. Li WT, Albrecht S, Yang LQ et al (2014) Mobility-controlled performance of thick solar cells based on fluorinated copolymers. J Am Chem Soc 136(44):15566–15576
17. Li WT, Yan L, Zhou HX et al (2015) A general approach toward electron deficient triazole units to construct conjugated polymers for solar cells. Chem Mater 27(18):6470–6566
18. Liu YQ, Weng B, Razal JM et al (2015) High-performance flexible all-solid-state supercapacitor from large free-standing graphene-PEDOT/PSS films. Sci Rep 5(1):17045
19. Lota K, Khomenko V, Frackowiak E (2004) Capacitance properties of poly(3, 4-ethylenedioxythiophene)/carbon nanotubes composites. J Phys Chem Solids 65(2–3):295–301
20. Mastragostino M, Arbizzani C, Soavi F (2002) Conducting polymers as electrode materials in supercapacitors. Solid State Ionics 148(3–4):493–498
21. Meng CZ, Liu CH, Chen LZ et al (2010) Highly flexible and all-solid-state paperlike polymer supercapacitors. Nano lett 10(10):4025–4031
22. Meng QF, Cai KF, Chen YX et al (2017) Research progress on conducting polymer based supercapacitor electrode materials. Nano Energy 36:268–285
23. Pan LJ, Qiu H, Dou CM et al (2010) Conducting polymer nanostructures: template synthesis and applications in energy storage. Int J Mol Sci 11(7):2636–2657
24. Shi Y, Peng LL, Ding Y et al (2015) Nanostructured conductive polymers for advanced energy storage. Chem Soc Rev 44(19):6684–6696
25. Shown I, Ganguly A, Chen LC et al (2015) Conducting polymer-based flexible supercapacitor. Energy Sci Eng 3(1):2–26
26. Snook GA, Chen GZ (2008) The measurement of specific capacitances of conducting polymers using the quartz crystal microbalance. J Electroanal Chem 612(1):140–146
27. Snook GA, Kao P, Best AS (2011) Conducting-polymer-based supercapacitor devices and electrodes. J Power Sources 196(1):1–12
28. Stuart AC, Tumbleston JR, Zhou HX et al (2013) Fluorine substituents reduce charge recombination and drive structure and morphology development in polymer solar cells. J Am Chem Soc 135(5):1806–1815
29. Ti D, Gao K, Zhang ZP et al (2020) Conjugated polymers as hole transporting materials for solar cells. Chin J Polym Sci 38(5):449–458
30. Uy RL, Yan L, Li WT et al (2014) Tuning fluorinated benzotriazole polymers through alkylthio substitution and selenophene incorporation for bulk heterojunction solar cells. Macromolecules 47(7):2289–2295
31. Wang L, Feng X, Ren LT et al (2015) Flexible solid-state supercapacitor based on a metal–organic framework interwoven by electrochemically-deposited PANI. J Am Chem Soc 137(15):4920–4923
32. Wang M, Xu YX (2016) Design and construction of three-dimensional graphene/conducting polymer for supercapacitors. Chin Chem Lett 27(8):1437–1444

33. Xiao SQ, Zhang QQ, You W (2017) Molecular engineering of conjugated polymers for solar cells: an updated report. Adv Mater 29(20):1601391
34. Xu BW, Hou JH (2018) Solution-Processable Conjugated Polymers as Anode Interfacial Layer Materials for Organic Solar Cells. Adv Energy Mater 8(20):1800022
35. Yang LQ, Tumbleston JR, Zhou HX et al (2013) Disentangling the impact of side chains and fluorine substituents of conjugated donor polymers on the performance of photovoltaic blends. Energy Environ Sci 6(1):316–326
36. Yu DS, Qian QH, Wei L et al (2015) Emergence of fiber supercapacitors. Chem Soc Rev 44(3):647–662
37. Zhao C, Jia X, Shu K, Yu C, Wallace GG, Wang C (2020) Conducting polymer composites for unconventional solid-state supercapacitors. J Mater Chem A 8:4677
38. Zhang ZT, Chen XL, Chen PN et al (2014) Integrated polymer solar cell and electrochemical supercapacitor in a flexible and stable fiber format. Adv Mater 26(3):466–470
39. Zhang ZT, Liao M, Lou HQ et al (2018) Conjugated polymers for flexible energy harvesting and storage. Adv Mater 30(13):1704261
40. Zhang ZT, Li XY, Guan GZ et al (2014) A lightweight polymer solar cell textile that functions when illuminated from either side. Angew Chem Int 53(43):11571–11574
41. Zhang ZT, Yang ZB, Deng J et al (2015) Stretchable polymer solar cell fibers. Small 11(6):675–680
42. Zhang ZT, Yang ZB, Wu ZW et al (2014) Weaving efficient polymer solar cell wires into flexible power textiles. Adv Energy Mater 4(11):1301750
43. Zhou HX, Yang LQ, You W (2012) Rational design of high performance conjugated polymers for organic solar cells. Macromolecules 45(2):607–632

Chapter 7
Conductive Polymers-Based Sensors

Conductive polymers (CPs) such as polypyrrole (PPy), polyaniline (PANI), and polythiophene (PTh) are particularly appealing among sensor materials because they have a flexible molecular chain, intrinsically high electrical conductivity, and a simple manufacturing procedure. Owing to the existence of a large conjugation structure with sp^2 hybridized carbons, this form of polymer, in particular, allows intermolecular charge transfer. CPs have been widely employed as sensing medium for a variety of purposes to date.

Because of their ability to sense physicochemical conditions in the surrounding environment and convert the chemical information into detectable electrical and/or visual signals, conductive polymers offer significant promise in sensor systems. Conductive polymers are excellent candidates for sensor applications owing to the properties: (i) several types of conductive polymers are available, (ii) they are sensitive to a wide range of analytes, (iii) the cost of preparation and processing is relatively low, (iv) polymer molecular engineering with recognition sites (host–guest) is possible for extremely selective and sensitive sensors, (v) a platform is provided for immobilization of biologically active species to fabricate biosensors, and (vi) reasonable performance features include low hysteresis, good reproducibility, and a relatively fast recovery and short response time.

Conductive polymers have been widely used as active materials in sensor applications due to their remarkable characteristics. Conductive polymer-based sensors, for example, may translate chemical information like concentration, activity, or PH into electrical or visual signals. Furthermore, conductive polymers can be a cost-effective and practical tool for designing sensors capable of detecting a wide range of chemical and biological substances in applications ranging from food control in the beverage and pharmaceutical industries to clinical diagnosis for the detection of hazardous substances.

Sensors based on conductive polymers are typically made up of a sensoactive layer (thin polymeric film) placed on a suitable transducer. The sensoactive layer interacts with the surrounding environment, and the electronic unit detects changes induced by analyte–polymer interactions. The power generation/sensor interrogation and pattern

Y. Wang and W. Feng, *Conductive Polymers and Their Composites*,
https://doi.org/10.1007/978-981-19-5363-7_7

recognition components are part of the electronic unit. The power generation/sensor interrogation subunit is in charge of the input signal, for example, interrogating the sensoactive layer and capturing opto- or electronic signals caused by analyte–polymer interactions. To boost sensor performance, this subunit may also do signal amplification. The pattern recognition subunit is in charge of using an appropriate pattern recognition approach to process the output signal. The majority of conductive polymer sensor signal transduction concepts are based on changes in their electrical and optical characteristics.

Conductive polymer sensors are derived from the change of their capacity, optical characteristics, redox properties, and so on. The difference in a certain attribute before and after exposure to a test target molecule determines the sensitivity of conductive polymer-based sensors. The electrochemical sensor detects the variation of the CP charge transport characteristics when exposed to certain analytes, and the change may be quantitatively connected to the analyte concentration. Conductive polymers reaction mechanisms include swelling, oxidation/reduction, charge transfer/screening, structural changes, and so on. The majority of CP sensors use electrochemical detection techniques such as potentiometric, amperometric, and conductometric approaches. Therefore, various sensing mechanisms for conductive polymer-based sensors have been developed in the form of many schemes. The change in electrical conductivity of conductive polymers following contact with analytes is monitored in conductometric sensors. Doping/dedoping, redox processes, and structural change can all result in conductivity change. Changes in the chemical potential of a system in potentiometric sensors are mostly generated by alterations in anion equilibrium inside the polymer film induced by doping/dedoping or redox activities.

7.1 Sensors Based on Conductive Polymers

Because of their low cost of synthesis, great sensitivity, and wide detection range of volatiles, CPs can be employed as sensitive layers. When CPs are exposed to an analyte, several apparent changes consist of backbone conformation changes, solvation effects on the polymer chain, and the attraction of dopant counter ions or electron transfer. Electron mobility changes and polymer matrix swelling can be turned into electrical and/or mechanical signals.

High-sensitivity CPs, mainly including PANI, PPy, and poly(3,4-ethylenedioxythiophene) (PEDOT), are utilized in a variety of sensors, including temperature sensors, gas sensors, strain sensors, humidity sensors, biosensors, and so on. Furthermore, due to a synergistic effect, the combination of CPs with other sensing materials such as carbonaceous materials, metals, metal oxides, metal sulfide, and so on generates diverse composites with enhanced sensing performance. Figure 7.1 depicts the structure of the primary coverage on sensors based on CPs and their composites.

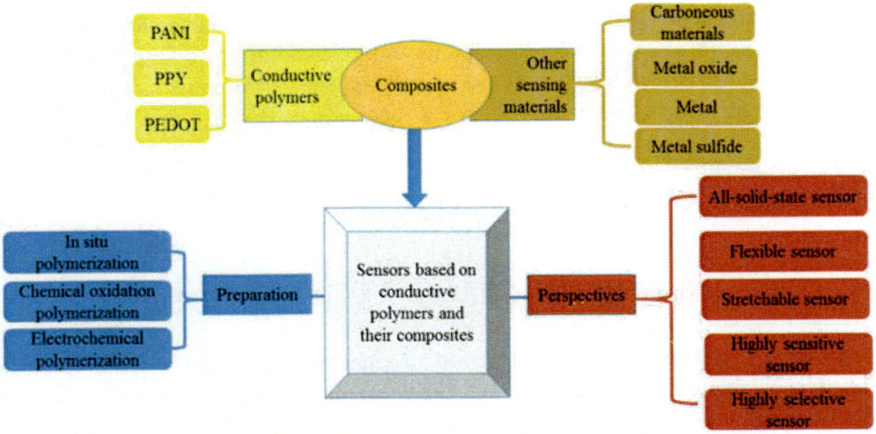

Fig. 7.1 Structure of the main coverage about sensors based on conductive polymers (CPs) and their composites. Reprinted from ref. [25], copyright 2019, with permission from John Wiley and Sons

7.1.1 PANI-Based Sensors

PANI has gained popularity due to the advantages such as low-cost monomer, easy synthesis, thermal stability, processability, and the best electrical conductivity of any CP. PANI is employed in the field of sensors because of its great sensitivity to gases, short reaction time, high stability, and inexpensive cost.

7.1.1.1 PANI Thin Film-Based Sensors

When PANI is exposed to a gas, its electrical conductivity changes dramatically. As a result, sensors based on PANI can create a variety of reactions when in contact with various types of gases. A sensor based on sensitive PANI film for ammonia (NH_3) gas detection possesses benefits such as compact size, cheap cost, flexibility, and low power consumption and exhibits a quick and steady linear response to NH_3 gas (t = 40 s) at concentrations ranging from 50 to 150 ppm.

Theoretical Investigation

Using the UB3LYP/6-31G(d) level, the detection of dangerous carbonyl gases such as acetone, formaldehyde, and phosgene by PANI ES was theoretically studied. In comparison with bare PANI ES, reduced gaps between the highest occupied molecular orbital and the lowest unoccupied molecular orbital (HOMO–LUMO gaps) of PANI ES demonstrate greater electron delocalization and conductivity along the backbone of PANI ES complexes in response to hazardous carbonyl chemicals. PANI has the maximum conductivity in response to phosgene, followed by formaldehyde and acetone, respectively, whereas the interaction energy demonstrates the opposite trend. The quantity of blue shift observed for acetone sensing is clearly more than that observed for formaldehyde and phosgene. As a result, the conductivity of PANI ES sensing with dangerous carbonyl gases is increased, and optical measurements may be utilized to identify these toxic gases.

Gas Sensor for Aroma Detection

Aroma is a key quality factor for food product approval. For volatile chemical fragrance detection, an array of gas sensors can be integrated into an electronic nose. PANI contains a range of functional groups that allow it to interact with various types of aroma molecules and has been utilized as sensing layers in gas sensors to detect aromas.

PANI placed on the interdigitated region of graphite electrodes by in situ polymerization was employed as a sensor to detect various artificial fruit scents (apple, grape and strawberry). During three aroma-analysis cycles, the gas sensor demonstrated good sensitivity and reversibility, as well as a detection limit of less than 0.62 ppm. For all aromas examined, resistance increased linearly with concentration in the range of 40–200 ppm. Under constant relative humidity (RH) ($55\% \pm 2.00\%$), the sensors demonstrated reversibility levels of $93\% \pm 1.23\%$, $90\% \pm 1.37\%$, and $89.5\% \pm 2.03\%$ for apple, grape, and strawberry aromas, respectively. As a result, electronic nose systems based on low-cost gas sensors using PANI film have the ability to evaluate aromas.

Doping PANI with various acids (hydrochloric acid (HCl), camphorsulfonic acid (CSA), and dodecylbenzenesulfonic acid) influences the detecting ability of chemical sensors based on PANI films in a sensor array to differentiate three artificial aromas: strawberry, grape, and apple. The HCl-doped PANI film possesses the best superficial regularity with smaller and better particle dispersion and demonstrates the highest and quickest reaction and best sensitivity.

Furthermore, the shape and sensing capabilities of gas sensors based on PANI layers to detect aroma are affected by different manufacturing processes. For films prepared by layer-by-layer and direct in situ adsorption polymerization, PANI granular aggregates and PANI nanoparticles of 70–150 nm were produced, respectively. The layer-by-layer approach produces rougher films because the already-formed nuclei grow as the bilayers increase. Sensor responses were significantly influenced by the substrate, with printed graphite interdigitated electrodes (PGIEs) sensors exhibiting the highest sensitivity, but sensors fabricated on gold interdigitated electrode (IDE) sensors presenting the best limit of detection and short response times and recovery times for various aromas. The reaction and recovery times are dependent on the rate of aroma penetration into the polymer as well as the roughness of the film. The IDE sensors had a faster reaction and recovery time than the PGIE sensors

Table 7.1 Response time (s) and recovery time (s) obtained from a PANI sensor considering different electrodes (IDEs and PGIEs) exposed to apple, strawberry, and grape aromas

PANI sensor		Apple		Strawberry		Grape	
		Response time (s)	Recovery time (s)	Response time (s)	Recovery time (s)	Response time (s)	Recovery time (s)
PGIEs	LbL in situ	68.00	14.66	67.30	14.66	61.66	12.66
	LbL interfacial	81.33	12.66	64.00	17.33	76.00	14.66
	In situ	71.14	6.66	91.33	7.33	75.33	5.33
IDEs	LbL in situ	20.00	47.33	10.00	9.33	6.66	28.00
	LbL interfacial	10.66	21.33	6.00	20.66	6.66	22.66
	In situ	15.33	4.00	3.33	27.33	6.00	20.00

Reprinted from ref. [25], copyright 2019, with permission from John Wiley and Sons
LbL, layer-by-layer

(Table 7.1), possibly due to a fast response originating from the decreased roughness and smoother morphology of the IDE substrate.

Optical Sensor for Chloroform Detection

The effectiveness of optical sensors based on PANI film for chloroform detection was affected by different synthesis factors, such as aniline (Ani) and dioctyl sodium sulfosuccinate (AOT) dopant ratios of 5:3, 5:5, and 5:7, and polymerization temperatures (−5, 0 and 25 °C). With a reaction time of 7 s, PANI with an Ani:AOT ratio of 5:5 and a polymerization temperature of 0 °C displayed the greatest conductivity (1.627×10^{-2} S cm^{-1}) and efficiency (sensitivity 0.0015). At the polymerization temperature of 0 °C, the polar on peak of PANI with an Ani:AOT ratio of 5:5 shows a red shift in wavelength from about 750 nm (before immersion) to approximately 780 nm (after immersion) in chloroform solution. Before and after immersion in chloroform, the quantity of benzenoid and quinoid structures throughout the PANI chain stays constant. This clearly shows that during sensor application, only a physical contact (dipole–dipole moment) occurs between the chloroform and PANI. The physical contact between the partial negative charge (Cl^-) of chloroform molecules and the partial positive charge (NH^+) on the PANI chain is the mechanism for the interaction between PANI and chloroform.

7.1.1.2 PANI Nanofiber-Based Sensors

PANI nanofibers for sensor applications offer higher sensitivity and faster reaction times than typical bulk materials due to a larger effective specific surface area and shorter penetration depth for target analyte molecules. Temperature-controlled polymerization was used to customize the diameter of PANI nanofibers. The conductivities of doped PANI nanofibers rise as the diameter decreases. Because of their enormous surface area and strong intrinsic conductivities, PANI nanofibers with lower diameters respond considerably more quickly than larger-sized fibers or bulk PANIs.

The concentration of the HCl solution determined the morphologies and electrical characteristics of PANI nanofibers generated using HCl-assisted interfacial polymerization. The sensors responded linearly to the target gas. When the sensors are subjected to NH_3, their gas detecting response increases rapidly and then instantly declines upon the removal of NH_3. The sensor doped with $[HCl] = 1$ mol/L demonstrated the greatest response and superior sensitivity. The sensors react almost linear to the rising NH_3 concentrations. Furthermore, when the sensor is exposed to NH_3, the reaction diminishes after the peak arrives, and the reduction gets steeper as the concentration of NH_3 increases. The NH_3 binds to the protons on the polymer chains, resulting in a reduction of conductivity. In comparison with HCl, the smaller molecule of NH_3, on the other hand, shortens the distance between the polymer chains, increasing conductivity.

7.1.2 PEDOT-Based Sensors

PEDOT, a significant polythiophene derivative, has several benefits for use in chemosensors and biosensors, including high stability, strong electrical conductivity, low oxidation potential, good biocompatibility, and electrochemical catalytic activity. PEDOT can easily produce a strong-willed film, is particularly eco-friendly and provides a quick electron transfer. Because PSS (through an insulator) connects PEDOT to water, doping it with PSS enhanced its solubility and processability. When compared to carbon black, PEDOT:PSS has a higher radiation transparency. However, it appears to be less prone to oxidation and have greater stability and deterioration resistance than other conductive materials. PEDOT:PSS is an intriguing sensing material due to its excellent conductivity and solubility in water, as well as its low redox potential. A synergistic blend of PEDOT:PSS with water-soluble electrocatalytic compounds or colloidal metal nanoparticles is a great way for altering sensing capabilities.

7.2 Processing of Conductive Polymers for Sensor Applications

Some characteristics are very desirable when building conductive polymer-based sensors, such as cheap synthesis and purification costs, high sensitivity, superior thermal, chemical, and electrical stability, and film-forming capabilities from solution or melt. Many CPs solubility has been increased by synthesizing derivatives as well as utilizing various combinations of dopants and solvents. These technologies have made it possible to create conductive polymer thin films using various ways. Thin films are required to produce sensoactive layers, which can be either pure or modified with various immobilized species on their surface. Furthermore, sensor sensitivity was significantly improved by using nanostructured films. Langmuir–Blodgett (LB) and layer-by-layer (LbL) or self-assembly methods have been effectively used to make nanostructured films with high molecular organization, along with sensitivity and morphological control.

7.2.1 Langmuir–Blodgett (LB) Films

The LB approach has distinct advantages, such as the capacity to construct nanostructured films with manipulable molecular architecture. Monolayer and multilayer films may be created with a great degree of shape and film thickness control.

7.2.2 Layer-by-Layer (LbL) Self-Assembly Technique

In comparison to the elegant LB approach, the LbL technique is more appealing owing to its lower relative cost, simple experimental setup and manipulation, and the potential of greater scale manufacturing. LbL films are generated as a result of the spontaneously adsorbing molecules from a solution onto a solid substrate via different adsorption mechanisms. The two basic processes that cause the formation of LbL films are chemical adsorption and physical adsorption. Decher presented an alternate strategy that used physical contact to adsorb amphiphilic compounds and polyelectrolytes. Rubner's group pioneered manufacturing LbL films of conductive polymers using Decher's approach. Because doped conductive polymers behave as polyelectrolytes, electrostatic attraction might be employed to create LbL films. Furthermore, LbL films may be created using other forms of physical interactions, such as hydrogen bonding. LbL films have already been made from many types of conductive polymers, such as PANI and its derivatives, PPy, and poly (p-phenylene vinylene). Each layer adsorption generally adds the same amount of material to the film; hence, the number of deposited layers may be used to regulate the film thickness. Furthermore, the layer thickness may be tuned by adjusting polymeric solution parameters such as pH, ionic strength, and polymeric concentration. The morphology of conductive polymers LbL films with low roughness may be modified by the introduction of various polyanions, such as polystyrene sulfonate, poly(vinyl pyrrolidinone), lignin sulfonate, poly(vinyl pyrrolidinone), and poly(allylamine hydrochloride).

7.2.3 Other Processing Techniques

Conductive polymers have also been processed into thin films directly on the electrode surface by other processes, such as electrochemical polymerization. However, the electropolymerized films are substantially thicker than that achieved with LB and LbL processes, and the film architecture is difficult to manage. To deposit polymeric films on surfaces, spin coating and casting are used mostly for soluble conductive polymers and their mixes with conventional polymers. These processes are less difficult and less expensive, and thin films may be easily manufactured. However, the thickness of the films produced by these processes is greater than the LB and LbL films. Sensor applications prefer very thin films, typically a few hundred nanometers thick, to provide high sensitivity and reversibility of sensor signals, hence preventing permanent adsorption of analyte molecules. As another intriguing method for producing ordered multilayer polymeric nanostructured films, spin self-assembly involves dumping a solution onto a spinning substrate. This technology has shown to be extremely efficient in creating nanostructured films with perfect morphological control and good electrical and optical characteristics.

7.3 Various Types of CP-Based Sensors

7.3.1 Conductometric Sensors

Electrical conductivity measurements of the sensor element in contact with the analyte under investigation are the foundation of conductometric sensors. Many sensor devices based on conductive polymers assess changes in conductivity caused by chemical species adsorbed on the surface or within the polymer layer. The interactions between the sensoactive layer (polymer, dopant counter ions, etc.) and the chemical species present in the analyte lead to the redox and acid–base reactions, charge transfer, and polarization processes, as induced the variations in conductivity.

Polyaniline, polypyrrole, polythiophene, poly(3,4-ethylenedioxythiophene), and other conductive polymers were often used in conductometric sensors. Various conductive polymers (or doped with different species) are coated onto transducers such as interdigitated microelectrodes in this type of sensor. The sensors are interrogated using measures of conductivity, resistivity, impedance, or admittance change. As previously stated, conductive polymers are very non-selective, responding to all sorts of substances at varying intensities. As a result, arrays of sensors constructed by various conductive polymer films rely on the generation of electrical patterns or fingerprints for a specific analyte or sample. Employing appropriated multivariate pattern recognition methods, they are subsequently processed to provide qualitative and quantitative information about the analyte.

Polyanilines doped with acid dopants have been employed as sensors detecting pH, ammonia, CO_2, ethylene, methanol, ethanol, and acetone, as well as various harmful volatilities such as H_2S, NO_x, SO_2, and so on. Several analytes, including ethanol, toluene, ammonia, water vapor, and other volatile chemicals, have already been detected using polypyrrol-based sensors. Polythiophene-based sensors have already been utilized to detect a wide range of chemical substances including ethylisovalarate, methane, chloroform, and ethanol.

Based on interdigitated microelectrodes, layer-by-layer (LbL) films of conductive polymers and other organic molecules were deposited to analyze natural and commercial coconut water, distilled, ultrapure, tap, and mineral water, various brands of red wines, sugar cane juice, various types and brands of commercial coffees, tea, and substances that comprise the four basic tastes: HCl (sourness), quinine (bitterness), NaCl (saltiness), and sucrose (sweetness). In the last scenario, the sensor array was capable of detecting differences in these four final compounds at the parts per billion (ppb) level. This method has also been used to identify off-flavor substances such as geosmine and methyl-isoborneol, which are algal exudates and give water a disagreeable taste.

The chemical stability of the sensing units against organic solvents was improved by taking nanostructuring of conductive polymers a step further. The crosslinked PANI-based enhanced sensing units reacted linearly to the concentration of adulterants in gasoline and ethanol samples. When compared to conventional films, nanostructured polymeric films in sensors give quicker response and increased sensitivity.

7.3.2 Gravimetric Sensors

For gravimetric sensors, the variations in the sorption/desorption of chemical species in the sensoactive layer result in mass changes or wave propagation properties that can be analytically quantified and regarded to the target analyte. Quartz crystal microbalance (QCM) and surface acoustic wave (SAW) sensors are the two most used gravimetric sensors. Although polypyrrole is the most commonly used polymer in this type of sensor, other polymers such as polyaniline, poly(vinylpyridines), and polypyrrole derivatives have also been utilized effectively in QCM sensors. Sensors for chromo ions, methanol, hexane, and other chemicals utilize polypyrrole doped with a variety of dopants and deposited via chemical and electrochemical polymerization processes. Polythiophene derivatives, poly(octylthiophene), poly(propanoate thiophene), and poly(decanoate thiophene), alternated with octadecylamine, a surface-active substance, were effectively employed as sensoactive coating in QCM sensors for NO_2. Functionalized polythiophenes have been employed effectively as selective coating materials in QCM sensors to detect heavy metal in solution. In the pH range of 2–12, a polyaniline-coated QCM sensor for detecting pH demonstrated high selectivity, sensitivity, and stability, and the good sensibility attributable to protonation/deprotonation by the hydrated proton (H_3O^+) rather than H^+. QCM sensors for a variety of organic compounds utilize molecularly imprinted polymers and other polymers able to perform host–guest interactions.

7.3.3 Optical Sensors

Optical sensors take advantage of changes in the optical characteristics of the sensoactive material upon exposure to an analyte of interest. The sensoactive material can be a conductive polymer or a conductive polymer incorporating immobilized molecules. Its optical property changes are employed for signal transduction, such as measurements of absorbance, fluorescence, reflectance, scattering, chemiluminescence, refractive index, and spectral shape. Surface plasmon resonance (SPR) sensors based on conductive polymers such as polyaniline and polypyrrole have already been proposed for the detection of biological species in biosensors and gas sensors. The use of molecularly imprinted polymers and nanoparticles in SPR sensors can successfully improve selectivity and sensitivity to a wide variety of analytes.

7.3.4 Fluorescence-Based Sensors

Fluorescent sensors rely on measuring the fluorescence emission of chemical species, as have been stimulated by specified excitation wavelengths. Luminescence sensors for metal ion detection employ conductive polymers such as polythiophenes, polypyrroles, and polyphenylenes functionalized with crown ethers, to improve sensitivity and selectivity. In another approach, molecularly imprinted polymers in fluorescence sensors impart selectivity toward specific analytes. Fluorescent conductive polymers with hyperbranched globular structures have also been employed as chemosensors and in explosive detection systems. When compared to linear fluorescent chemosensors, hyperbranched globular structure polymers provide significantly more diffusional channels for excitons to move through, allowing them to be rapidly destroyed by quenchers. Changes in Vis/NIR optical absorbance were utilized to detect ozone levels ranging from 2.5 to 100 ppm.

A biological element immobilized or entrapped in a matrix or membrane coupled to a transducer is often used in biosensors. These biologically active compounds play an important role in providing recognition surface sites for the analyte of interest, resulting in selectivity and sensitivity for the sensor. The matrix serves two crucial functions: (i) as an immobilizing agent for biological material, preventing its loss and degradation, and (ii) as a redox mediator, facilitating electron flow between biological species and the electrode. Conductive polymers can be deposited in situ in tiny areas of microdevices. Moreover, they provide additional interesting features such as stability, compatibility with many biological substances, and the ability to tailor properties such as functionalization, doping level, and dopant type. Although poly(pyrrole) and poly(3,4-ethylenedioxythiophene) (PEDOT) have both been used effectively as electroactive components in (bio)sensing, PEDOT showed higher electrochemical stability. Another method for detecting glucose used a nitrogen-doped electrode functionalized with polyaniline [2].

7.4 Properties of Conductive Polymer Hydrogels and Their Application in Sensors

7.4.1 Conductive Polymer Hydrogels (CPHs) for Sensors

CPHs have been widely used in sophisticated sensor technologies that can turn gathered information into electrical signals or other needed forms as a class of unique smart materials or responsive materials. When subjected to changes in environmental circumstances such as temperature, light, pressure, moisture, gas, pH, and chemical compounds, CPHs undergo physical/chemical transformations. For next-generation wearable, implantable, and portable applications, improved sensors must have strong mechanical flexibility, better biocompatibility, self-repairing capabilities, and a simple preparation procedure. As a result, significant efforts have been undertaken to provide CPHs with stretchability and flexibility, self-healing properties, and biocompatibility. Intelligent systems and materials necessitate accurate and efficient sensing. CPHs are regarded as smart responsive materials since they undergo chemical/physical changes in response to external stimulation, allowing for wider use in the field of sensors. Table 7.2 exhibits a full list of CPH-based sensors.

7.4.2 Application of CP Hydrogel on Sensors

7.4.2.1 Strain Sensor

Human skin is the greatest sensory organ in the human body, allowing us to precisely feel our surroundings and live a safe existence. In recent years, electronic skin and wearable strain sensors inspired by human skin have been widely investigated to replicate sensory capabilities comparable to those of human organs and enable applications for human health monitoring and illness detection. Because of their flexibility and stretchability, CPHs are a promising choice for wearable strain sensors with outstanding sensing capability.

Sensitivity is the most important consideration for sensors. In general, the high sensitivity of CPHs-based piezoresistive strain sensors may be accomplished by modifying the morphological design of the CPHs, such as by including porous structures into the CPHs. The deformation ability of the piezoresistive strain sensor under specified force determines its sensitivity, and the porous construction can boost the sensor deformation. Gu et al. created a durable macroporous conductive hydrogel-based strain sensor with little performance loss after 1200 cycles of operation. The PPy was dispersed consistently across the hydrogel network, resulting in an excellent electrical characteristic. This strain sensor is capable of detecting all types of human motions, demonstrating its immense potential in wearable systems.

A highly sensitive resistive pressure sensor based on CPH film with a hollow spherical microstructure is also achieved. PPy hydrogel is used to create a microstructured

Table 7.2 Applications of CPHs in various types of sensors

	CPHs/CPHs composites	Application	Sensing performances
Strain sensor	PPy/2-hydroxyethyl methacrylate	Piezoresistive sensor	Resistance response: 60 at 50% strain
	PPy	Piezoresistive sensor	Sensitivity: 56–113 kPa^{-1} Limit of detection: 0.8 Pa
	PPy/CNT	Piezoresistive sensor	Response time: 70 ms Limit of detection: <0.1 N
	PANI/poly(N-isopropylacrylamide)	Piezoresistive sensor	Gauge factor: 3.92 Response time: 0.4 s
	PANI/P(AAm-co-HEMA)	Piezoresistive sensor	Gauge factor: 11
	PAM	Piezoresistive sensor	Gauge factor: 0.693 at 1000% strain
	PANI/PAA	Piezoresistive sensor	Sensitivity: 37.6 kPa^{-1} Response time: 50 ms
Biosensor	PANI/PtNPs	Glucose biosensor	Sensitivity: 96.1 μA mM^{-1} cm^{-2} Response time: 3 s Limit of detection: 0.7 μM
	PPy/AuNPs	Carcinoembryonic antigen biosensor	Linear range: 1 fg mL^{-1} to 200 ng mL^{-1} Limit of detection of 0.16 fg mL^{-1}
	PPy/Nafion/CNTs	Glucose biosensor	Sensitivity: 2860.3 μA mM^{-1} cm^{-2} Limit of detection: 5 μM
	PPy/copper oxide/reduced graphene oxide	Glucose biosensor	Linear range: 0.1–100 mM Limit of detection: 0.03 μM
	PANI/PtNPs	Metabolite biosensor	Sensitivity: 322.8 μA mM^{-1} cm^{-2} (uric acid), 0.88 μA mM^{-1} cm^{-2}(cholesterol), 0.57 μA mM^{-1} cm^{-2}(triglyceride)
	PANI/PtNPs	Metabolite biosensor	Sensitivity: 7.49 μA mM^{-1} cm^{-2} (triglycerides), 3.94 μA mM^{-1} cm^{-2}(lactate), 5.03 μA mM^{-1} cm^{-2}(glucose)
Gas sensor	PANI/CeO_2	NH_3 sensor	Response time: 57.6 s (50 ppm)
	PAM/carrageenan	NO_2 and NH_3 sensor	Sensitivity: 1.3 ppm^{-1} (NH_3), 78.5 ppm^{-1} (NO_2) Limit of detection: 2 ppm (NH_3), 1.2 ppb (NO_2)

(continued)

Table 7.2 (continued)

	CPHs/CPHs composites	Application	Sensing performances
	PANI	Amine sensor	Sensitivity: $\Delta R/R_0 = 225\%$ (5 ppm NH_3), 46% (5 ppm putrescine), 17% (5 ppm cadaverine)

Reprinted from ref. [8], copyright 2019, with permission from John Wiley and Sons

film via a multiphase reaction, which results in hollow spherical microstructures and a low effective modulus of elasticity (Fig. 7.2a) [18]. The sensor created by sandwiching this film between copper foil and flexible poly(ethyleneterephthalate) has extremely high sensitivity in the low pressure region. It is worth noting that the sensor sensitivity may be increased further by micropatterning the film with a triangle cross-sectional profile (Fig. 7.2b), and the sensitivity eventually reached ~56.0–133.1 kPa^{-1} under 30 Pa (Fig. 7.2c). Another typical way for increasing sensitivity is to include conductive fillers into CPHs. Conductive fillers dispersed over the 3D hierarchical structure of CPHs can generate a low-density percolated network that benefits conductivity and sensitivity. Khalili et al. created a pressure sensor out of PPy hydrogel with carbon nanotubes (CNTs) and graphene nanoplatelets (GNPs) embedded (Fig. 7.2d, e) [11]. It can detect pressure changes of less than 0.1 N and has a reaction time of 70 ms (Fig. 7.2f).

Aside from sensitivity, certain other important characteristics such as flexibility, linearity, biological adhesion, stability, and so on must be considered in order to broaden the use of strain sensors, particularly in wearable electronic devices.

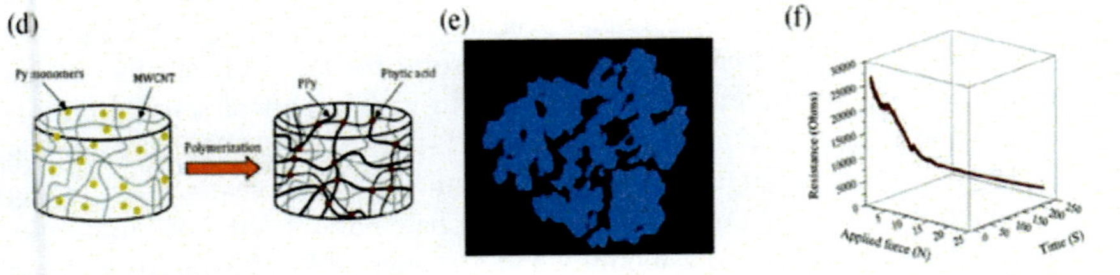

Fig. 7.2 **a** Structure of sensor based on hollow-sphere PPy. **b** Fabrication of micropatterned PPy film. **c** Improved sensitivity of the micropatterned film pressure sensor. **d** The polymerization process of the hydrogels. **e** The interconnected network of the hydrogels. **f** Time and pressure response of the PPy/CNT hydrogel. Reprinted from ref. [8], copyright 2019, with permission from John Wiley and Sons

Excellent mechanical qualities are critical for consistent sensor performance and sturdy applications. Wang et al. used a dual-synergistic network with two components of poly(N-isopropylacrylamide) (PNIPAAm) and PANI mechanically linked to resistive-type strain sensors (Fig. 7.3a–c). The sensor resistance does not change significantly throughout the stretching-releasing cycles test, and it may still function as a sensitive sensor after 350 tests. The linked network topology enables effective energy dissipation, which accounts for the development of outstanding mechanics and stability. Linearity of resistance change during applied strain is required for reliable sensor measurement and calibration. Wang et al. produced a PANI network with acrylamide and hydroxyethyl methyl acrylate copolymer P(AAm-co-HEMA) hydrogel for strain sensor with exceptional linearity in the 0–300% strain range (Fig. 7.3d, e) [26], which can be explained by network deformation being strain adaptable. People's experiences with direct touch wearable gadgets can be improved by using biocompatible and sticky materials. To achieve both self-healing and stretchy qualities, Xu et al. created a composite hydrogel containing PANI, PAA, and phytic acid (Fig. 7.3h), which demonstrated great stretchability (500%) and outstanding self-healing capabilities (>99% healing efficiency in 24 h). The composite may be made into a strain sensor with great sensitivity in both static and dynamic pressure, even for non-planar pressure sensing (Fig. 7.3i) [24].

7.4.2.2 Gas Sensor

With the fast expansion of civilization, a great number of harmful gases, such ascarbon oxides (CO_x), nitrogen oxides (NO_x), sulfur dioxide (SO_2), and ammonia (NH_3), are being created by industry and agriculture, posing a major threat to the environment and human health. Furthermore, several gases are thought to represent biomarkers for human disorders (e.g., monitoring NO_x can be used for invasive detection of lung infections and bowel disease). As a result, sensitive gas sensing has been given a high priority for human healthcare, hazardous monitoring, and food deterioration detection. However, most commercially available gas sensors are metal oxide-based, with disadvantages such as high operating temperatures and limited sensitivity. CPHs-based gas sensors offer the following benefits over them: they can operate at room temperature, have a quick reaction time and a low detection limit.

Chemiresistive PANI-based gas sensors are useful for fast and sensitive amine detection, and they have been widely used due to their excellent sensing performances and ease of fabrication and measurement. PANI reversible doping/dedoping process serves as the basis for the sensing mechanism. During the gas-induced doping process, polarons are transferred from the PANI backbone to the NH_3 molecule, resulting in a reduction in carrier mobility and a change of PANI from its emerald base (EB) form to its emerald salt (form). Wang et al. suggested a CeO_2@PANI hydrogel-based NH_3 gas sensor for room temperature (Fig. 7.4a). When subjected to 50 ppm of NH_3, the nanocomposites displayed an improved response and a reduced detection time (Fig. 7.4b), which was attributed to the p–n junction core–shell structure generated by the intimate contact between the CeO_2 core and the PANI shell

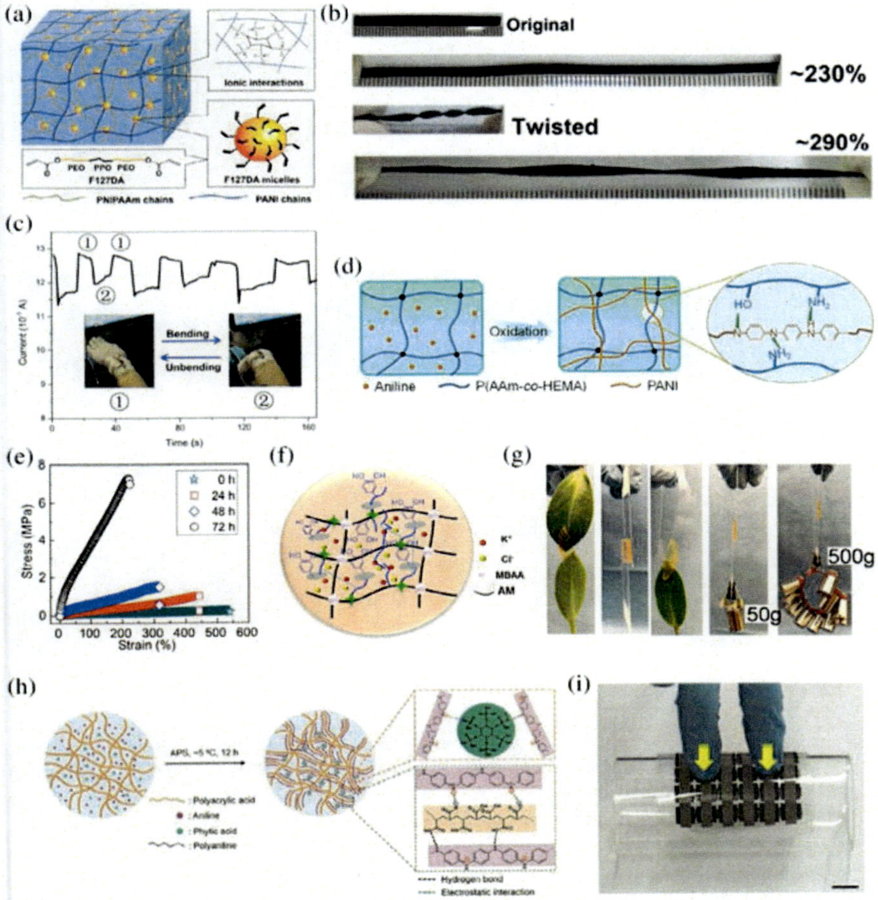

Fig. 7.3 **a** Network structure of F-PNIPAAm/PANI hydrogels. **b** Hydrogel deformations under different condition. **c** Current response of wrist joint with the sensor. **d** Schematic illustration to the synthesis of PANI/P(AAm-co-HEMA) hydrogels. **e** Stress–strain curve. **f** Schematic network structure of PAM hydrogel. **g** The adhesion measurements of the PAM hydrogel [10] **h** Schematic illustration synthetic process of composite hydrogel. **i** Non-planar pressure sensing of the composite hydrogel sensor. Reprinted from ref. [8], copyright 2019, with permission from John Wiley and Sons

(Fig. 7.4c) [23]. Furthermore, the gas sensor demonstrated external environment stability due to the use of phytic acid as a gelator, which boosts the mechanical strength and chemical stability of the sensor.

Wu et al. have developed a double-network (DN) hydrogel sensor based on PAM and carrageenan that exhibits good sensitivity to NH_3 and NO_2 as well as mechanical resilience and stretchability (up to 1200% strain). In addition to its high sensitivity and stretchability, the sensor may increase water retention, which is critical for high-performance gas sensors. The movability of K^+ and Cl^- ions is hampered

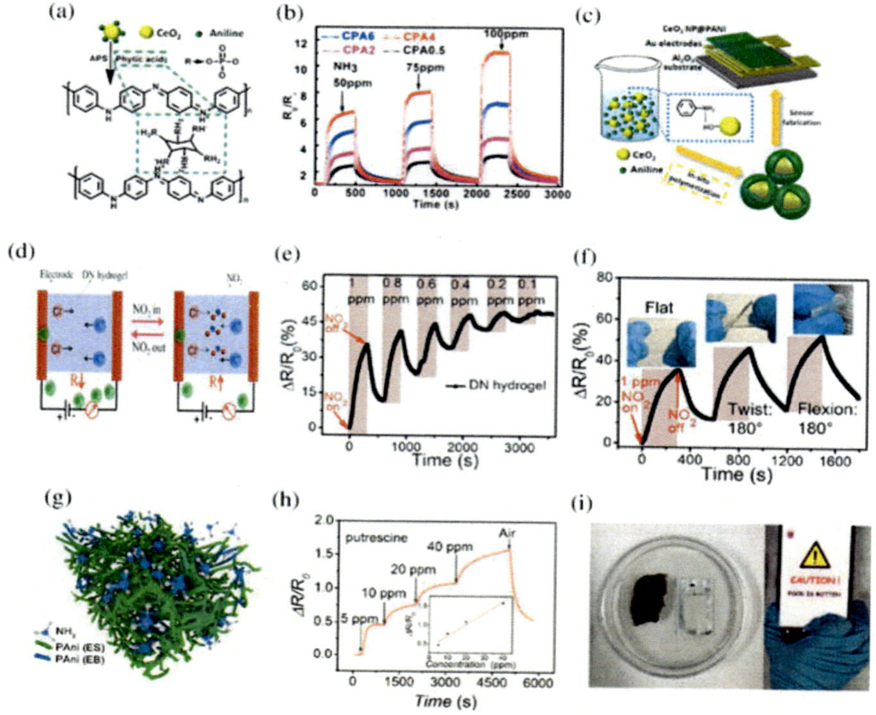

Fig. 7.4 **a** Schematic illustration of the preparation of CeO_2@PANI hydrogel. **b** Response curves of different concentrations of ammonia. **c** Core/shell structure of CeO_2@PANI hydrogel. **d** Schematic illustration the working mechanism of the NO_2 sensor. **e** Dynamic response of NO_2 at different concentration by sensor. **f** Dynamic response of sensor to 1 ppm NO_2 at flat, 180 twisted, and 180°bent states, respectively. **g** The general sensing mechanism of PTS–PANI. **h** Sensor responses in different concentrations of putrescine. **i** Application of NFC tag in meat spoilage detection of smart phones. Reprinted from ref. [8], copyright 2019, with permission from John Wiley and Sons

by gas diffusing into the hydrogel, resulting in an increase in resistance (Fig. 7.4d–f) [27]. The sensor has exceptionally low detection limits of 2 ppm for NH_3 and 1.2 ppb for NO_2. The equivalent sensitivity values are 1.3 ppM^{-1} and 78.5 ppM^{-1}, respectively. The sensor high sensitivity is attributed to the improved adsorption of NH_3 and NO_2 molecules on the hydrogel by functional groups (e.g., –OH, $-NH_2$, and SO_3^-). As smart responsive materials, conductive polymer cryogel composites with super and interconnected porosity and rapid reaction to stimuli are frequently favored. They can be employed in very sensitive gas sensor applications due to the change in gas-sensitive conductivity. Sahiner et al., for example, created poly(2-hydroxyethylmethacrylate)/graphene oxide (p(HEMA)/GO) cryogel composites and their reduced versions, p(HEMA)/rGO cryogel composites. The conductivity of p(HEMA)/rGO cryogel composites rose thrice and reduced fourfold after 15 min of exposure to HCl and NH_3 gases, respectively.

The excellent sensitivity and processability of the CPH-based gas sensor allow it to perform advanced innovative functionalities. Ma et al., for example, described a highly sensitive PANI-based gas sensor for detecting food deterioration. The p-toluene sulfonate hexahydrate (PTS)-doped PANI film has a higher surface-to-volume ratio, which means that there are more active sites for gas adsorption and reaction (Fig. 7.4g). The sensor has strong sensitivity to 5 ppm of NH_3, putrescine, and cadaverine, with $\Delta R/R_0$ of 225%, 46%, and 17%, respectively (Fig. 7.4h). With its superior sensing capabilities, PANI may be paired with near field communication (NFC) tags via inkjet printing to create a wireless gas sensor. The presence/absence of gas produces a resistive change in PANI and impendence dismatch/rematch of the tags, altering mobile phone reading and enabling food spoiling detection (Fig. 7.4i) [16].

7.5 Conductive Polymer Composites for Sensing Applications

7.5.1 *PANI Composites for Sensors*

7.5.1.1 PANI and Carbonaceous Material Composites for Sensors

Carbonaceous materials, such as carbon nanotubes (CNTs) and graphene oxide (GO), have been identified as suitable options for gas sensing due to their unique electrical, physical, and chemical characteristics, as well as their high chemical stability. CNTs are an active material for gas sensing because of their high surface-to-volume ratio, large specific surface area, and extensive conjugation of π-electrons along the surface. Owing to its two unique oxidation states (insulating EB and metallic ES), PANI is sensitive to a variety of gases. PANI, on the other hand, suffers a volumetric change during oxidation/reduction (doping/dedoping), making it mechanically unstable. PANI mechanical stability can be improved by including appropriate fillers like as CNTs or graphene. CNTs boosted the flexibility and resilience of PANI, as well as the number of active sites, by increasing intra and interchain charge mobility along the polymer chain in the presence of electron donor or acceptor gases. For hydrogen detection, sensors based on multiwalled CNT (MWCNT) and single-walled CNT-doped PANI composite thin films outperform pure PANI.

PANI/MWCNT Composites

Highly sensitive gas sensors based on PANI/MWCNT nanocomposites for detection of trace-level NH_3 gas (up to 2 ppm NH_3) revealed a reaction time of about 6 s. A PANI/MWCNT-modified screen-printed carbon electrode (SPCE)-based electrochemical sensor demonstrated a linear response ($R_2 = 0.99902$), a lower detection limit, and greater selectivity and sensitivity than previously reported biosensors. The sensor detection limit was 10 μmol^{-1} L, and its sensitivity was 0.38 $mAmmol^{-1}$ L cm^{-2} in the urea concentration range of 10–50 μmol^{-1} L. Compared to bare indium tin oxide (ITO), carboxylic acid functionalized MWCNTs ((MWCNTs-COOH)/PANI/ITO composite reduced the oxidation potential of nifedipine (NIF) by about 170 mV, demonstrating a synergistic impact on NIF electrochemical oxidation and a lower detection limit of NIF. The substantial electrochemical performance variation of the produced films was attributed to the excellent conjugation of PANI nanoparticles and MWCNTs-COOH.

PANI/graphene Composite

Graphene (*G*) with high electron mobility and large surface-to-volume ratio has strong hydrophilic nature and high optical absorbance. With rising NH_3 concentration, the absorbance spectrum of an optical sensor based on PANI/G nanocomposite grows linearly. The sensor outperforms that coated only with PANI, with sensitivity relative to absorbance unit (AU), response and recovery times of 132.8 AU percent, 112 s, and 185 s, respectively.

Because the strong quantum confinement and edge effects result in a good electron transfer rate and chemical stability, graphene quantum dots (GQDs) have gained a lot of attention as an excellent sensor material. GQDs have a significant impact on the specific surface area and analytical performance of electrochemical sensors. In situ polymerization was used to effectively construct a PANI/(functionalized GQDs) composite with outstanding performance, which was then used to manufacture electrochemical sensors for the measurement of calycosin in complicated samples of traditional Chinese medicine. The electron transfer rate on the sensor surface was greatly boosted by a glassy carbon electrode (GCE) modified with PANI/(functionalized GQDs) composite. Under optimum circumstances, the constructed sensor has adequate repeatability, long-term stability, strong selectivity and sensitivity, and a wide detection range (1.1×10^{-5} –3.52×10^{-4} mol/L) with a detection limit of 9.8×10^{-6} mol/L [signal–noise ratio (S/N) = 3].

Figure 7.5 depicts the gas response of flexible pure PANI and S and N co-doped GQDs (S, N: GQDs)/PANI hybrid gas sensors to NH_3 concentrations ranging from 1 to 100 ppm [6]. The flexible pure PANI and (S, N: GQDs)/PANI hybrid gas sensors have a response of around 10.1% and 42% and a limit of 1 ppm and 500 ppb, respectively, to detect NH_3 at 25 °C in 57% RH. The flexible (S, N: GQDs)/PANI hybrid outperforms previously reported hybrids in terms of NH_3 sensing performance. The enhanced NH_3 sensing properties of the flexible (S, N: GQDs)/PANI hybrid gas sensor can be attributed to the following factors: (i) the reversible acid–base doping/dedoping process of PANI; (ii) the intrinsic sensing properties of S, N: GQDs; and (iii) The swelling process of the (S, N: GQDs)/PANI hybrid on the diffusion of NH_3 molecules.

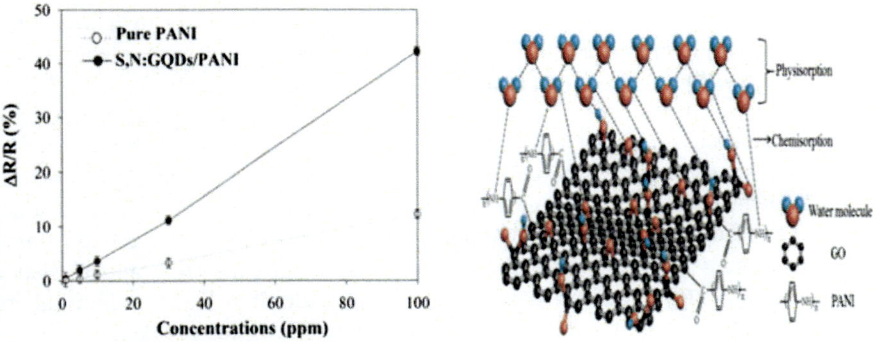

Fig. 7.5 **a** Gas response of flexible pure PANI and (S, N: GQDs)/PANI hybrid gas sensors toward 1–100 ppm of NH_3 at 25 °C and 57% RH. **b** Scheme for humidity sensing for PANI/GO films. Reprinted from ref. [25], copyright 2019, with permission from John Wiley and Sons

PANI/GO Composite

Because of its numerous oxygen-rich functional groups, GO, as one of the most significant graphene derivatives, has a high hydrophilicity. Because of its hybrid nanostructure and the hydrophilic groups on GO and PANI, the described PANI/GO nanocomposite humidity sensor offers greater sensitivity, better repeatability, and faster response/recovery characteristics than previously reported equivalents. Water molecule adsorption is sensitive to the PANI/GO nanocomposite film. The pattern of water molecule adsorption at the PANI/GO film is depicted in Fig. 7.5b [31]. Water molecules bind to the PANI/GO hybrid via the hydroxyl, carboxyl, and epoxy functional groups linked to the GO nanosheets and the amino groups on PANI. The water layer absorbed through chemisorption is created initially, followed by first-layer physisorption of water molecules due to double hydrogen bonding. Second-layer physisorption of water molecules occurs at high RH due to the action of single hydrogen bonding. A Grotthuss chain reaction generates hydronium ions as charge carriers ($H_2O + H_3O^+ \rightarrow H_3O^+ + H_2O$), and free water penetrates the PANI/GO film interlayers.

7.5.1.2 PANI/Metal Oxide Composite

As sensitive materials for gas detection, many inorganic semiconducting metal oxides such as undoped or doped titanium dioxide (TiO_2), tin(IV) oxide (SnO_2), and zinc oxide (ZnO) have been extensively researched. PANI/inorganic nanocomposites for gas sensing applications have been proposed to solve the disadvantages of PANI as a sensing material, such as low mechanical strength and lack of selectivity. PANI composites containing metal oxides including ZnO, TiO_2, copper(II) oxide (CuO), iron(III) oxide (Fe_2O_3), SnO_2, cerium(IV) oxide (CeO_2), tungsten oxide (WO_3), and others have good sensing characteristics.

ZnO has been used to make PANI/ZnO composites in order to increase the sensing performance of gas sensors at room temperature. The synergistic impact of organic and inorganic components boosted the sensing response of PANI/ZnO hybrid film-based sensors for NH_3 detection substantially. ZnO nanorod arrays can produce nanoscale gaps for gas diffusion while also providing a large number of adsorption sites to improve responsiveness. The PANI/ZnO nanocomposite demonstrated far superior gas sensing characteristics to either ZnO or PANI alone. A ZnO nanosheet composite coated with vapor phase polymerized aniline demonstrated optimal sensing performance with high response magnitude (relevant resistance change of *ca* 2150% toward NH_3 as low as 10 ppm), ultralow detection limit (about 5 ppb), good repeatability, and excellent selectivity. The extremely desired sensing characteristics of the nanocomposites are attributed to the shape and interactions between p-type PANI and n-type ZnO.

A PANI/TiO_2 composite with superficial PANI deposition on TiO_2 has an aggregated granular porous morphology and an average size of 25 nm. The composite sensor has a humidity detecting response of 84.21% in the range 25%–95% RH and response and recovery times of 60 s and 100 s, respectively. At ambient temperature, PANI/niobium-doped TiO_2 (Nb-TiO_2) nanocomposites with varying weight percentages of Nb-TiO_2 nanoparticles exhibited high sensitivity and rapid reaction to NH_3 gas. PANI@flower-like WO_3 nanocomposites with varying WO_3 mole percentages (2–50%) were placed on flexible polyethylene terephthalate substrates to create NH_3 sensors with fast response and recovery rates, selectivity, high repeatability, and minor humidity effects to 10 ppm NH_3.

7.5.1.3 PANI/Au Composite

Noble metal Au particles are utilized in glucose sensors due to their high dispersion over a polymer matrix, antipoisoning properties, improved electrocatalytic action, superior biocompatibility, and wide oxidation potential range. To create a highly sensitive glucose sensor, the Au particles were electrodeposited on PANI film under optimal growth conditions for maximal dispersion. The constructed sensor presented great stability, reproducibility, and repeatability, a wide linear range of 0.01–8 $mmol^{-1}$ L, a quick reaction time of less than 5 s, a high sensitivity of 113.6 $\mu A\ mmol^{-1}\ L\ cm^{-2}$, and a low detection limit of 0.5 μmol^{-1} L. Furthermore, in the presence of interferents such as sucrose, uric acid, and ascorbic acid, the sensor exhibited strong selectivity toward glucose.

The response time of a PANI/Au hybrid optical sensor for detecting the pH of saliva samples was short. When the pH in solution was adjusted, the absorption at certain wavelengths of 665 nm and 785 nm in the visible-near-infrared spectra of this sensor varied. The sensitivities were 0.0299 a.u. pH^{-1} (a.u., arbitrary unit) over a pH range of 5–8 and 0.0234 a.u. pH^{-1} with a linear pH range of 2–8.

7.5.1.4 PANI/polymer Composite for Sensors

Elastomer or thermoplastic elastomer polymer matrices, which do not require vulcanization, combine the mechanical qualities of rubbers with the processability and recyclability of thermoplastics. Thermoplastic elastomer styrene–butadiene–styrene (SBS) triblock copolymer is an intriguing thermoplastic elastomer for the creation of stress/strain sensors for big strain, robotic, and industrial automation applications. For piezoresistive sensors, a composite of conductive PANI and thermoplastic elastomer SBS was produced using standard polymer extrusion. The material has good processability, electrical conductivity, mechanical characteristics, and piezoresistive responsiveness. The composite with 40% PANI component has a piezoresistive gauge factor close to 1 for deformation measurements between 0.1 and 3 mm in bending cycles. For the detection of hexamethylphosphoramide, sensor devices using PANI covered polyurethane membranes exhibit good sensitivity. The sensing performance was likely attained because PANI resistance varies significantly when exposed to a hexamethylphosphoramide environment due to deprotonation. When employed as NH_3 sensors, CSA-doped PANI/polystyrene composites with high conductivity show good sensitivity and reversibility.

7.5.2 PPy Composites for Sensors

7.5.2.1 PPy/CNT Composite

As a nitrite sensor in aqueous medium, a nanohybrid composite synthesized by electrochemical-assisted synthesis of PPy with sodium dodecyl sulfate (SDS) as a surfactant and supported by functionalized MWCNTs (f-MWCNTs) was employed. When f-MWCNTs are added to the PPy/SDS matrix, the resulting PPy/SDS/f-MWCNT composite films have lower impedance and greater capacitance than pure PPy/SDS films. The capacitance of the composites steadily rose as the fraction of f-MWCNTs increased. A GCE modified with PPy/SDS/f-MWCNT composite demonstrated a sensitive response to nitrite electrochemical oxidation with a low detection limit of 0.11 mol/L (110 nmol^{-1} L)($S/N = 3$).

7.5.2.2 PPy/G Composite

The wide effective area of PPy functionalized graphene nanosheets (GNs) in a PPy-GNs-modified electrode increased the sensitivity for detecting atorvastatin due to their porous structure. Atorvastatin peak currents are considerably greater at the PPy-GNs/GCE than at the bare GCE and rise linearly with atorvastatin concentration. Pressure/temperature/strain sensors made of polydimethylsiloxane-coated microporous PPy/graphene foam composite exhibit great sensitivity, quick response/recovery, and long durability after 10,000 pressure loading cycles.

The synergistic interaction of the PPy layer and G resulted in an exceptional performance of a hybrid sensor consisting of an ultrathin PPy layer over chemical vapor deposition grown *G*. In the selective sensing of NH_3 at room temperature, a high, quick, and reversible response was produced, along with relatively good stability, repeatability, and humidity resistance. Following NH_3 absorption on the PPy layer, electron transport from NH_3 to the PPy layer and subsequently to *G* via the PPy layer reduced the resistance of the PPy layer and *G*, respectively, and increased sensor sensitivity. Furthermore, the ultrathin PPy layer with a porous nature has a significant impact on sensor selectivity, responsiveness, and humidity resistance.

7.5.2.3 PPy/Metal Oxide Composite

The most sensitive voltammetric sensor for the electroanalysis of clofazimine is a PPy/bismuth oxide (Bi_2O_3) nanocomposite with a detection limit of 0.97 μmol mL^{-1} and a quantification limit of 3.23 μmol mL^{-1}. With a correlation value (R_2) of 0.99, its well-defined reduction peak changed linearly. The improved sensor also has excellent good recovery, repeatability, and long-term stability. When compared to PPy/GCE, TiO_2/GCE, and bare GCE, a PPy/TiO_2/GCE voltammetric sensor presented superior electroanalytical performance. The improved sensor has a linear response for sulfamoxole oxidation over a concentration range of 1.25 to 12.50 μgmL^{-1}, a correlation coefficient (R_2) of 0.9980, a detection limit of 1.24 ngmL^{-1}, and a quantification limit of 4.15 ngmL^{-1}.

Based on PPy-dodecyl benzene sulfonate-cerium oxide nanocomposite (PPy-DBS-CeO_2), an electrochemiluminescence H_2O_2 sensor with a detection limit of 1.2×10^{-16} mol/L demonstrates high repeatability and stability. The addition of CeO_2 nanoparticles in the polymeric matrix of the composite improved the electrochemiluminescence signal. Under ideal circumstances, the signal and the logarithm of H_2O_2 concentration exhibited a linear relationship ranging from 5.0×10^{-16} to 5.0×10^{-5} mol/L ($R_2 = 0.9922$).

7.5.2.4 PPy/metal Composite

Template-free and template-assisted procedures were used to create PPy nanosphere/Ag nanocomposite and PPy fiber/Ag nanocomposite, respectively. The PPy fiber/Ag nanocomposite sensor outperforms the PPy nanosphere/Ag nanocomposite sensor in terms of sensitivity, response time, and recovery time. Even in the presence of 250 ppm NH_3 gas, the PPy fiber/Ag nanocomposite sensor was not saturated, but the PPy nanosphere/Ag nanocomposite sensor was saturated at 100 ppm NH_3 gas concentration. The sensitivity of Au-nanoparticle-decorated PPy nanowires to arsenic was measured to be 0.22 and 4.38 μmol^{-1} L for two sets of Au concentration ranges, respectively, while the detection limit of the sensor was measured to be 0.32 μmol^{-1} L.

7.5.3 PEDOT:PSS Composite

7.5.3.1 PEDOT:PSS/Metal Composite

The incorporation of optimum concentrations of silver nanowires (AgNWs) into PEDOT:PSS film considerably increased the sensitivity of flexible NH_3 sensors using PEDOT:PSS/AgNW composite film as the active layer. The sensor demonstrated good sensing capability for detecting very low NH_3 concentrations below 500 ppb, as well as significant selectivity for water and common organic vapors. To check the freshness of pork, the sensor was combined with a self-designed portable data gathering device.

7.5.3.2 PEDOT:PSS/Metal Oxide Composite

ZnO nanorods (NRs)/PEDOT:PSS hybrid heterojunctions for highly efficient UV sensors demonstrated diode-like behavior and a good sensitivity of 37.65 when illuminated with UV (=256 nm, 130 W). The responsivity and sensitivity of the ZnO NR/polymer heterojunction device were increased by a factor of 10 and 8, respectively, as compared to bare ZnO NR arrays. The improved performance might be attributed to efficient charge separation guided by the built-in potential at the interface between the ZnO NRs and the PEDOT:PSS.

7.5.3.3 PEDOT:PSS/Carbonaceous Materials Composite

During the electrochemical experiments, a reduced graphene oxide (rGO)-PEDOT:PSS/GCE sensor demonstrated good stability, repeatability, and reproducibility, with a high peak current and low working potential (electrocatalytic activity). It gave piroxicam and nimesulide analytical curves with detection limits of 1.0×10^{-7} and 2.4×10^{-9} mol/L, respectively. To assess overall antioxidant capability, a G/PEDOT:PSS/SPCE nanocomposite was created. The electrosprayed G/PEDOT:PSS can improve current responsiveness simply by increasing the SPCE surface area. The significant reduction in the peak potential difference (ΔEp) between the anodic and cathodic peak potentials of G/PEDOT:PSS/SPCE shows that electron transport kinetics are quicker than in the unmodified SPCE. With the addition of N-GQDs, the sensing response of the resistive NH_3 gas sensor based on PEDOT:PSS and N-doped GQD dopant improves significantly. In comparison with the sensor without N-GQDs, the reaction time of the N-GQDs-doped sensor was 6.8 min, and the stability was improved.

7.5.4 Refined Chemical Sensor Systems Based on Conductive Polymer/Cyclodextrin Hybrids

With growing concern about the impact of environmental toxins on human health and the environment, chemical sensors have received a lot of interest as a way of identifying and monitoring possible environmental risks. Chemical and biological sensors with high sensitivity, selectivity, flexibility, and reusability have been produced employing a variety of attractive materials and architectures. Cyclodextrins (CDs) are a novel type of cyclic oligosaccharides with six (α-CD), seven (β-CD), and eight (γ-CD) glucopyranose units that have been widely employed as effective and ecologically friendly sensing medium in chemical sensors. Furthermore, due to their natural nature, CDs have higher compatibility with diverse organic compounds and lower toxicity.

The synergistic coupling of CPs and CDs is projected to yield new sensing materials, since the resulting CP/CD hybrids offer the advantages of both CPs and CDs. Signal transduction can be aided by CPs, which is crucial for reliable sensor operation. Many organic compounds, such as potentially dangerous substances, can be accommodated on CDs. Even though the CP/CD hybrid has a limited sensitivity to certain molecules, this may be addressed by adding other components, such as a photocatalyst.

The most widely utilized CPs, such as PPy and PANI, have been extensively employed for hybrid creation with CDs. CDs, for example, have been used as a catalyst in the production of PPy. Using CD molecules as intermediates, morphology control in PPy materials is possible. CD addition has resulted in the formation of novel structures like as nanowires, microprisms, and microsheets. PANI has also been used in the production of hybrids with CDs. CDs naturally produce a wide range of adaptable hybrids with other CPs, including PTh, as well as those with more complex molecular structures.

7.5.4.1 Sensors Using Polypyrrole/Cyclodextrin Hybrids

PPy is more attractive for the creation of hybrid materials with CDs than other CPs because it is generally stable under ambient conditions and easily manufactured utilizing basic methods such as electrochemical polymerization and vapor deposition. To create and improve the performance of innovative sensors, a number of strategies such as shape control, surface modification, and the use of additive have been presented.

A research on the analytical performance of PPy/CD revealed that the electrochemically produced PPy/CD hybrid had better sensitivity and selectivity to numerous organic neurotransmitters. In another work, an electrochemical impedance technique employing PPy and β-CD as sensing materials was devised to detect the muscle relaxant rocuronium bromide, with a sensitivity approaching 1 ppm. Using a PPy/β-CD hybrid, a simple formaldehyde sensor was created. Breslin's research group has been conducting considerable research on PPy/β-CD sensors since 2010. They created a PPy film doped with sulfonated β-CD and used cyclic voltammetry to determine the sensing capability of PPy/β-CD. In the presence of a competing chemical, ascorbic acid, this combination allowed for the selective detection of dopamine. They created a urea sensor with an extremely high sensitivity of 0.1 nM using a fundamentally comparable technique.

Several beautiful nanostructures have been built to improve the efficiency of the PPy/β-CD sensor. The sensing behavior of the hexagonal PPy/β-CD plate was demonstrated. The signal strength strikingly depends on analyte (dopamine) concentration. Despite the unique shape of the sensing material, the sensor dopamine sensitivity did not increase much. This work was notable because a systematic technique to revealing the structure–performance relationship in the PPy/β-CD hybrid sensors was provided.

Reverse microemulsion polymerization is used to create PPy nanotubes, and the surface of the resultant PPy nanotube is changed using β-CD. Because surfactants with a somewhat longer chain are added, producing more stable micelles, the reverse microemulsion technique is thought to be effective for controlling dimensional regularity such as diameter and size. To boost sensor performance, the resulting sensing medium is embedded in a hydrogel matrix to improve compatibility with other components.

To improve sensor performance, a third component can be introduced to PPy/CD hybrid composites. Carbon nanostructures are often employed to increase the fidelity and performance of several sensors. Using a PPy adorned graphene/β-CD hybrid at a screen-printed electrode, hazardous heavy metal ions such as Hg^{2+} were identified. The detecting profile of the hybrid sensor was influenced by a variety of factors, and it attained a Hg^{2+} detection sensitivity of 50 mM. To detect catechol, a CD-modified pyrrole monomer was polymerized to produce PPy on reduced graphene oxide (RGO). A hybrid polymer composite for dopamine detection was created by combining another polymer, CD-doped poly(N-acetyltriamine), with a PPy composite film. An ultrasensitive immunosensor for celiac disease diagnostics was created by attaching an adamantane-appended polymer earing antigen to a PPy/CD surface. Table 7.3 contains further information on the PPy/CD hybrid sensors discussed previously. Because of their adaptability, PPy/CD hybrids will find usage in a variety of disciplines, including electrical energy storage/conversion and optoelectronic devices.

7.5.4.2 Sensors Using Polyaniline/Cyclodextrin Hybrids

A PANI/CNT/β-CD combination on a glass carbon electrode (GCE) offers a dependable way for measuring dopamine. The results revealed that the response of the PANI/CNT/β-CD hybrid is superior to that of the PANI/CNT, PANI, and plain GCE. A comparable PANI/CNT/β-CD hybrid film was added to a carbon electrode for detection of 1-phenylalanine, an amino acid with a detection limit of 1 nM. Garrido et al. reported several intriguing themes including the use of innovative film-type

Table 7.3 Performance summary of sensors based on PPy/CD hybrids

Sensing material	Analyte	Performance (Limit, Linearity)
PPy/β-CD	Catechol, pyrogallol	4, 1 μM
PPy/CD	Rocuronium	1–10,000 ppm
PPy/β-CD	Formaldehyde	–
PPy/sulfonated β-CD	Dopamine	3.2 μM
PPy/sulfonated β-CD	Urea	0.1 M
PPy/β-CD	Dopamine	–
PPy/graphene/β-CD	Hg(II)	0.47 nM
Poly(pyrrole-CD)/RGO@electrode	Dopamine	–
Molecular imprinted PPy/β-CD/Au/graphene	Quercetin	0.1 nM
β-CD/poly(N-acetyltryramine)/PPy film	Dopamine	5 nM
PPy/CD	Antibody	–

PANI/CNT/β-CD hybrids for the management of hazardous compounds such as herbicides, fungicides, antibiotics such as citrofloxacin, and cocaine.

A new immunosensor for antigen detection was developed, which used PANI/Au nanoparticles as labels and an interdigitated electrode with connected carboxymethyl-CDs as a platform. A simple interfacial process was used to create a PANI/Au hybrid material, and antibodies were added to the hybrid materials. The electrode surface was treated with carboxymethyl-CDs and antibodies. Antigens (CEA) and antibodies coupled to the hybrids were then injected into the system. PANI/Au-labeled antibodies were collected by the immunosensor following the reaction between the antibodies and antigen. This resulted in a decrease in electrical resistance, which had a linear relationship with antigen concentration.

PAcANI (poly(N-acetylaniline)) has been widely employed in the surface engineering of several electrodes, including GCEs and CPEs. PAcANI/β-CD was coated on a GCE surface for uric acid sensing and chichonine monitoring utilizing 1,4-hydroquinone as a probe. Methylene blue (MB) molecules were added to PAcANI/β-CD to detect amantadine. A PAcANI/β-CD hybrid on a CPE were used to identify carbidopa and levodopa.

Ternary PAcANI/β-CD/carbon nanomaterial hybrids have been widely researched for sensing applications. PAcANI and CNT synergistic action has resulted in improved performance. Serotonin and dopamine levels were measured using a PAcANI/β-CD/CNT hybrid. Graphene, on the other hand, is a promising candidate for ternary hybrid creation. Graphene has been used to create more beautiful and complex sensor structures. Organic pollutant chemicals such as 1-aminopyrene (1-AP), for example, have been discovered utilizing a PAcANI/β-CD/graphene hybrid on a GCE. Methylene blue was used as a signal probe to test cholesterol using a

Table 7.4 Performance summary for the PANI/CD hybrid-based sensors

Sensing material	Analyte	Performance (Limit, Linearity)
PANI/MWCNT/β-CD@GCE	Dopamine	12 nM, 0.1 μM-1 mM
PANI/MWCNT/β-CD@CE	1-phenylalanine	1 nM
PANI film/MWCNT/β-CD@GCE	Herbicide	0.99 μM, 10–100 μM
PANI film/MWCNT/β-CD@GCE	Benzazone	1.6 μM,10–80 μM
PANI film/MWCNT/β-CD@GCE	Fungicide	1.04 μM
PANI film/MWCNT/β-CD@GCE	Antibiotic	50 nM, 10-80 μM
PANI film/MWCNT/β-CD@GCE	Cocaine	1.02 μM
PANI/carboxymethyl β-CD/Au	Carcinoembryonic antigen	7 pg/mL
PAcANI/β-CD@GCE	Uric acid	10–200 μM
Poly(sulfanilic acid-co–N-acetylaniline)/β-CD@GCE	Uric acid	2.7 μM, 10–350 μM
PAcANI/β-CD@GCE	Cinchonine	2.0 μM, 4–80 μM
MB/ PAcANI/β-CD@Electrode	Amantadine	-
PAcANI/β-CD@CPE	Levodopa Carbidopa	0.2, 0.8 μM
PAcANI/β-CD@CPE	Guanine,adenine	0.05 μM
PAcANI/β-CD/CNT@CPE	Serotonin,dopamine	Sub- μM
PAcANI/β-CD/MWNT@GCE	Hydroquinone	0.8 μM
PAcANI/β-CD/CNT@SCE	Biosensor	5 nmol/mL
PAcANI/β-CD/graphene@GCE	1-aminophenol	–
PAcANI/β-CD/graphene@Electrode	Cholesterol	0.5 μM, 1–50 μM
PAcANI/β-CD/RGO@Pt electrode	antigen	6.7 pg/mL

PAcANI/β-CD/graphene sensor. Recently, a prostate-specific antigen biosensor was described. Table 7.4 contains more information on the aforementioned PANI/CD hybrid sensors.

7.5.4.3 Sensors Using Polythiophene and Polythiophene Derivatives/Cyclodextrin Hybrids

Polythiophene (PTh) is a fascinating new material with potential uses in sophisticated sensing technology. PTh, on the other hand, is constrained by its poor solubility in most solvents and its comparatively complicated synthesis. As a result, its derivatives, such as poly (alkylthiophenes) (alkyl-PTh) and poly(3,4-ethylenedioxythiophene) (PEDOT), are thought to be more helpful than pure PTh. Furthermore, a specific polyelectrolyte, PEDOT doped with poly (styrenesulfonate) (PSS), is widely used in a variety of applications, including solar cells.

Cho et al. reported the effect of a CD molecule, hydroxypropyl-β-CD, on the electrochemical oxidation of PEDOT as a solubilizer. Their findings revealed that the addition of hydroxypropyl-β-CD in conjunction with an anionic surfactant had no effect on the characteristics of the resultant PEDOT film. This conclusion is helpful since the majority of PTh and PEDOT polymers are made using electrochemical polymerization. A theoretical analysis on the inclusion compound between a α-CD dimer and oligothiophene was provided in order to promote both theoretical pursuit and experimental demonstration of relevant PTh/CD complexes. Bouchta et al. investigated the feasibility of employing a PTh/CD combination for sensor systems, reporting the production of a stable poly-3-methylthiophene/γ-CD film for the analyses of neurotransmitter, chlorpromazine, levodopa, and dopamine. A detection limit of less than 1 mM was reached. Fukuhara et al. developed a highly intriguing sensor in which peptide chirality was assessed by a PTh/CD conjugated structure, the molecular conjugate being formed by adding permethyl-α-CD to PTh. When enantiomeric amino acids were introduced to the backbone of PTh/CD, a hypochromic alteration in the UV spectra occurred, resulting in enantioselectivity discrimination.

PEDOT/CD hybrids are being extensively researched for sensing applications. At a GCE surface, an appealing enzyme-free biosensor for quantitative creatinine analysis with PEDOT/β-CD was shown. A remarkable reaction time of about 60 s was obtained, demonstrating the prospect of real-time diagnosis. Even in the presence of competing chemicals, the sensor proved dependable. PEDOT/β-CD has also been utilized to detect plant compounds such as the medicinally active hyperoside and the natural color shikonin. These findings imply that the configuration in which a PEDOT/β-CD hybrid is coupled with CNTs is particularly sensitive to changes in electrode surfaces. In addition, the PEDOT/β-CD hybrid has been used as a Nafion membrane for an Au electrode. To increase sensor performance, PEDOT worked as an electron mediator. Nafion may also contribute to the overall conductivity of the sensing material, while CD molecules promote compatibility among the three key components. PEDOT/PSS has also been a common sensor component. Due to the benefits of PEDOT/PSS, it has been used for ultrasensitive detection of dopamine without the requirement for costly new metals or ITO. For cholinergic monitoring, an all-solid-state nanowire-type sensor was developed. This enzyme-free sensor displayed efficient acetylcholine detection in rat brains. Table 7.5 summarizes sensors based on PTh and its derivatives with CDs.

7.5.4.4 Sensors Using Other Conductive Polymer/Cyclodextrin Hybrids

For sensor applications, CPs with more complex chemical structures are also available. Despite the problems associated with its synthesis, the resultant sensor systems are seen to be useful. On a graphite electrode, a new CP poly(2-(2-octyldodecyl)-4,7-di(selenoph-2-yl) -2H-benzo [1,2,3 [triazone]) was combined with β-CD to form a cholesterol sensor. For sensor applications, CPs with more complex chemical structures are also available. The polymer was made in such a way that alkyl chains

Table 7.5 Performance summary of sensors using PTh and its derivatives with CDs

Sensing material	Analyte	Performance (Limit, Linearity)
Poly(3-methylthiophene-γ-CD)film	Neurotransmitters	0.1–1.0 μM
PTh-CD conjugate	Peptide chirality	–
PEDOT/β-CD	Creatine	50 μM
PEDOT/β-CD	Hyperoside, shikonin	–
PEDOT/β-CD/CNT @Electrode	Folic acid, uric acid	0.8, 0.07 μM
PEDOT/β-CD/Nafion@Au electrode	Paracetamol	36.1 nM, 3–300 μM
PEDOT-PSS/β-CD	Dopamine, catechol	9.6, 27.5 nM
PEDOT-PSS/β-CD	Acetylcholine	10 μM–0.1 M

were introduced into the polymer molecular structure, resulting in effective immobilization. It is worth noting that this procedure was used on genuine blood samples. The experimental results of this work were backed by theoretical reasoning utilizing density functional theory (DFT), which suggested that hierarchical self-assembly of CD and N,N-dimethyl-4(phenylethynyl)aniline on N-doped graphene might give a high-performance sensor for dihydroxybenzene isomers. The experimental aspect of the study was difficult since the authors employed real water samples from the Nanjing River.

To create a bioassay for cocaine detection, poly(p-phenylene)-β-CD was combined with poly(ethylene glycol) to build a coating on a microwell. At a GCE treated with β-CD-doped poly(2,5-diaminobenzenesulfonic acid), electrochemical behavior changed dramatically. When levodopa was added, there was a significant change in oxidation potential. β-CD contributes to more stable current profiles in the system. Bisphenol A, an ecologically harmful substance, was detected using molecularly imprinted CD-poly(L-lysine) gel layers on polyterthiophene film. This study uncovered an intriguing phenomenon: CD molecules may be placed in ideal places for chemical recognition. This technology introduces a novel approach to monitoring harmful substances in natural surroundings.

7.6 Chemical Sensors Based on Conductive Polymers

The fundamental difficulty in constructing a chemical sensor with extremely sensitive, quick, and reversible response at room temperature and under convenient working conditions is to produce a sensitive material with high sensitivity and also compatible with sensor production procedures. The reversible redox process facilitates anion uptake and release. Similarly, the primary sensing method in conductive polymers is often based on the variation in doping level as a consequence of redox interaction with analyst molecules, resulting in a change in conductivity and other features, such as optical capabilities. As a result, conductive polymers have a wide range of applications in chemical sensors and biosensors.

7.6.1 Chemiresistors

Chemiresistor is the most popular type of sensor that employs CPs. They can be made quite readily and cheaply. They also utilize most famous attribute of a CP, its conductivity. When the CP interacts with gaseous species, it might behave as an electron giver or acceptor. The hole conductivity of a p-type CP improves as it contributes electrons to the gas. When the same CP is used as an electron acceptor, its conductivity falls. Aside from the change in carrier numbers, there may be a change in their bulk mobility. It is generally caused by changes in the polymer backbone conformation. The response originating in the majority of the CP has a relatively long time constant (tens of seconds to minutes), which is frequently accompanied by hysteresis. The sluggish entry of gases into the CP causes these consequences. Because secondary doping modifies both mobility and carrier number, their relative contribution to the total change in conductivity varies.

7.6.2 Chemically Sensitive FETs

Based on work function modulation, three types of microfabricated devices are classified as sensors: chemically sensitive FETs (CHEMFETs), chemically sensitive capacitors, and chemically sensitive diodes. CHEMFETs are available in a variety of forms for both gas and liquid applications. It is critical to differentiate whether the current is flowing via the CP or the silicon. In this regard, they are further classified as thin-film transistors (TFTs) and insulated gate FETs (IGFET).

In a TFT, current flows through the CP, whose conductivity is regulated by the analyte contact and/or the electric field. The interpretation of the chemical response of TFTs becomes complicated for the same reasons as that of chemiresistors, namely the separation of the various kinds of modulation of conductivity and of the work function. Localized energy states can change the work function value but not the conductivity of the CP. However, if the CP is merely employed as the gate conductor in a typical, silicon-based transistor, the issue is substantially simplified. In such situation, the current passes through the silicon channel with no regard for the conductivity of the CP. The response of such a sensor is solely determined by chemical modulation of the CP work function. The ability to build multisensing microfabricated arrays is a realistic result of work function tuning. A generic array platform has previously been designed.

7.6.3 Capacitors and Diodes

Chemically sensitive capacitors and CHEMFETs are closely linked. This is not unexpected given that the "heart" of a CHEMFET is a capacitor, which is frequently made from the same materials and has the same size. The signal in CHEMFETs is acquired by measuring the channel transverse conductivity (transconductance) in the silicon plate. Chemically sensitive capacitors, on the other hand, are activated by an alternating current voltage, the voltage-dependent capacitance is measured using an impedance bridge, and the analytical information is retrieved from the shift in the capacitance–voltage curve. This shift has the same significance as the gate voltage shift, V_G. Although capacitors are plainly easier to manufacture, data gathering is significantly more difficult than with the comparable CHEMFET.

Particular care must be made to the quality of the contacts between the CP and the metal in CHEMFETs and Schottky diodes. They are capacitive and cannot be chemically modified if they are blocked, for example, by thick insulating oxide. This can happen when the CP is coated on less noble metals like aluminum. In such situation, a capacitive divider is formed by connecting the blocking contact capacitor C_C to the insulating gate capacitor C_G. This is an unfavorable design in any potentiometric sensor because the voltage drop across such a divider causes the signal to wander in an unexpected manner. It is feasible to choose the metal/CP pair so that a Schottky barrier forms between them. In such situation, the capacitor is purely fictitious (that is, without a dielectric). The junction possesses rectifying qualities, and the direct current flowing through it is proportional to the work function of the CP and the applied voltage. Because most CP are p-type semiconductors with very high work functions, the metals that may be used to make Schottky contacts are restricted. Nonetheless, low work function metals such as Al, Ca, In, or Na have been employed [9].

7.6.4 Future Trends

Chemical sensing is currently used to address a variety of daily demands, including clinical diagnosis, environmental monitoring, food and beverage quality evaluation, and homeland security. Conductive polymers play an important role in this situation as sensing components with high sensitivity to a wide variety of chemical compounds as well as changes in the physicochemical conditions of the surrounding environment. Furthermore, conductive polymers are often less costly than traditional inorganic semiconductors used in chemical sensors, and they may be made with customized characteristics using well-established molecular engineering processes. These exceptional characteristics are supported by a rising number of scientific articles and patents published during the 1990s, which include both scientific inquiries and technology applications involving conductive polymers in sensors. However, extensive use of conductive polymers in sensors is fraught with difficulties, including the creation of more stable, selective, and sensitive devices. Furthermore, the low repeatability of sensor signals remains a hurdle. Sensor fabrication, as well as ambient factors, can have a significant impact on sensor response (sensor poisoning). However, innovative breakthroughs in nanotechnology enable the manufacture of nanoscale materials and devices to make a significant contribution to the development of tiny devices with well-controllable features based on conductive polymers. Thin film deposition techniques, such as layer-by-layer (LbL) assembly, allow less expensive production pathways that are subject to scaling-up, resulting in sensors with more repeatable characteristics. If conductive polymer-based sensors are not long-lasting or maintain a consistent response over time, nanotechnology technologies for sensor production may favor a trade-off between performance and cost. Furthermore, nanofillers such as nanoparticles, nanofibers, carbon nanotubes, graphene, and other molecular elements capable of performing host–guest interactions have provided unprecedented levels of sensitivity and selectivity to conductive polymers, improving sensor performance and broadening the range of potential applications. Furthermore, molecular imprinting has shown its usefulness, particularly in terms of selectivity and detection limit, which are often limited in conductive polymer-based sensors.

7.7 Application of Conductive Polymers on Biosensors

Biosensors with great precision and accuracy are regarded as the most unique and cutting-edge instrument for monitoring bioanylate concentrations in clinical diagnostics, food industry, and toxicant detection. A biosensor is a piece of analytical equipment that turns a biological reaction into a readable signal. A biosensor is made up of three parts: a bioreceptor, biomolecule immobilization, and a transducer. The bioreceptor identifies a specific analyte, whereas biomolecules are immobilized on the matrix for stability, and the transducer translates a biological signal into a detectable electrical signal.

The biocompatibility and unique electrical features of CPs, which may translate biological information into electrical impulses, account for their vast range of applications. Furthermore, CPs feature numerous functional groups that give maximal enzyme loading via interactions between the enzyme molecules and the functional groups of the polymers, allowing for well-organized scaffold biosensors.

From the standpoint of biosensor application, CP has the most distinguishing features, including (i) tunable sensitivities by adjusting synthetic variables such as incorporated counter ions and polymerization temperature, and (ii) fabrication of sensor arrays via electrochemical deposition or solution casting, which allows miniaturization and mass production of biosensor devices.

7.7.1 Biosensors Utilizing the Effective Energy Transfer of CPs

CPs are good light harvesters, and the captured energy may be effectively transmitted to a neighboring acceptor through the conjugated chain. Swager and Zhou demonstrated in 1995 how CPs may be employed as molecular wires to connect synthetic receptors attached to the CP backbone. The attachment of a quencher to the receptor dramatically reduced the intensity of the CP emission, and the poly-receptor system demonstrated a 65-fold increase in sensitivity when compared to a mono-receptor fluorescent chemosensor model. The efficient energy migration along the polymeric backbone resulted in the effect. An electron is stimulated from the highest occupied energy band (the π-band) to the lowest unoccupied energy band (the π^*-band) when it is excited by a photon. An exciton is made up of an excited electron and the associated "hole." The CP works as a conduit for the exciton, allowing it to move quickly up the chain. When the exciton reaches a unit with a quencher-occupied receptor, it is quenched. The exciton migrates efficiently across the polymer due to the molecular wire method. This means that a single quencher molecule can quench multiple repeat units in the chain, resulting in an enhanced quenching effect. Biosensors based on CP quenching have been constructed by coupling a ligand of interest to a quencher.

FRET can also transmit energy to a nearby chromophore, resulting in optical amplification. This phenomenon has been employed in a variety of biosensor applications, primarily leveraging the effect of proteins on FRET, and has been beneficial for sensitive detection of genetic material. Single nucleotide polymorphisms (SNPs) identification is critical in fields such as forensic science and medicine. Biosensor devices based on CPs have been demonstrated to have the specificity required to discriminate SNPs, and numerous applications have been developed. Duan et al. employed Taq DNA polymerase and fluorescein-labeled dGTP at the guanine nucleotide for probe extension following complementary binding to the SNP target. FRET from an additional CP to the fluorescein molecule enabled for the detection of SNP in a solution where the concentration of the wild-type sequence was 49 times greater than that of the mutant. This single nucleotide base extension process and optical amplification of CPs has also been used to evaluate the DNA methylation status of CpG sites.

7.7.2 Biosensors Utilizing the Conformational Changes of CPs

Because of their use in quick detection and monitoring in clinical and food diagnostics, bioprocess monitoring, and environmental defense monitoring, biosensors have recently sparked a lot of attention. A biosensor attaches an immobilized biospecific recognition or catalytic component, such as protein (enzymes, antibodies, and antigens), ribonucleic acid, deoxyribonucleic acid, polypeptides, cells, or tissue slices, to the surface of a transducer. Then a biological recognition event is converted into a measurable electrical signal, thereby detecting the presence of the target analyte.

A CP conjugated backbone is flexible, and any change in its conformation may be monitored optically since it affects the CP absorption and/or emission characteristics. As a result, if a CP interacts with a biological target to create a change in the backbone structure, it may be easily identified due to the impacts on fluorescence released by the CP. These assumptions are based on the fact that the backbone conformation is tightly related to the CP electrical characteristics. Any action that causes the main chain to twist from planar to non-planar conformation (rod-to-coil transition) reduces the effective conjugation length, resulting in a blue shift in the absorption spectra. The overall chain configuration has an effect on the light radiated by the CPs. When the polymeric backbones are split, the emission spectra shifts blue, whereas backbone aggregation shifts the emission spectra red. Furthermore, the near proximity of chains encourages self-quenching, which is visible as a reduction in fluorescence upon aggregation. The optical transition of CPs has been observed for a variety of stimuli including heat, ions, and solvents, allowing for the development of a variety of innovative sensing devices [12].

7.7.3 Organic Biosensors

Biosensors are analytical instruments that detect, record, and communicate data regarding a biological process (e.g., blood pressure) or the presence of a certain chemical molecule (the analyte). As a sensing device in a biosensor monitors the biological species, quantifiable electrical signals proportional to the concentration of the analyte are created.

The CPs have been used in a variety of biosensor applications, including medical and environmental monitoring. Blood glucose biosensors, which have been extensively researched, are a good example of medical uses. They employ enzymes to oxidize glucose in a series of stages, and the concentration of glucose is measured by a CP-based transducer. Typical CPs used in glucose biosensors include PPy, PANI, and polyaminobenzoic acid (PAB). Cholesterol biosensors assess the amount of cholesterol in the blood, which is a significant parameter in clinical diagnosis. PPy can be employed as a transducer in cholesterol biosensors to physically immobilize the enzyme cholesterol oxidase. The oxidation of cholesterol in PPy films produces amperometric responses in cholesterol biosensors.

Environmental hazards can also be detected using biosensors. When organic vapor or moisture is detected in gas sensors for environmental monitoring, the CPs demonstrate a rapid change in electrical conductivity due to polymer degradation. PANI films have been shown to be effective in ammonia gas sensors, which is useful for food packaging. CP films such as poly(3,4-ethylenedioxythiophene)/polystyrene sulfonate (PEDOT/PSS) and poly(anilinesulfonic acid) have been demonstrated to be capable of detecting a change in a wide range of humidity in a humidity sensor.

The major problem of biosensors is to integrate physiologically reactive materials like antibodies, cells, and enzymes with CPs to serve as an electrical signal generator. Because of their biocompatibility, sensitivity, and endurance, CPs are excellent matrices for bioreceptor immobilization. Furthermore, they are adaptable enough to withstand biomolecule embedment. Physical adsorption, chemical adsorption, covalent bonding, crosslinking, and trapping are all methods for immobilization.

Physical adsorption takes use of physical interactions between bioreceptors and CPs, such as Van der Waals forces and electrostatic forces. Electrostatic forces hold the anionic bioreceptors to the cationic CPs. It should be noted that physical adsorption does not necessitate the functionalization of monomers. However, due to the weak forces of physical adsorption, bioreceptor leaching may occur over time. Chemical adsorption, which includes covalent interaction between bioreceptors and CPs, is introduced in this context. Despite its simplicity, perfect control of biomolecule orientation on the surface layer is still difficult. Such random orientations may obstruct active spots on the CP surfaces. As a targeted coupling technique, crosslinking agents and functional groups such as $-NH_2$, $-COOH$, and $-SH$ are selectively placed onto biomolecules. These types of tailored coupling technologies allow for exact control of biomolecule orientation. However, the rigorous treatment by hazardous chemicals involved in the crosslinking process makes it challenging for biomolecules to sustain enzymatic function.

Electrochemical polymerization often employs entrapment techniques in which monomers, dopants, and biomolecules are combined together in a solution. Polymerization of the CPs and immobilization of the biomolecules occur concurrently in this method. However, because a high concentration of biomolecules is required throughout the process, it is not cost-effective. The entrapment technique makes sensing components less accessible to analytes. Furthermore, the hydrophobic CPs disrupt the quaternary structure of proteins, reducing their biological function as sensing components.

Despite the benefits of low cost and ease of manufacture, CPs in biosensors have drawbacks in that they are more susceptible to surrounding chemical conditions such as oxygen, acids, and bases. Furthermore, the charge carrier mobility is inferior to that of other conducting materials such as carbon nanotubes and silicon nanowires.

CPs can be employed in conjunction with a biorecognition element or can combine biorecognition and transduction, resulting in the selective and continuous detection of biomolecular interactions. Surface-biomolecule interactions in biosensors need that the CP create a stable film in aqueous solution. This can be accomplished by either (i) electrochemical or (ii) chemical polymerization. As long as the support material is conducting, the former approach may be used to almost any arbitrary shape. However, it often results in surface coatings with significant roughness from fractal growth processes, which may reduce batch-to-batch repeatability. Chemical polymerization can produce a highly conducting material without the use of an underlying support and offers a variety of processing methods (some of which are suitable for large-scale production), such as spin coating, dip coating, drop casting, inkjet printing, electrospinning, freeze-drying, and so on, each with their own geometric constraints and material formats (e.g., thin films, gels, and fibers). To avoid film breakdown and delamination in aqueous solutions, additives (such as crosslinking agents and adhesion boosters) may be necessary.

Electrical readout methods are used in electrochemical biosensors to evaluate changes in the electrical characteristics of the CP film caused by specific binding between the biorecognition element and the analyte. CP films were initially utilized as simple electrode coatings in electrochemical biosensors to facilitate charge transfer from the biorecognition element to the underlying electrode. Signal transduction in this situation is accomplished using electrochemical methods such as electrochemical impedance spectroscopy (EIS), cyclic voltammetry (CV), and amperometry. Further advancements in the field result in the integration of the biorecognition element with an organic electronic device based on a CP film as the primary electronic component (e.g., a transistor). Organic thin film transistors (OTFTs), as opposed to CP-coating-based biosensors, rely on field-effect or electrochemical doping for signal transduction and have various benefits, including inexpensive cost of fabrication, quick response, and signal amplification.

7.7.4 Doped CP-Based Biosensors

Doping CPs with acids is typically recommended because it increases electrical conductivity while also acting as a functional material with functional groups that tend to connect with multifunctional proteins for the building of biosensors. Kaneto et al. created a urea biosensor by covalently immobilizing urease on poly(N-3-aminopropyl pyrrole-co-pyrrole) doped with para-toluene sulfonate (PTS). The NH_2 group on the PTS-doped CP covalently linked with the urease enzyme on the surface through a carbodiimide anchoring process. Aside from the NH_2 functional group, the CP displayed porous shape as a consequence of the large PTS dopant, which resulted in high enzyme loading and thus good enzyme electrode performance in terms of long life stability, dynamic range, and rapid response time. Luo et al. created a 3D macroporous structure of poly(sodium 4-styrene sulfonate) (PSS)-doped polyaniline for alphafetoprotein recognition. The sensitive and selective biosensor had excellent conductivity, a large surface area, and several functional groups, which assisted in efficient antibody immobilization with a detection limit of 3.7 fg mL^{-1}. Because the lone polyaniline composite exhibits redox activities under acidic conditions, which reduces the sensitivity of the majority of the biosensing system, polyaniline was doped with PSS.

Mathiyarasu et al. employed camphorsulfonic acid to dope polyaniline for use as a potentiometric glucose biosensor by covalently immobilizing glucose oxidase. The resultant CP had a uniform and smooth microporous structure, resulting in a homogeneous composite with constant enzyme loading and a more conductive platform for electron transport. The redesigned CP electrode demonstrated improved selectivity, sensitivity, and a good linear concentration range of 1–50 mm. Wernet et al. created a flexible polypyrrole doped with sulfated poly(β-hydroxy ethers) in order to determine its properties. The resultant biosensor had a greater H_2O_2 oxidation potential at a lower applied current, and a linear range of up to 100 mm. Marisa et al. created a bi-enzymatic biosensor for gluconic acid identification based on PAAMPSA-doped polyaniline by co-immobilizing 6-phospod-gluconate dehydrogenase and gluconate kinase (6PGDH and GK, respectively). Gluconic acid works as a biomarker for the quantitative measurement of Botrytis cinerea infection in grapes, and the resulting biosensor demonstrated excellent sensitivity, linear concentration range, low limit of detection (5 μm), and acceptable storage stability.

Nicholas et al. used poly(2-acrylamide2-methyl propane sulfonic acid) (PAAMPSA) as a dopant for doping polyaniline for bovine serum albumin and lysozyme recognition investigations. They also conducted studies to examine the range of pH over which the doping mechanism is active by examining the reaction of ICP to proteins at physiological pH and measuring the conductivity variations of PANI/PAAMPSA with respect to altering pH. Furthermore, the use of PAAMPSA as an acid dopant led in water-soluble CP with good conductivity.

Using tosylate-doped PEDOT, Noemi et al. created an aptasensor for label-free detection of kanamycin A and ampicillin. With a dynamic concentration range of 10 nm–1 mm for kanamycin A and 100 pm–1 μm for ampicillin, the resulting biosensor was highly selective and sensitive to the investigated target. Genuine samples of ampicillin-spiked low-fat milk were successfully tested to evaluate the biosensor functioning using real samples. Erslan et al. used an amperometric biosensor to determine acetylcholine using polyvinylsulfonate (PVS)-doped PEDOT film. Crosslinking acetylcholinesterase-choline oxidase with bovin serum albumin and glutaraldehyde was used for the immobilization. When such big dopants are utilized, PEDOT is doped with PVS, which boosts conductivity, improves environmental stability, preserves charge neutrality throughout the reduction step, and augments the polymerization rate. Because of these characteristics, the ICP demonstrated a broad linear range and a low limit of detection of 5.0×10^{-9} mol/L with a reaction time of 200 s [19].

7.7.5 Conductive Polymer-Based Electrochemical Biosensors

Electrochemical biosensors are a potential analytical approach for detecting biomolecules in a sensitive and selective manner. The unique properties of conductive polymers, as well as the ability to entrap enzymes during electrochemical polymerization, have been used to create amperometric biosensors. The majority of biosensors reported in the literature employ an enzyme as a biological recognition element. Clark and Lyons demonstrated the first biosensor, known as the "enzyme electrode," in 1962 for the measurement of glucose.

The covalent immobilization of choline oxidase and glutamate oxidase enzymes onto poly-5,2′:5′,2″-terthiophene-3′-carboxylic acid, poly-TTCA layers on a glassy carbon electrode resulted in amperometric enzymatic biosensors for choline and glutamate. The electrochemically manufactured conductive polymer layer oxidized the enzymatically generated H_2O_2 at a lower voltage and was the foundation of detecting choline and glutamate. Other interfering anions were detected by the biosensors with high selectivity. A xanthine biosensor was created by covalently immobilizing xanthine oxidase (XO) onto a functionalized conductive polymer (poly-5,2′:5′, 2″-terthiophine-3-carboxylic acid), poly-TTCA, by forming an amide link between the carboxylic acid groups of poly-TTCA and the amine groups of the enzyme. The linear dynamic ranges of anodic and cathodic xanthine detections were between 5.0×10^{-6}–1.0×10^{-4} M and 5.0×10^{-7} to 1.0×10^{-4} M, respectively. The detection limits for anodic and cathodic processes were determined to be 1.0×10^{-6} M and 9.0×10^{-8} M, respectively.

The biosensor applicability was evaluated by detecting xanthine in blood serum and urine real-world samples. A cytochrome c3 (Cyt c3)-based biosensor for detecting and quantifying superoxide radicals has been developed (O_2^-). The mutant of Cyt c3 from the sulfate-reducing bacterium (Desulfovibrio vulgaris Miyazaki F. strain) were covalently bound to a conductive polymer (poly-TTCA) covered electrode. The bioelectrocatalytic reaction to O_2^- was monitored to assess the possible applicability of the Cyt c3-modified electrode. The hydrodynamic range was 0.2–2.7 $\times 10^{-6}$, while the detection limit was 5.0×10^{-8} M. In another study, Cyt c was immobilized onto lipid-bonded conductive polymer layers of model biomembranes (poly(3,4-diamino-2,2:5,2-terthiophene) (polyDATT) for the creation of another O^{2-} biosensor.

For the detection of H_2O_2, a new amperometric biosensor was constructed by depositing horseradish peroxidase/polypyrrole (PPy) onto the surface of ferrocenecraboxylic acid functionalized sol–gel generated composite carbon electrode. The detection limit was determined to be 5×10^{-5} M, and after 1 week of storage at 4 °C, the biosensor preserved 90% of its initial sensitivity. The biosensor has a quick reaction time and a relatively broad linear range, making it suited for H_2O_2 detection. The influence of pH on the co-electropolymerization of glucose oxidase and polypyrrole (PPy) and the properties of a glucose oxidase/PPy biosensor has been investigated. This glucose biosensor has a linearity of 0 to 10 mM, a sensitivity of 7 $nAmM^{-1}$, and a long-term usability of roughly two weeks. Despite its limited linear range, the biosensor demonstrated great sensitivity.

Seker and Becerik described the electrochemical co-entrapment of glucose oxidase and *p*-benzoquinone into a polypyrrole matrix for the creation of a glucose biosensor. An amperomteric tyrosinase biosensor based on conducting poly (3,4-ethylenedioxythiophene) (PEDOT) for herbicide and phenolic chemical quantification has been reported. The detection limits for monophenol and di-phenol varied from 5×10^{-9} M to 0.5×10^{-6} M, whereas the detection limits for diuron and attrazine were 2.14×10^{-6} M and 4.6×10^{-6} M, respectively. Although this biosensor can detect a variety of phenolic chemicals, selectivity is a key issue with it.

Jiang et al. demonstrated that adding polyvinylalcohol (PVA) to a conductive polymer film increased sensitivity over a pure PPy sensor. Kan et al. showed greater stability of this polyaniline-uricase biosensor due to higher affinity between uricase and polyaniline generated via template technique. Haccoun et al. created a reagentless lactate biosensor based on an electrocopolymerized copolymer film of poly(5-hydroxy-1,4-naphthoquinone-co-5-hydroxy-3-acetic acid-1,4naphthoquinone).

Lactate oxidase (LOD) and lactate dehydrogenase (LDH) were co-immobilized on electrochemically produced polyaniline (PANI) films using a physical adsorption approach to make a lactate biosensor. Asberg and Inganas used poly-4-vinylpyridine to crosslink horseradish peroxidase in a highly conducting PEDOT/PSS dispersion for hydrogen peroxide estimation. Morrin et al. electrochemically applied nanoparticulate polyaniline (PANI) doped with dodecylbenzenesulfonic acid (DBSA) on a glassy carbon electrode surface for horseradish peroxidase physical adsorption.

Horseradish peroxidase was electrostatically immobilized on the surface of a polyaniline film electrochemically formed onto a platinum disk electrode by Mathebe et al. The correlation coefficient and linear range of the polyaniline/peroxidase-based biosensor were determined to be 0.995 and 2.5×10^{-4} to 5×10^{-3} M, respectively. Brahim et al. created a 2-hydroxyethyl methacrylate/polypyrrole system that contains glucose, cholesterol, and galactose oxidase. The nine-month durability of this technology suggests that it may be utilized subcutaneously to monitor glucose, cholesterol, and galactose. Ivanov et al. employed a crosslinking approach to immobilize cholinesterase on glassy carbon electrodes modified with polyaniline.

Langer et al. immobilized choline oxidase in nanostructured polyaniline layers with regulated nano and micro porosity for choline quantification in food. The polyaniline biosensor based on choline oxidase remained stable for around 30 days. Nanostructured polyaniline layers enhance sensor performance and stability. Qu et al. have published a choline amperomteric biosensor based on a polyaniline multilayer screen and a layer-by-layer constructed functionalized carbon nanotube. The biosensor sensitivity was improved by layering carbon nanotubes. Shim et al. created a disposable and mediatorless immunosensor based on a conductive polymer covered screen-printed carbon electrode employing a separation-free homogenous approach for the detection of rabbit IgG as a model analyte.

Ac impedimetric electrochemical interrogation was employed to create a label-free and reagent-free immunosensor based on direct integration of antibodies into conductive polymer films at the surface of the screen-printed electrode (SPE). It has been proven that the real component of the impedimetric response operates as a dominating component of the ac impedimetric response of an anti-BSA-loaded conducting polypyrrole (PPy) film when exposed to various concentrations of BSA. BSA was detectable with a linear response ranging from 0 to 1.136×10^{-6} M. The observed faradic reaction current is the result of antibody–antigen interaction. The use of SPE may benefit the downsizing of this immunosensor.

Tahir et al. studied the properties of polyaniline compounds in various protonic acids for use in diarrhea virus detection using polyclonal and monoclonal BVDV antibodies. They demonstrated that polyaniline with perchloric acid had the maximum conductivity at pH 6.6, and the biosensor sensitivity varied from 10^3 to 10^4 cell culture infective dose (CCDI) mL^{-1}. Lillie et al. created simple immunosensor formats by polymerizing pyrrole loaded with avidin or antibody to luteinizing hormone (LH) on a gold interdigitated electrode and proved that impedance spectroscopy can detect LH at concentrations ranging from 1 and 800 IU L^{-1}. Farace et al. used impedance spectroscopy to build a reagentless immunosensor for the detection of luteinizing hormone based on antibody entrapped in a conducting polypyrrole matrix.

Electrochemical techniques for detecting DNA hybridization offer several advantages, including the fact that they are very rapid to detect and may be readily applied to a portable DNA sensor. Shim et al. described an electrochemical technique for directly detecting DNA hybridization based on a conductive polymer polymerized on a glassy carbon electrode using a carboxyl group terthiophene monomer. A conductive polymer-based DNA probe was also used to detect protein-DNA interactions. A sensitive electrochemical test for DNA and proteins was devised, which used electrocatalytic reduction of hydrogen peroxide by labeled hydrazine on probe immobilized surfaces. A conductive polymer, poly-5,2′:5′,2″-terthiophene-3′-carboxylic acid (poly-TTCA), is covalently bonded to the dendrimer (DEN) and hydrazine in the process. The linear dynamic ranges for the electrocatalytic detection of DNA and proteins were discovered to be from 1.0 fM to 10 μM and 10 fg/mL to 10 ng/mL, respectively, with detection limits of 450 aM (2700 DNA mols. in a 10-μL sample) and 4.0 fg/mL.

The conducting polypyrrole/polyvinyl sulfonate films were electrochemically produced and utilized to trap calf thymus DNA for the creation of a DNA biosensor capable of detecting DNA hybridization. It has also been reported that conducting poly (thiophen-3-yl-acetic acid-1,3-dioxo-1,3-dihydroisoindol-2-ylester) (PTAE) may be used as a DNA hybridization electrochemical sensor. To creat a DNA electrochemical sensor, Youssufi and Makrouf prepared a conducting polypyrrole modified with ferrocenyl groups. The detection limit of this sensor was less than 1.0×10^{-12} M of DNA. Poly(aniline-aniline boronic acid) wires with redox functionalities at neutral pH solutions were created using ds-DNA templates. These wires provided direct electrical contact between the biocatalyst and the electrode. According to early research, glucose oxidase may be similarly connected with negatively charged polyelectrolyte templates such as polystyrene sulfonate.

A biosensor for measuring DNA hybridization based on polyaniline intercalated graphite oxide nanocomposite (PAI/GO) has been disclosed. Zhu et al. created a unique and sensitive electrochemical DNA biosensor for detecting DNA hybridization using an electrochemically synthesized polyaniline nanowire and methylene blue. For hybridization tests, single-stranded DNA was mounted on a self-assembling polymer based on polyallyamine that had been treated with thioctic acid. Lori et al. created a conductive polymer surface for the immobilization of his-tagged biomolecules by doping nitriloacetic acid (NTA) into electropolymerized polyvinylsulfonate-doped polyaniline (PANI/PVS) at a screen-printed carbon electrode. The resultant NTA-PANI-/PVS film exhibited intriguing electrochemical characteristics.

A conducting copolymer electrosynthesis employing pyrrole and pyrrole-oligonucleotide (ODN) at a platinum disk electrode has been described. Cadmium sulfide (CdS) nanoparticles were mixed into the polymer layer to boost sensitivity. As a basis for a reagent-free technique, electrochemical impedance spectroscopy may be utilized to observe charge transfer via conducting polypyrrole films loaded with oligonucleotide probes generated on carbon nanotube-modified electrodes. For complementary and mismatched target sequences, this approach offers various advantages, including good selectivity and shortened reaction time without the need of mediators or fluorescent materials [20].

7.7.6 Conductive Polymer Nanostructures for Biosensors

Due to the sluggish penetration of the target molecules into the conductive polymer, the reaction time is relatively long when bulk polymer is utilized to make a sensor. There is sometimes a concomitant hysterical reaction. In contrast, due to the porous nature of the nanotubes or nanofibers, the reaction time of a sensor composed of conductive polymer nanostructures is projected to be substantially quicker. For example, a nanofiber thin film of polyaniline (PANI) generated using interfacial polymerization responds much faster to acid doping and dedoping than a conventional film. Nanofiber films, in particular, exhibit no thickness dependency on performance in the range of 0.2–2.5 lm. Conductive polymer nanostructures with a high specific surface area and a porous structure are so expected to be good sensing materials. A variety of conductive polymer-based sensors, including gas sensors, hydrogen sensors, and acetic acid sensors, have been described to date. However, the creation of conductive polymer nanostructure-based sensors with high sensitivity, cycle number, and quick reaction time remains desirable.

Because of the capacity to adjust the sizes and structures, and hence the characteristics, of these nanomaterials, studies of conductive polymer-based sensors have indicated a tendency toward the creation of nanostructured conductive polymer-based sensors. Conductive polymer nanostructures such as nanotubes, nanowires, and nanoparticles provide new possibilities for biological and environmental sensing platforms. Actually, a suitable biosensor may benefit from a number of enhancing elements (nanostructure, composites, specific molecular design, etc.). On the one hand, nanostructures can be used to improve a nanocomposites-based sensor, and vice versa. Furthermore, the processability of conductive polymer materials is frequently critical for the creation of certain nanostructures or nanocomposites.

Because of their unique qualities that combine with those of CPs (biocompatibility, direct electrochemical synthesis) and nanomaterials, nanostructured CPs were an attractive building block for producing very sensitive biosensors (e.g., large surface area, flexibility for the immobilization of biomolecules and quantum effect). Because of their larger active surface area and lower penetration depth for target molecules, PANI NSs offer better sensitivity and quicker temporal response than typical bulk counterparts. The large surface area and porous structure allow chemicals to diffuse quickly into the framework, allowing them to be used as biosensors. Nanoparticles (NPs), nanotubes (NTs), core–shell nanomaterials, and hollow nanospheres are examples of PPy nanomaterials that have been utilized as possible components in electrical devices and chemical sensors. However, due to their ease of functionalization and environmental stability, PPy nanoparticles have lately been employed for biosensors.

The growing technology for controlling matter patterns on the nanoscale length scale is likely to result in completely new sorts of biological sensors. Constrained polymerization of these conductive polymers is an important method for giving them regulated structure and size. This results in the production of nanostructured conductive polymers (NSCP) with quick and discrete electrochemical switching properties, as well as a significant enhancement in conductivity and sensitivity for biosensor applications. Because of the high surface-to-volume ratio, NSCP can produce superior sensitivity through improved interaction between the CP and analytes. They enable analytes to have quick adsorption/desorption kinetics, resulting in fast response/recovery times even at room temperature. Furthermore, they can exhibit higher signal strength due to variations in charge transport behavior throughout the bulk of the materials, rather than only the surface region.

NSCP has historically been used to detect a wide range of analytes, including hazardous gases, volatile chemical compounds, and biological organisms. NSCP has evolved as a new research and development field focused on the development of novel smart materials for application in contemporary and future generation biosensors. Despite the fact that conductive polymers are successful matrices for serving as a transducer for biosensor applications, there are various obstacles to address, with nanoconcepts appearing in related research arenas as downsized sensors with excellent performance that are in demand. The NSCP-based electrodes have a strong chance of meeting this criteria. The NSCP having a large interfacial area performed better in sensing applications. Because of its larger effective surface area and shorter penetration depth for target molecules, the NSCP has better sensitivity and a quicker reaction time than its typical bulk equivalent. The large surface area and porous structure also allow for quick diffusion of molecules into the framework, which improves loading of biomolecules without compromising their biological activity, allowing them to be used as biosensors. Furthermore, the dimensional compatibility of NSCP with biomolecules aids in the fabrication of nanoconducting devices inside the sensor framework and facilitates electron transport.

The strong qualities of nanodimension CPs, including as high surface energy, quick and discrete electrochemical switching, and dimensional compatibility with biomolecules, have made them excellent transduction matrices in the field of biosensing devices. Nanowires, nanotubes, and nanoparticles formed of conductive polymers have been used successfully in biosensors. NSCP has several uses in biosensor technology. Catalytic biosensors, DNA chips, aptasensors, field-effect transistor (FET) biosensors, and immunosensors based on CP nanoparticles have demonstrated high-performance sensing characteristics toward target biological species.

7.7.6.1 Catalytic Biosensors

Glucose Detection

One of the primary clinical uses of NSCP in biosensors is the development of glucose concentration monitoring devices for diabetic patients. NSCP are commonly used as efficient transduction matrices in glucose biosensors to give suitable specificity and repeatability, and they can be conveniently made in high amounts at cheap cost. Ramanavičius et al. use the application of GO_X-PPy nanoparticles in the creation of a basic biosensor for glucose detection as proof of GO_X self-immobilization inside a produced polymeric layer. During polymerization, Pt sputtered nanoporous alumina disks were used as base electrodes to induce nanoporosity into PPy sheets. The resultant NS PPy has aided in improving the performance of biosensors by increasing the effective surface area for detecting electron flow. In the instance of physical adsorption, the sensor was shown to be more robust against H_2O_2 and to have improved glucose detecting capabilities. This sensor has produced quick, easy, and precise readings with high sensitivity (3.7AM^{-1} cm^{-2} for H_2O_2 and 3mAM^{-1} cm^{-2} for glucose, respectively) and very fast reaction times (3 s in each case).

H_2O_2 Detection

Biosensors based on nanostructured conductive polymers are potential options for electronic biological species detection. The measurement of hydrogen peroxide is critical for ensuring the safety and quality of pharmaceutical and cosmetic compositions. The HRP-modified PANI nanoparticles-based biosensor for H_2O_2 detection has increased enzyme deposition as well as signal-to-noise ratio. Nanodimension and biosensing performance have a close link. In terms of H_2O_2 detection, nanoPANI/DBSA film is a more efficient transducer matrix than bulk PANI/PVS film. The nanoPANI/DBSA film has a high protein loading capacity and requires a protein concentration six times lower than the PANI/PVS layer to achieve monolayer coverage, a greater signal-to-background ratio, and a shorter reaction time.

Choline Detection

Choline serves as a substrate for choline oxidase, which is immobilized on the surface of a sensor active element such as NSCP. Langer et al. created a choline nanosensor by immobilizing choline oxidase on NS PANI layers with variable thickness and porosity. The sensitivity of this biosensor is 5 mA/mM, and the storage stability is 30 days. The scientists discovered a link between the thickness of NS PANI and biosensor performance; however, they did not demonstrate how the biosensor performance fluctuates with thickness.

7.7.6.2 Bioaffinity Sensor

DNA Sensor

As an electrochemical transducer, NSCP can provide transfer energy as well as direct label-free, electrical detection of DNA hybridization. The NS PANI and PPy are the most commonly utilized in the manufacture of electrochemical DNA sensors, primarily for single-stranded DNA immobilization. Zhu et al. showed the fabrication of physiologically functionalized PANI nanowires and their application to DNA sensing. They demonstrated that this DNA sensor can efficiently distinguish complementary sequences from non-complementary and three-mismatched sequences. The sensor can successfully distinguish complementary from non-complementary DNA sequences, with a detection limit of 1.0×10^{-12} mol/L of complementary target.

Immunosensor

A single PPy nanowire-based nanoimmunosensor with cancer antigen (CA 125) antibody has been described for CA 125 biomarker detection and quantification. The immunosensor was very sensitive, with a detection limit of 1U/mL CA 125 and a dynamic range of up to 1000 U/mL in 10 mM phosphate buffer. Tolani et al. created poly(pyrrolepropylic) nanowire-based quick and extremely sensitive immunosensors for real-time monitoring of human serum albumin (HSA), which is used to detect early stages of renal illness. The detection is based on the specific binding of HSA to anti-HSA that is covalently bonded to a nanowire. Mulchandani et al. created high density PPy nanowire arrays for pathogen detection via the extraordinarily sensitive regulation of the electrical conductance of the nanowires, produced by the change in electrostatic charges from the pathogen attachment to antibodies [4, 5].

7.7.6.3 NSCP for Different Biosensor Applications

Bioelectronic Noses and Tongues

Human sense mimicry is a key strategy for developing unique smelling and tasting gadgets. Owing to their ease of functionalization, CP nanoparticles were examined to increase the performance of the B-nose. Jang et al. used reverse microemulsion polymerization in an apolar solvent to create PPy NTs. Co-polymerization simply changed the surface of the PPy NTs with carboxylic groups, and the carboxylated PPy NTs (CPNTs) were employed for the B-nose. Yoon et al. presented the integration of hORs and CPNTs into a field-effect transistor (FET) platform controlled by a gating effect, resulting in a CPNT-based B-nose. Covalent bonds existed between the hORs and transistors in the CPNT B-noses. As a result, they were more stable in liquid than the standard B-nose system and worked better in terms of odorant detection.

The FET B-nose under a liquid-ion gating mechanism was described as p-type, with holes as charge carriers, because of its aqueous-stable covalent bonding. The liquid-ion gated FET B-nose demonstrated extraordinarily sensitive real-time responses to target molecules, with a minimum detectable level (MDL) of around 400 fM. Surprisingly, the covalent functionalization procedure restricted the quantity of hORs loaded on the CPNTs, resulting in an increased MDL. This finding suggests that the surface immobilization of hORs is an important factor in the B-nose performance. In contrast to the previous study, Lee et al. demonstrated a human nose-like nanobioelectronic nose (NBE-nose) that responds to gaseous odorants. The NBE-nose had a sensing specificity response similar to that in the cellular signal transduction pathway. The NBE-nose also exhibited antagonistic behavior, similar to a human nose.

By combining CPNTs with human taste receptors, CPNTs were also used to mimic a nanobioelectronic tongue (NBE-tongue) (hTRs: hTAS2R38s). Song et al. used the aforementioned technology to create a FET NBE-tongue to show the mechanics of human taste. The sensing abilities of the NBE-tongue in terms of sensitivity and selectivity were strikingly similar to those of the human nose. Furthermore, the NBE-tongues responded differently to target bitter compounds depending on the taster and non-taster types of hTRs, and their MDL for phenylthiocarbamide (PTC) was around 1 fM. The bitter-taste perception of NBE-tongues differed for chemicals containing thiourea moieties in genuine vegetable samples, as caused a concentration-dependent increase in drain-to-source current (ISD).

Aptasensors

Aptamers, which are made up of oligonucleotide or peptide molecules that bind to a specific target biomolecule, are promising bioprobes for biomolecule detection due to their small size, chemical and environmental stability, and ability to be produced on a large scale. Aptasensor nanomaterials include carbon-based nanomaterials, metal nanomaterials, ceramic nanomaterials, and CP nanomaterials.

A direct electrochemical deposition approach can be used to insert the aptamer into a PPy nanowire (PPy-NW). Huang et al. showed the inclusion of an aptamer and PPy-NW-grown gold electrodes in a single process. The single PPy-NW with the aptamer was integrated into the fluidic substrate and demonstrated excellent specificity and rapid stabilization times toward IgE protein solutions in the range of 0.01 to 100 nM. The PPy-NW aptasensor sensing technique was a conductance change under a fixed voltage. The specific binding of the aptamer and IgE protein can increase the negative charge density on the PPy-NW because the IgE protein (with an isoelectric point of $pI = 5.2–5.8$) has a negative charge in pH 7.4 phosphate buffered saline (PBS) solution, resulting in an increase in conductance. Under the fluidic system, the PPy-NW with a MUC1 aptamer demonstrated highly specific MUC1 biomarker detection.

7.7.7 *Biosensors Based on Conductive Polymer Films with Nanofeatures*

Conductive polymer films are versatile materials for biosensors because they can be produced quickly with controlled thickness and porosity, and they can also be selectively immobilized by electropolymerization over very small areas of various surface morphologies, providing an easy route for functionalizing microelectrodes. The nanostructured conductive polymer films that increase biosensor sensing performance may be classified into three aspects: nanoscale film thickness, porosity, and surface shape.

7.7.7.1 Biosensors Based on Conductive Polymer Nanothin Films

The sensitivity and reaction time of conductive polymer biosensors are strongly influenced by the thickness of the film. The target analyte, on the other hand, can adsorb onto and diffuse into a thinner film more easily than a thick film. Thinner films also provide a bigger specific area for interaction with the target analyte. As a result, when the films get thinner, increased sensitivity and faster reaction time are predicted. Several investigations have been conducted to examine nanothin conductive polymer film-based biosensors. Electrochemical polymerization, spin coating, and layer-by-layer (LBL) processes might be used to create nanothin films of conductive polymers into which bioactive molecules could be placed. For example, a simple electropolymerization procedure was used to create an ultrathin (55 nm) PPy/glucose oxidase (GOD) film in a supportive electrolyte-free monomer solution for potentiometric glucose biosensing. The use of a lower current density for a longer length of time to obtain the same amount of charge transferred allowed for the creation of a more uniform and permeable ultrathin film, which resulted in improved glucose sensitivity.

Although the spin-coating approach is a simple way for producing nanothin films, it is restricted to the use of soluble conductive polymers. A glucose sensor based on an organic thin film transistor (OTFT) with a channel consisting of glucose oxidase (GOx) immobilized by a spin-coating process on a poly (3,4-ethylenedioxythiophene-poly (styrenesulfonate) (PEDOT-PSS) conductive polymer film has been disclosed. Under optimum testing settings, a quick reaction time (within 20 s) and high sensitivity (~1.65 μA/mM) were achieved, as well as a linear connection between the drain current and glucose concentration (range from 1.1 to 16.5 mM).

Because it can regulate the thickness of the film at the one molecule-layer level, the layer-by-layer (LBL) approach is well suited for the fabrication of nanothin films. The LBL process was used to create ultrathin PANI and PANI-PAA hybrid films. PANI and polystyrene sulfuric acid (PSSA) composite films produced by a self-assembly process showed increased sensitivity and response time. The reaction time fell from 10 to 8 s and the recovery time decreased from 4 to 2 min when the film thickness of PANI–PSSA was reduced from three layers to two layers. Furthermore, the LBL approach has been used to create glucose biosensors based on LBL self-assembled ion-sensitive field-effect transistors (ISFET). The glucose biosensor was constructed with a self-assembled PANI thin film as the channel, and its sensitivity to glucose was 0.738 μA/mM. These findings show that nanoparticles and thin polymer films are particularly promising materials for the development of low-cost biosensors.

7.7.7.2 Biosensors Based on Conductive Polymer Films with Nanoscale Roughness

Because changing the surface roughness of the conductive polymer film increases the specific surface area, it can also improve the performance of the accompanying sensor devices. A rougher nanoscale surface, for example, resulted in more binding sites accessible for probes in a conductive polymer-based impedimetric aptamer biosensor. Sensitivity for particular targets was obtained at 10 ng/ml. Nanopillar-enhanced electrodes (NEEs) functionalized with glucose oxidase (GOx) and PPy have also been used to improve glucose sensing capability. NEEs with roughness factors around 60 were ideal, resulting in good sensing performance for glucose detection with sensitivity as high as 36 $\mu A\ cm^{-2}\ mM^{-1}$. Furthermore, a glutamate micro-biosensor based on glutamate oxidase mounted onto nanostructured conductive polymer layers was disclosed for in vivo monitoring of glutamate release triggered by cocaine. The electrochemically generated nanostructured conductive polymer films allowed the immobilization of significant quantities of enzyme by covalent attachment, resulting in an extremely durable and sensitive micro-biosensor for in vivo measurements. The glutamate micro-biosensor has a detection limit of 0.1 M, a reaction time of less than 10 s, and a linear range of 0.2 to 100 M.

7.7.7.3 Biosensors Based on Micro- or Nanoporous Conductive Polymer Films

Because of its porous character, the response and associated sensitivity of nanoporous film were generally independent of thickness, facilitating the interaction of target molecules with all exposed regions, both outside and within the film. The reaction time of porous films made of nanofibers is not affected by thickness. The nanoporous film may be produced not only by hard template methods, but also by soft-template methods like nanofiber-based films.

With colloidal particles as hard templates, macroporous conductive polymers such as PPy, PANI, and PTh may be produced. In one example of nanoporous conductive polymer films-based biosensors, 3-methylthiophene (MTH) and thiophene-3-acetic acid (TAA) were electrochemically copolymerized onto ITO-coated glass plates containing various sizes of polystyrene (PS) template particles, and then glucose oxidase was immobilized onto the copolymer films. It was proposed that the amperometric response of this enzyme electrode was caused by the diffusion of the enzyme and electron-mediator molecules. The interconnected channel structure of copolymer films, with its long distance regularity and adequate size, allowed enzyme and electron-mediator molecules to permeate into the interior pores of the copolymer film. These geometric variables should have a large impact on both enzyme immobilization and the electron transfer reaction of mediator molecules on the electrode surface. However, no increase in the quantity of immobilized enzyme was found in the macroporous structure of MTH/TAA copolymer films.

A hard template approach was used to create a PANI-uricase biosensor. A template for PANI film with considerable holes for immobilizing enzymes resulted in higher active uricase doped onto this template and a significant reaction current. The template technique improved the biosensor stability and increased the affinity between uricase and substrate. A soft-template approach might likewise be used to create micro- or nanoporous conductive polymer films. A PANI-uricase biosensor, for example, was created by employing uricase as a soft-template. The PANI-uricase electrode was created via electrochemical polymerization, and it was subsequently hydrolyzed in 6.0 mol/dm^3 hydrochloric acid solution to eliminate any uricase that had been impacted by the aniline monomer from the PANI film. Finally, active uricase was immobilized in the PANI film using the conductive polymer doping and dedoping approach. The affinity between uricase and substrate was higher in the PANI-uricase biosensor created using the template method than that prepared using the two-step method. The determination of biosensors shows that the response current of a biosensor manufactured using the template approach declined by only about 18% over 60 days, but the response current of a biosensor prepared using the two-step procedure fell by around 39% in just 40 h. In addition, for the detection of DNA hybridization, an electrochemical DNA biosensor based on an electrochemically produced PANI nanowire screen has been created. The DNA biosensor detection limit for distinguishing complementary and non-complementary DNA sequences was determined to be 1×10^{-12} mol/L.

Molecular imprinting is an essential and complex process for creating conductive polymer films with corresponding nanocavities to the imprint molecules. For the identification of chiral amino acids, molecularly imprinted PPy nanowire-based films were created. Electrochemical polymerization effectively produced PPy films made of nanowires with an average diameter of around 100 nm and a few micrometers in length, in which enantiomeric camphorsulfonic acid (D- or L-CSA) molecules operated as both the dopant and the molecular imprinting pseudotemplate. Electrochemical impedance spectra (EIS) and CD spectra were used to illustrate the enantioselective interactions between PPy nanowires imprinted by enantiomeric CSA and phenylalanine enantiomers. Sensors based on molecularly imprinted PPy for ascorbic acid and saccharide detection have also been described.

7.7.8 *Biosensors Based on Conductive Polymer Nanostructures*

7.7.8.1 Biosensors Based on Conductive Polymer Nanoparticles

Nanoparticles can be utilized to modify electrode surfaces as well as biological receptor molecules such as enzymes, antibodies, or oligonucleotides. Because of its nanoscale, zero-dimensional nanostructures have unique physical and chemical features. For the following reasons, nanoparticles are an appropriate structure for the use of biosensors: (1) minimal diffusion limitations; nanoparticles are readily dispersed in aqueous media and over the electrode surface during the modification process; (2) maximum surface-to-volume ratio; this provides a larger active area for the design of a biosensor; (3) high efficiency of bioreagent loading; biomolecule immobilization can be realized through physical entrapment or surface modification; and (4) enhanced process ability; they can be well-dispersed in aqueous media.

In general, conductive polymer nanoparticles are scattered on the electrode surface to raise the area/volume ratio and promote biomolecule adsorption. A chemical technique was used to encapsulate a redox enzyme-glucose oxidase (GOx) into the conductive polymer PPy. Their research demonstrated the feasibility of using PPy/GOx nanoparticles in the design of amperometric biosensors. Furthermore, an amperometric enzyme biosensor made on PANI nanoparticles was disclosed. Electrodeposited polymer nanoparticles produced a highly ordered conductive nanostructured electrode surface, allowing uniform electrostatic adsorption of protein and resulting in higher signal-to-background ratios (61 ± 3) and shorter response times (0.62 s) than electrochemically prepared PANI/polyvinylsulfonate films.

Enough enzyme was securely immobilized during the creation of a phosphate biosensor using conductive polymer nanoparticles. The biosensors have a reaction time of roughly 6 s. Between 1.0 M and 100 M, a linear response was observed, and the detection limit was determined to be at 0.3 M. In addition, an ascorbic acid sensor was created by drop-casting PANI nanoparticles onto a screen-printed carbon-paste electrode. The PANI nanoparticles not only aided the catalytic process but also enabled for the detection of ascorbate at a low applied voltage of 0 V and operating at neutral pH, so bypassing the problem of sample interference.

7.7.8.2 Biosensors Based on Conductive Polymer Nanoarrays

Highly organized nanostructures made from (1D) nanowires or nanotubes, which are named as nanoforests or nanoarrays, have piqued the curiosity of biosensor designers. The benefit of properly aligned nanostructures is the combination of 1D nanowires/nanotubes and a highly ordered alignment. To optimize the value of employing nanostructured electrodes as improved elements, researchers studied the function of reaction kinetics and mass transfer in glucose sensing with nanopillar array electrodes. A conducting PANI nanotube array-based ultrasensitive biosensor for detecting DNA hybridization has been created. Even at the ultralow concentration, it displayed a very high sensitivity of 1.0 fM and remarkable hybridization specificity. This ultrasensitive feature was discovered to originate from the collective action of PANI nanotubes, increased PANI conductivity, and quicker hybridization kinetics of the orientated nanostructure. A PPy nanotube array was also employed to create an amperometric biosensor. The nanotube array has significantly improved this unique sensor adsorption property. By tailoring the mean pore diameter of the alumina disk to match the target enzyme, this nanotube array sensor may be employed with any of these enzymes.

7.7.8.3 Biosensors Based on Individual Addressable Conductive Polymer Nanostructures

Individual addressable one-dimensional (1D) nanostructures, such as nanowires and nanotubes, have aroused the interest of the researchers in biological sensing. The large contact area between analyzers and nanotubes (or nanowires) might result in a significant density of polarons and bipolarons in conductive polymer backbones, affecting the electrical conductivity of the nanotubes (or nanowires). As a result, using these nanostructures as transducers in sensors would result in increased and amplified sensitivity. Despite extensive research on 1D metallic and semiconductor nanomaterials-based sensor applications, the reversible electrical conductivity of conductive polymers, together with their ease of processing/modification and rich chemistry, makes them particularly appealing as 1D sensing materials.

Conductive polymer nanowires can be electropolymerized between pre-patterned micro- or nanoelectrodes, and bioactive analyzers can be implanted in situ as well. The built-in electrical connections and the one-step insertion of analyst into the conducting-polymer nanowire during polymerization are significant advantages over previously reported silicon nanowire and carbon nanotube biosensors that need post-synthesis modification and placement. Individually addressable PANI nanoframework electrode junctions (PNEJ) for sensing applications, for example, have been deposited in a parallel-oriented array. The PNEJ arrays displayed good performance in terms of high sensitivity and rapid reaction time in detecting HCl and NH_3 gases, ethanol vapor, and the pH of NaCl solutions. Furthermore, using target-guided deposition of PANI nanowires as in situ signal amplification, a nanogapped microelectrode-based biosensor array for ultrasensitive electrical detection of microribonucleic acids (miRNAs) has been accomplished. With a detection limit of 5.0 fM, the target miRNA may be measured in the 10 fM–20 pM range. However, because PANI nanoframeworks are made up of several interconnected PANI nanowires, their sensitivity is lower than that of a single semiconductor nanowire. These approaches reveal the capacity to create individually addressable nanowires between electrodes spaced tens of nanometers to a few micrometers apart, implying the possibility for high-density biosensor nanoarray fabrication [28].

7.7.9 Biosensor Based on Conductive Polymer Hydrogel

Human bodily fluids (such as blood, sweat, and tears) contain a significant quantity of chemical information, indicating the deep condition of biological molecules for human health. Biosensors can detect analytes in biological fluids non-invasively, giving a window into the overall dynamic state of biological molecules in the human body. CPHs have demonstrated remarkable benefits among several biosensor interface materials due to their flexibility to immobilize biological recognition species and strong biocompatibility stemming from their resemblance with extracellular matrix. CPHs with large surface areas and linked nanostructures are thus appealing platforms for biological sensing.

The majority of human metabolite biosensors rely on enzymes, which allow for intrinsic selective detection of various analytes. Because the loss of enzyme and the decline in enzyme immobilization stability cause the biosensor response signal to fade, it is critical for the enzyme-based sensor to preserve the stability of the enzyme immobilization. Because excess phosphate groups on phytic acid can react with amino groups on the enzyme, a porous PANI hydrogel utilizing phytic acid as a dopant and gelator demonstrates improved enzyme immobilization stability. Rong et al. also created a biosensing platform based on PPy hydrogel/Au nanocomposites using carcinoembryonic antigen (CEA) as a model analyte. This generic and straightforward approach for introducing conductive filler into the matrix of CPHs may be used to a wide range of different nanoscale materials.

Shrestha et al. developed a hybrid material for high-performance glucose biosensors made of PPy, Nafion, and multiwalled carbon nanotubes (Fig. 7.6a, b). The addition of other nanoscale materials improves the electrocatalytic activity of the enzyme and increases the rate of electron transfer. Moozarm Nia et al. developed a copper oxide/PPy nanofiber/reduced graphene oxide composite-based glucose biosensor with good sensitivity and a low detection limit (0.03 M). In practical applications, it is required to create a biosensor platform capable of detecting many metabolites at a low cost and with increased efficiency. Zhai et al. developed a glucose enzyme biosensor using Pt nanoparticles (PtNPs)-modified PANI hydrogel heterostructure. The sensor superior performance is ascribed to the 3D porous nanostructures, which facilitate the diffusion of glucose and hydrogen peroxide. PtNPs operate as catalysts to accelerate hydrogen peroxide oxidation by efficiently lowering the diffusion length of the molecules, and the PANI hydrogel framework promotes fast charge transfer between the electrode and enzyme. PtNPs-modified PANI hydrogel was presented by Li et al. as a biosensor platform for immobilization of distinct enzymes (uricase (UOx), cholesterol esterase/cholesterol oxidase (ChEt/ChOx), and lipase/glycerol kinase/glycerol3-phosphate oxidase (LIP/GK/GPO)) for detecting of three respective analytes (uric acid, cholesterol, and triglycerides). Because of the superior wettability of the enzyme solution on hydrophilic hydrogel, the enzymes were evenly deposited at high density on the PANI hydrogel matrix. The biosensor platform demonstrated excellent metabolite sensing performance with a wide linear range covering the entire range of metabolites in human serum (Fig. 7.6c), a 3 s fast response time, and

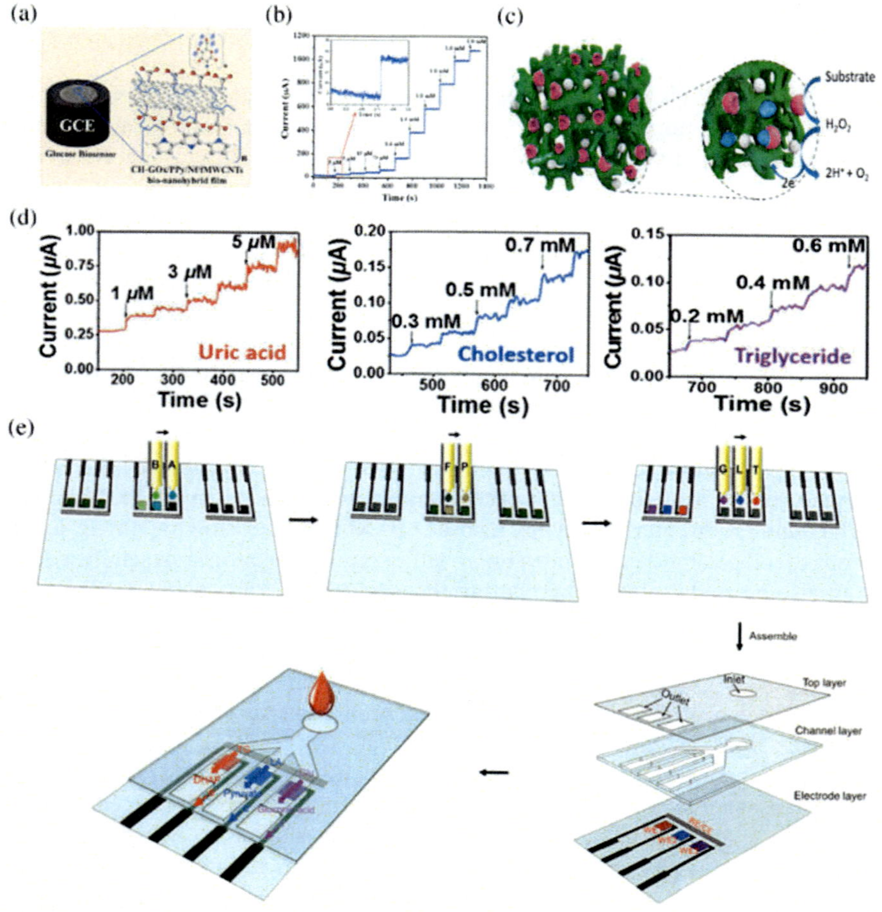

Fig. 7.6 **a** Schematic illustration of glucose biosensor based on chitosan-glucose oxidase immobilizedpolypyrrole/Nafion/functionalized multiwalled carbon nanotubes bio-nanohybrid film. **b** Current response of the electrode with successive addition of different concentrations of glucose. **c** Schematic illustration of the general sensing mechanism of the CPH-based electrode platform. **d** Current–time response curves of metabolites. **e** Schematic illustration of the inkjet-printed multiplexed biosensor based on conductive hydrogels. Reprinted from ref. [8], copyright 2019, with permission from John Wiley and Sons

excellent sensitivity (322.8 $\mu A\ mM^{-1}\ cm^{-2}$ for uric acid, 0.88 $\mu A\ mM^{-1}\ cm^{-2}$ for cholesterol, and 0.57 $\mu A\ mM^{-1}\ cm^{-2}$ for triglycerides). Furthermore, the solution-based synthesis and high water content of CPHs offer innovative biosensor platform building approaches. Inkjet printing is an appealing additive manufacturing process with no material waste, cheap cost, and quick fabrication, which opens up new possibilities for large-scale device realization. Li et al. demonstrated a multiplexed biosensor that can detect diverse samples on a chip using the "drop-on-demand" inkjet printing approach, in which polymer, enzyme, and metal nanoparticle monomers may be directly and effectively printed onto the electrode (Fig. 7.6d, e) [13, 14]. The as-fabricated sensor array displayed an exceptional sensing capacity to detect several chemicals at the same time (triglyceride, lactate, and glucose with sensitivities of 7.49, 3.94, and 5.03 $\mu A\ mM^{-1}\ cm^{-2}$, respectively).

7.7.10 *Conductive Polymer Composites to Electrochemical Biosensors*

7.7.10.1 Polyaniline and Composite Sensors

Based on the PANI-modified electrode, a chemical sensor for detecting dissolved oxygen was developed. However, the most promising use of these composites was discovered in electrochemical applications. The manufacture of electrochemical sensors based on PANI and its composites, in particular, has been a prominent method in many electroanalytical experiments, offering great sensitivity and selectivity to detect target molecules.

Glucose detection was reported utilizing PANI and $NiCo_2O_4$ NP composites. In glucose analysis, the non-enzymatic glucose sensor demonstrated excellent electrocatalytic performance. The catalytic active sites in $NiCo_2O_4$ with conductive PANI worked synergistically to oxidize glucose, exhibiting high sensitivity and a low detection limit. Electrochemical deposition of poly(3,4-ethylenedioxythiophene)/graphene oxide (PEDOT/GO) nanocomposite followed by electrochemical reduction of nanocomposite resulted in the development of a selective dopamine sensor. A dual polymer system comprising catalytic molecularly imprinted polymers on a PANI layer was utilized as a recognition layer for the electrochemical detection of catechol, as illustrated in Fig. 7.7a [22].

PANI composites containing metal NPs and carbon materials were created, and they demonstrated exceptional electrochemical activity in a variety of applications. A very sensitive glucose sensor, for example, was constructed using a composite PtNP/PANI hydrogel electrode. The nanostructured PtNPs and the PANI aero-gel on which they sit have a synergistic effect that results in the immobilization of a high density of GOx, allowing for efficient glucose oxidation. As illustrated in Fig. 7.7b, a comparable PtNP/PANI hydrogel-based electrode was utilized to detect human metabolites, and the sensor probe performed admirably [29].

MoS_2 was coupled with PANI by easy aniline monomer oxidation on the MoS_2 matrix. This composite material functioned as a good conductive skeleton with a large electrocatalytic surface area, as well as a direct electron transfer channel. The hybrid PANI/MoS_2 material outperformed pure PANI and MoS_2 in electrochemical performance. As a result, this substance was employed to detect DNA molecules electrochemically.

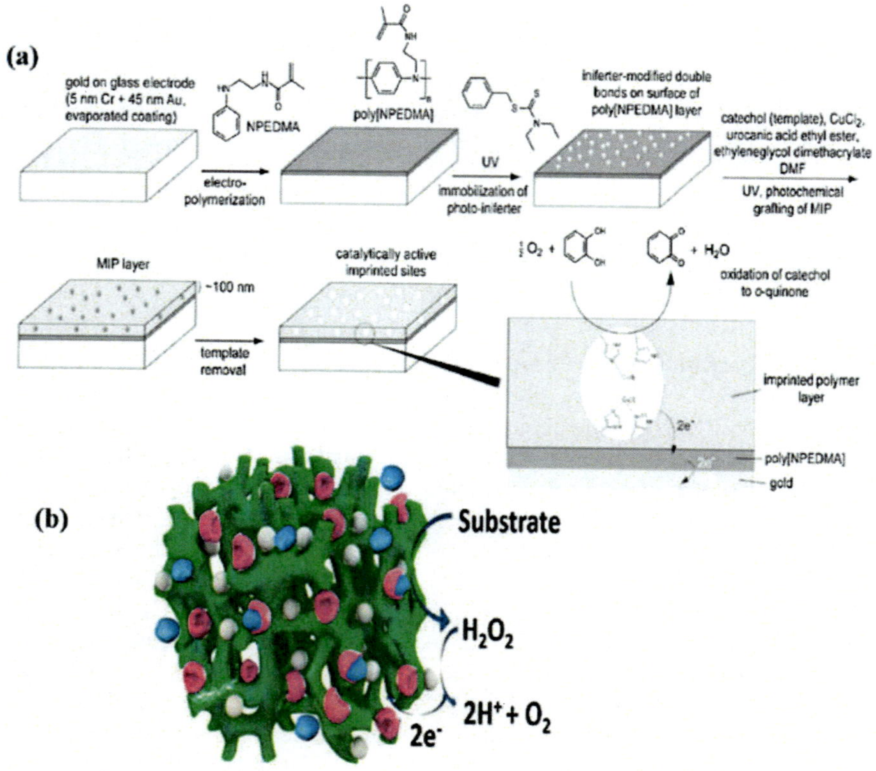

Fig. 7.7 **a** Schematic of the MIP-conductive polymer hybrid electrode imprinted with catechol. **b** Schematic of the PtNPs and enzymes-loaded 3D porous PANI hydrogel hydride electrode. Reprinted from ref. [17], copyright 2017, with permission from Elsevier

Fouling of electrode surfaces is a serious issue that researchers encounter while fabricating biosensors. Polyethyleneglycol (PEG) was grafted onto PANI nanofibers to solve this challenge. This composite material was conductive and had a wide surface area, and it demonstrated outstanding antifouling efficacy in both single protein solutions and complicated human serum samples. The antifouling effect was achieved by making the surface hydrophilic, which prevented interactions with non-specific proteins in complicated biological samples. This combination was utilized by the researchers to create an electrochemical biosensor for detecting the breast cancer susceptibility gene BRCA1. Many CP composites were created in place of metal or metal oxide electrode materials to detect nucleic acids in biological samples.

7.7.10.2 Polypyrrole and Composite Sensors

As electrochemical electrode materials, PPy and its composites with metal and metal oxide NPs, MWCNTs, graphene, enzymes, and electron transfer mediators have been explored, notably for electrochemical biosensor applications. PPy has been employed in electrochemical sensing materials because of its advantageous and distinctive chemical, physical, and electrical features. H_2O_2 was detected using an AuNP-PPy/Prussian blue nanocomposite sensor. This H_2O_2 sensor probe demonstrated great sensitivity, rapid responses, and good stability as a result of the PPy, with a detection limit of 8.3×10^{-10} M.

It has been reported that graphene and overoxidized PPy composites may detect adenine and guanine at the same time. The sensor probe was made by electroplating PPy and GO together using a potential cycling approach, followed by electrochemical reduction of GO. The resultant PPy/graphene was subsequently oxidized at +1.8 V, yielding an oxidized surface of a PPy/graphene composite layer (PPyox-gra). The oxidized form of PPy is permeable to cationic species but repellent to anionic and neutral species, which aids in interference avoidance and confers selectivity to the sensor probe. An enzymatic glucose sensor was created by covalently attaching PPy nanowires GOx to PPyCOOH/PPy composite nanowires. This sensor probe demonstrated not only good sensitivity with a large linear dynamic range and a low detection limit, but also high stability and glucose specificity.

7.7.10.3 Polythiophene and Composite Sensors

Organic CPs based on thiophene have been shown to have excellent conductive characteristics. As semiconducting material, they exhibit extraordinary stability and improved conductivity when conjugated with metal NPs, graphene, and carbon compounds. They may also be engineered with certain functional groups that allow them to be chemically bonded to transducer molecules in a biosensor, making them a very dependable substrate material group.

Hyaluronic acid was doped onto PEDOT to provide a more selective sensor surface. This substance is very porous, highly hydrophobic, and has a huge number of carboxylic acid functional groups. These functional groups were extremely helpful in immobilizing CEA antibodies, resulting in a highly sensitive, selectively binding sensor surface with a detection limit of 0.3 pg/mL. PEDOT:PSS, which is commercially available, was electrospun into nanofibers and then shaped into conducting paper. As an alternative to the usual indium tin oxide, gold, and glassy carbon electrodes, this conducting paper-based biosensor was proved to detect CEA. Many different material combinations using thiophene-based polymers have been investigated for use as glucose biosensors. For example, to obtain strong covalent interaction between the biomolecule and the immobilization substrate, GOx was immobilized on a polymerized, freshly synthesized thiophene conducting monomer.

Using a hard template approach, three-dimensional nanoporous PEDOT was electrochemically produced. Cu NPs were electrodeposited deeper into the PEDOT nanostructure because it has a huge surface area and is very stable and conductive. The prepared porous CuNP/PEDOT nanocomposite was excellent at electrocatalyzing glucose oxidation due to its unique structure in which the electrodeposited CuNPs provided many active sites for glucose oxidation and the 3D porous structure can facilitate glucose molecule diffusion and accelerate electron transfer. In vivo glucose testing has shown to be a more reliable and accurate method of diagnosing diabetes than in vitro analysis of patient blood samples. Many polythiophene conjugates are employed in in vivo devices since CPs are non-toxic chemical compounds.

7.7.10.4 Gold-Based CP Composites

Gold nanoparticles (Au NPs) have distinct chemical and physical properties, making them excellent candidates for the creation of biological sensors. Due to their unique conductive property and high surface energy and surface-to-volume ratio, Au NPs provide persistent immobilization of a large amount of biomolecules while preserving their biological activity, as well as direct and fast electron transfer during biochemical reactions between the biomolecule and electrode.

Electropolymerization is one of the most generally acknowledged techniques for creating adequate immobilization matrix due to the acceptable qualities of CP, namely film thickness, charge transfer, and permselective capabilities. Though biomolecule immobilization onto different CPs matrix has received a lot of interest as a consequence of these unique properties, the main restriction is the enormous amount of monomers and biomolecules required for immobilization. Simultaneously, it provides improved electrochemical activity, eliminates biomolecule leakage, and allows for fast diffusion of substrates and products while using less biomolecule. Although Au/CP nanocomposites have been utilized to determine a variety of biological species, practical implementation of the Au/CP composite has been difficult due to the limited solubility of CPs and the different optical characteristics of CP and Au NPS.

One of the most intriguing uses of Au/CP-based nanocomposite in DNA sensing is the on-site determination of the gender of arowana fish (or dragon fish), which is required for optimal generation of this cultured species. Marugan et al. created a kappa-carrageenan-Au NP/polypyrrole-based DNA biosensor on which they covalently bonded the ssDNA probe of the dragon fish. Because polypyrrole is brittle, kappa-carrageenan was employed to create a mechanically robust CP composite for biosensing. Genomic DNA samples obtained from fish scales were extracted, and the resultant biosensor demonstrated a wide response range, high stability, and a low detection limit of 5×10^{-18} mol/L. Mikolaj et al. produced a smooth polypyrrole film with non-aggregated Au NPs for DNA detection. In comparison to typical thiol self-assembled monolayers, the modification of the electrode with this composite layer enabled for the deposition of a greater number of DNA probes (up to two orders of magnitude).

De et al. created a platform for biosensing three distinct biomolecules, namely glucose, DNA, and protein, by immobilizing glucose oxidase, single-stranded DNA, and Lamin A antibody utilizing polyaniline nanowire coated with gold nanoparticles, respectively. The detection and immobilization methods were different for each system, and the biosensor demonstrated superior specificity, improved sensitivity, greater stability, and a low detection limit, proving the Au/polyaniline composite appropriate as a generic biosensor platform. Denkbas et al. developed a disposable, stable, and cost-effective biosensor composed of self-assembled peptide nanotubes, Au NP, and polyaniline on a pencil graphite electrode for detecting prostate-specific antigen (PSA). The detection limit was determined to be 0.68 ng mL^{-1}. The biosensor use was demonstrated by effectively detecting PSA in blood serum samples using the modified biosensor.

Omidfar et al. created an ultrasensitive sensing platform for detecting E. coli by first coating a screen-printed carbon electrode with polyaniline film and then with Au NP. The avidin–biotin interaction was then used to immobilize covalently avidin, followed by biotinylated DNA probe, on the modified electrode. The reported electrochemical test was shown to detect 4×10^6 to 4 CFU of *E. coli*, implying that this geneosensor has the ability to provide accurate and speedy diagnosis of *E. coli*-imposed illnesses. Au/CP composites have also been widely employed in medical diagnostics to identify harmful cancer cells. Shim et al. created a new, easy-to-use, and biocompatible aptamer-based nanobiosensor for detecting non-small-cell lung cancer (NSCLC) cells. The scientists covalently attached a MUC1 aptamer probe on an ICP composite produced by self-assembly of a 4-([2,2′:5″,2″-terthiophen]-3′-yl) benzoic acid layer on Au NPs and reached a detection limit of 8 cell mL^{-1}.

Wang et al. presented an electrochemical nanobiosensor for early detection of breast cancer by direct identification of microRNA-21 in clinical samples without the need for RNA extraction or amplification. They described a label-free self-assembled polypyrrole/gold nanocomposite biosensor using toluidine blue as a signal amplifier with a detection limit of 78 am. The self-assembly of polypyrrole-coated gold nanoparticles produced a superlattice of the close-packed kind, resulting in maximal current generation. Brain derived neurotrophic factor (or BDNF) is a neurotrophin that has a role in a variety of mental and neurological illnesses. Recently, Chandra et al. developed a microfluidic immunosensor made of gold nanoparticles, functionalized polyterthiophene-based ICP, and an anti-BDNF probe mounted on a composite layer for detecting BDNF from neuronal cells extracellular matrix. The immunosensor has a dynamic linear range with a detection limit of 1.5 ± 0.012 pg mL^{-1}.

Gold-based CP composites have also been used in biological fluids to detect co-existing chemicals simultaneously. Folic acid (a water-soluble vitamin) and dopamine (a neurotransmitter) are essential for human metabolism and the central nervous system. By using in situ chemical oxidative polymerization, Stephen et al. created a poly(omethoxyaniline)/Au-based nanocomposites for unique simultaneous measurement of folic acid and dopamine. In pharmaceutical and human urine samples, the sensor proved stable, accurate, and sensitive to dopamine and folic acid, with limits of detection of 0.062 and 0.090 mμ, respectively. The simultaneous detection of uric acid (2×10^{-7} mol/L) and epinephrine (8×10^{-8} mol/L detection limit) in human blood serum was examined by depositing Au NP/polyaniline film made by Langmuir–Blodgett technique, which showed great stability, sensitivity, and broad linear range.

7.7.10.5 Carbon Nanotube-Based CP Composites

CNTs properties are susceptible to being altered by biomolecules, which has led to its investigation as a sensing element for biological sensors. They have been recognized as an ideal material for the fabrication of biosensors due to their exceptional surface-to-volume ratio, wide electrochemical window, efficient electron transfer, and large specific surface area, as it is possible to achieve ultrafast detection of biomolecules at very low concentrations. As a result, CNT-based biosensors are among the greatest candidates for ultrasensitive sensing methods. Though CNTs have several advantages, some constraints, such as poor processibility, restrict their applicability because the randomly placed CNTs on the electrode surface benefit easy peel-off, lowering the repeatability of the biosensor. As a result, numerous techniques, such as multifunctionalization and composite creation with CPs, have been developed to make CNTs processable while minimizing agglomeration. Wrapping CNTs in conjugated polymers has also been reported to make them processable. In summary, CNTs have a π-conjugated structure as well as a highly hydrophobic surface, which allows them to engage with aromatic CPs via π-π and hydrophobic interactions, resulting in novel structures known as composites that may be used for biosensing.

Kang et al. used chitosan-coupled CNT/polyaniline composite for glucose detection, which is one of the most popular applications for CNT-based biosensors in diabetes diagnosis. The polyaniline layer was created using oxidative graft polymerization, and chitosan linked CNTs were covalently bonded with the CP layer using glutaraldehyde as a bifunctional linker. Using a bifunctional linker, pendant hydroxyl groups of chitosan were employed to covalently immobilize glucose oxidase. Because the 3D network structure of the composite electrode boosted the biocatalytic activity of the enzyme, the electrode was able to offer a biocompatible environment. Basu et al. created a new bi-enzymatic (horseradish peroxidase and glucose oxidase) glucose biosensor by combining polypyrrole and carboxy-modified multiwalled carbon nanotubes to produce a composite, which outperformed single enzyme biosensors in terms of sensitivity, shelf-life, and linearity. Thus, the improved biosensor performance is related to the inclusion of horseradish peroxidase in addition to glucose oxidase, which improves the total biochemical reaction. They also discovered that the sensitivity of the bi-enzymatic electrode is three times that of the mono-enzymatic electrode.

Pan et al. immobilized glucose oxidase in a poly(o-aminophenol)/CNT composite-based biosensor. Co-polymerization of glucose oxidase, o-aminophenol, and CNT at the electrode surface in mildly acidic solution immobilized the enzyme. They compared the two biosensors with and without CNT and discovered that the CNT-based electrode had twice the sensitivity, 1.5 times the response current, and twice the detection limit, as well as fast amperometric response, good anti-interferent ability, good stability, and excellent reproducibility.

Hydrogen peroxide (H_2O_2) is the result of the enzymatic process that occurs during biosensing. Xia et al. first investigated hydrogen peroxide biosensing using a polyvinylpyridine/CNT/Prussian blue-based composite based on electrode synergy, which demonstrated an excellent detection limit of 25 nm. They expanded this approach for determining glucose by identifying the hydrogen peroxide generated during the enzymatic interaction between glucose and immobilized glucose oxidase. Another researcher detected hydrogen peroxide at a trace level using cytochrome c-immobilized on polyaniline/carboxylated MWCNT. The modified electrode was sensitive to H_2O_2 and had an excellent detection limit of 0.2 μmol L^{-1}.

Kidney dysfunction is often characterized by a rise in the concentration of urea in the blood and a reduction in the amount of urine. As a result, urea measurement in blood samples is of tremendous interest to the medical community. Hassan et al. used a poly(o-toluidine)/CNT nanocomposite to immobilize urease enzyme on the composite film to create a novel approach for continuous and direct monitoring of urea in blood samples. Electropolymerization and in situ preparation were used to modify the working electrode, which demonstrated strong electrochemical response with a detection limit of 0.03 mm for urea in blood samples. Shim et al. created an amperometric l-lactate biosensor in order to maintain the content of l-lactate in human blood, which would otherwise induce clinical illnesses such as hypoxia and acute heart disease. They created the biosensor by immobilizing lactate dehydrogenase and nicotinamide adenine dinucleotide (NAD) on functionalized carboxylic acid-based polythiophene and multiwalled carbon nanotubes (CNTs). Covalent bond formation resulted in immobilization of the acid group of the composite and the amine group of the enzyme. The resultant biosensor has a large number of active sites, a good response range of 5–90 μm, and a detection limit of roughly 1 μm, which was successfully shown in milk and human blood samples. Another researcher created the same biosensor by immobilizing lactate dehydrogenase on a poly(3-methylthiophene)/MWCNT-based hybrid. Under optimum circumstances, the resultant biosensor has a concentration range of 1×10^{-6} mol/L to 5×10^{-4} mol/L with a detection limit of 5.6×10^{-7} mol/L.

Estimating cholesterol or triglyceride levels in blood is critical since high levels can lead to hyperlipidemia. Malhotra et al. created a nanocomposite of polyaniline and single-walled carbon nanotubes by co-immobilizing lipase and glycerol dehydrogenase, and their applicability in triglyceride sensing was investigated. An excellent storage stability of 13 weeks and a decent reaction time of 12 s were recorded. They also created a nanocomposite electrode for cholesterol biosensing out of polypyrrole and carboxy functionalized MWCNT in their prior study. Since the majority of the study effort involves detecting 30% free cholesterol, cholesterol esterase and oxidase were immobilized on the composite surface for the detection of total cholesterol. The suggested biosensor has a fast response time (9 s) and a detection limit of 0.04 mm L^{-1}.

Another study created a polypyrrole and carboxylic-based MWCNT-based biosensor for DNA detection. This COOH alteration of CNT is required for establishing a bond between the amino group of ssDNA and the (–COOH) group of MWCNT, which aids in covalent immobilization. Because of the synergistic properties, both selectivity and sensitivity were increased, and the suggested biosensor identified a complementary DNA sequence of 5.0×10^{-12} mol L^{-1}.

Gonorrhea is another sexually transmitted illness that has been recognized as a cofactor in HIV transmission, making quick and accurate identification of gonorrhea a crucial goal. Malhotra et al. employed a polyaniline/CNT composite to detect *Neisseria Gonnorhoeae (NG)* by immobilizing its bio probe with glutaraldehyde as a crosslinker. This bioelectrode had an excellent reaction time of 60 s, a 75-day stability under refrigerated circumstances, a low limit of detection of 1.2×10^{-17} mol/L, and a negative response to non-NG species as well as gram-negative bacteria. Shim et al. developed an amperometric immunosensor for detecting IgG by covalently immobilizing anti-IgG on MWCNT-embedded carboxylic acid-based polyterthiophene composite. To reduce H_2O_2, they used another hydrazine-labeled secondary antibody combination. The suggested immunosensor has a linear range (0.1 to 10 ng mL^{-1}) with a low limit of detection (0.084 ± 0.004 ng mL^{-1}) and was tested in rabbit serum samples.

7.7.10.6 Graphene and Their Derivatives-Based CP Composites

Graphene (GR), a 2D honeycomb lattice with an atomic thickness of sp^2 linked carbon atoms, is a graphite monolayer that has received great interest because of its optical, mechanical, electrical, thermal, electrochemical, and sensing capabilities. The structure of GR, as well as performance features such as high charge mobility and surface-to-volume ratio, may be translated into extremely sensitive sensing applications.

Unlike CNTs, which have a tubular shape, GR is a 2D sheet with an open structure, hence both sides of GR sheets may be used to host biomolecules. However, it has been shown that the intrinsic van der Walls interactions of the 2D graphene layers result in restacking to form graphite or produce irreversible agglomeration. As a result, this disadvantage can be solved by the inclusion of another macromolecule or the usage of polymers. The unique properties of CP and GR in combination have recently attracted the interest of researchers as a consequence of the developing a novel composite material for use in next-generation biosensor devices. This GR-based CP composite acts as a biomolecule loading platform, amplifying or harnessing the system adsorption, redox reaction, catalytic reaction, and transport characteristics. As the grafted CP with GR affects the in-plane features of composite materials, CPs operate as conducting conduits that interact with GR and analytes to improve sensitivity, selectivity, response time, and detection limit, finally giving rise to remarkable biosensors. Because GR sheets with CP matrix exhibit weak adhesion between them, several ways to improving GR adhesion qualities by changing their structure have been devised. Furthermore, the chemical inertness of GR makes it difficult to create multifunctional and changed hybrid nanomaterials. Therefore, chemically modified graphene oxide (GO) and reduced graphene oxide (rGO) have been produced for multifunctionalization. Covalent immobilization of the biomolecule, for example, as well as a faulty plane and a rich edge reaction site, all contribute to the identification of the target analyte.

Ho et al. developed a new composite incorporating reduced GO/PEDOT nanotubes for electrochemical biosensing of hydrazine and hydrogen peroxide and observed a significant increase in electrochemical activity when compared to pristine PEDOT and rGO electrodes. The suggested biosensor demonstrated high selectivity and sensitivity, with detection limits of 2.2 μm for hydrazine and 3.2 μm for hydrogen peroxide. Another researcher, Wei et al., demonstrated a large-scale one-step electrochemical synthesis of multilayer GR/polyaniline composite for hydrogen peroxide identification via the reduction process. Horseradish peroxidase was immobilized on this produced composite electrode, which demonstrated an excellent limit of detection of 8×10^{-8} mol/L. An effective strategy for the advancement of a biosensing platform was demonstrated based on sulfonated GR/polypyrrole nanocomposite for glucose detection. The surface of the nanocomposite was changed with successive carboxyphenyl groups to immobilize glucose oxidase. Because of the permselective properties of the grafted carboxyphenyl and overoxidized polypyrrole layer, the proposed biosensor enabled good selectivity toward glucose in the presence of interferents (paracetamol, uric acid, ascorbic acid, and cysteine). This eliminated the need for additional membranes, simplifying the biosensor fabrication. A MIP was synthesized using the co-electropolymerization of o-phenylenediamine and aminophenyl boronic acid. For determining salbutamol, this MIP was built using a GR-PEDOT:PSS composite-modified electrode. The goal of employing composite was to increase the sensitivity of the current, which demonstrated an excellent detection limit of 100 pm with a wide linear range of 1 nm–1.2 μm. The biosensor demonstrated strong reusability of at least ten repeated measurements without significant loss of electrochemical response, as well as good recovery findings for tangible sample analysis (actual swine meat) without the requirement for pretreatment measures.

Levofloxacin is a quinolone antibiotic that is used to treat a variety of bacterial illnesses. However, excessive levofloxacin intake causes disorders such as tendon damage, heart disease, and muscle atrophy. Feng et al. used a new poly(sulfanilic acid)-rGO composite film to create a sensitive and selective levofloxacin sensor. The suggested biosensor demonstrated high selectivity and electrocatalytic activity, a broad linear range, and a low limit of detection (0.12 μm) based on the synergistic effect. The built sensor also showed high repeatability and reliability, and it was effectively employed to detect levofloxacin in human urine samples. By immobilizing the corresponding DNA sequence, Huang et al. validated a DNA biosensor based on oxidized GR and polyaniline nanowires. The resultant composite displayed good electrochemical behavior toward this complementary DNA, with a detection limit as low as 3.25×10^{-13} mol/L and high storage stability. Nasrabadi et al. predicted a label-free and enzyme-free electrochemical immunosensor made of poly(p-phenylene diamine) (PPD) and GR nanocomposite for detecting neuron-specific enolase biomarker in human serum. The suggested non-enzymatic biosensor improved not only stability but also sensitivity, with a broad linear range and a limit of detection as low as 0.3 ng mL^{-1} for biomarker detection.

Dopamine is a neurotransmitter that causes neurological problems when the levels of dopamine are abnormal. The interference induced by ascorbic and uric acid, which coexist in biological samples, is a serious challenge in its detection. As a result, Travas-Sejdic et al. created a highly selective PEDOT-modified laser scribed graphene biosensor for detecting dopamine in the presence of various interfering chemicals. This electrode, which was created by direct laser scribing on polyimide sheets followed by electrodeposition of PEDOT, displayed 3D porous morphology with outstanding electrochemical properties and a detection limit of 0.33 μm. Luo et al. created a PEDOT/GO nanocomposite for specifically detecting dopamine in the absence of common interferences such uric acid and ascorbic acid. Nanocomposite was electrodeposited on the electrode surface, followed by electrochemical reduction to produce a reduced nanocomposite-modified electrode with good electrocatalytic activity toward dopamine oxidation.

By immobilizing uricase on the polypyrrole/uricase/GR composite, Pereira et al. validated a simple electrochemical approach for identifying uric acid in human urine. The proposed biosensor demonstrated high repeatability, stability, and significant results over its large linear range (0.002 to 0.024 μmol L^{-1}), as well as a low limit of detection (0.541 nmol L^{-1}). Liu et al. used a MIP/reduced GO composite for the simultaneous identification of tyrosine and uric acid. The MIP layer was electropolymerized using 2-amino-5-mercapto-1,3,4thiadiazole monomer on a rGO-modified electrode while immobilizing the dual template of uric acid and tyrosine. They described the sensing process of the biosensor by first recognizing the target molecule and then catalyzing the oxidation reaction on the composite of MIP/rGO by measuring the electrochemical behavior of tyrosine and uric acid. Under ideal conditions, the resultant nanocomposite had an increased electron transfer rate, more recognized binding sites, a low limit of detection for tyrosine (0.046 μm) and uric acid (0.0032 μm), and a broad linear range.

Lysine is one of the eight essential amino acids required by the human body, but it is the most threatened amino acid and is readily destroyed. Thus, when food is stored for an extended period of time, it is critical to establish the nutritional content of the food. L-Esma et al. created an amperometric biosensor for measuring L-lysine by immobilizing L-lysine-oxidase (LyOx) on GR/polymer (vinyl ferrocene). The resultant biosensor demonstrated long-term stability due to the biosensor appropriate microenvironment, which aids in the preservation of LyOx bioactivity. The reported biosensor demonstrated excellent analytical qualities such as high sensitivity, a low limit of detection (2.3×10^{-7} mol/L), and a rapid reaction time (<5 s). Wen et al. developed an amperometric organophosphate pesticide biosensor by immobilizing acetylcholinesterase (AChE) on a graphene/polyaniline composite. The resulting composite had a high specific area, strong conductivity, an increase in AChE surface loading, and excellent encapsulated and layered structures with a detection limit of 20 ng L^{-1}.

7.7.10.7 Hybrid CP Composites

Because nanomaterials have one or more bottlenecks, either during biosensor manufacture or during sensing, hybrid CP composites employing more than one nanomaterial are required. The combination of graphene and CP produced a surface with higher electrical conductivity, resulting in a selective and sensitive biosensor, and Au NP gave strong affinity for DNA for immobilization. Mucin 1 is a sort of protein biomarker that assists in the identification of breast cancer, therefore Prabhakar et al. developed an aptasensor for the detection of mucin 1 utilizing a PEDOT/GO/AuNP composite. The aptasensor demonstrated high reusability (eight times), stability (14 days), and a low limit of detection (0.031 fm) when used in spiked human blood samples, achieving 85–93% recovery. A flatoxins B1 (AfB1) are another hepatocarcinogenic mycotoxin; hence, Raju et al. created a biosensing platform of PEDOT/GO composite coated with AuNPs for electrochemical detection of AfB1. The rationale for the AuNP decorating on GO was to improve the electrode sensitivity, as GO is known to have low sensitivity due to its weak conductivity. Furthermore, AuNP assists in improved immobilization, resulting in a benign microenvironment that preserves their activity and inhibits leaching. The resultant PEDOT/GO/AuNP composite-based biosensor has an outstanding linear range of 0.5–20 ng mL^{-1}.

The quantitative measurement of glucose is critical because of the rising need in diabetes diagnosis, the food sector, and biotechnology. So Luo et al. developed a two-step electrochemical approach for manufacturing PEDOT/RGO/nickel nanoparticles for non-enzymatic glucose biosensor fabrication. Because of their environmental friendliness, cheap cost, and outstanding electrocatalytic activity, nickel-based materials have emerged as an electrode active material for non-enzymatic glucose biosensors. Under ideal conditions, the glucose sensor had a linear range and a low detection limit of 0.8 μm.

Fisher et al. demonstrated a scalable nanostructured biosensor based on a multilayered graphene petal nanosheet (MGPNs)/platinum nanoparticle (PtNPs)/PEDOT composite using glucose oxidase as a biorecognition element. The sensor platform was developed using a combination of 0D NPs and 2D support arranged in third dimension to achieve exceptional performance. The research resulted in the development of a biosensor with excellent performance and heightened sensitivity toward glucose, with a detection limit of 0.3 μm. The glucose biosensor performed well, with consistent glucose sensitivity after five weeks of use, and its performance was benchmarked against a diverse assortment of glucose biosensors.

Tang et al. created an enzymatic glucose sensor based on a layer-by-layer assemblage of polyaniline/MWCNT/AuNP composite screen on which glucose oxidase was mounted. The resultant biosensor preserved 92% of its original activity after 30 days and had a detection limit of 0.19 μm. Traditional metal oxides, such as copper oxide, have gotten a lot of interest since they have a lot of advantages for biosensors, such as strong stability, easy synthesis, and impressive redox process at different potential ranges under appropriate reaction circumstances. Alias et al. created a nanocomposite of copper oxide/RGO/polypyrrole for use in glucose detection, and the resulting biosensor had a low limit of detection of 0.03 μm.

Dechtrirat et al. developed a one-pot simple and environmentally friendly method for the manufacture of PEDOT:PSS/AuNPs/graphene nanocomposites for the selective detection of uric acid and dopamine in the presence of ascorbic acid. The composite was created by electroexfoliating graphene in polystyrene sulfonate (PSS) and then forming AuNP and PEDOT in situ. The as-synthesized dispersion of PEDOT:PSS/AuNP/graphene was used to fabricate a biosensor on the electrode surface by drop casting, and the resulting biosensor revealed a linear response of (1 nm–300 mm) and (10–1 mm) with low limit of detection of 100 pm and 10 mm of dopamine and uric acid, respectively, and was measured in real pharmaceutical and human serum samples. Yang et al. used electrodeposition synthesis of a new PEDOT/CNT/nickel oxide composite to detect dopamine, serotonin, and tryptophan all at the same time. The composite was discovered to be a 1D co-axially structured nanotube of CNT and nickel oxide/PEDOT composite with high electrocatalytic activity and a linear response with low detection limits of 0.026, 0.063, and 0.210 μm for dopamine, serotonin, and tryptophan, respectively, which was then used for simultaneous detection in serum samples.

Zhang et al. developed a novel technique for building a sensitive levofloxacin (LEV) sensor by incorporating AuNPGR on MIP polypyrrole, achieving a detection limit of 0.53 μmol L^{-1} and a good linear range. In their investigation, GR-AuNP was used as an electrooxidation promoter for LEV on an electrode, while MIP polypyrrole was used as a recognition element. Another analgesic and antipyretic medicine that can induce acute renal and liver failure is paracetamol. Thus, Luo et al. created an Au/GR nanoparticle with a core–shell structure that has high stability and catalytic activity doped into PEDOT to form a nanocomposite by an electrodeposition method. The modified electrode showed satisfactory sensitivity to paracetamol across a wide range of concentrations, with a detection limit as low as 41 nm, and was capable of detecting paracetamol in actual samples with high accuracy. Orlando et al. used silver nanoparticles, carbon black, and PEDOT:PSS as an effective voltammetric biosensor to detect levofloxacin and paracetamol at the same time. Carbon black was chosen because it is less expensive than graphene and has catalytic and electrocatalytic qualities that improve sensor performance. The succeeding biosensor demonstrated high repeatability and stability, with detection limits of 1.2×10^{-8} and 1.4×10^{-8} mol/L, respectively, for paracetamol and levofloxacin.

7.7.11 Application of Conductive Polymer Biosensor

Conductive polymers are versatile materials for sensing applications since they not only have unique features but may also be employed as immobilization matrices, receptors, and transducers (redox systems for electrical charge transfer) in a biosensor. Conductive polymers have been utilized to make biosensors in a variety of sectors, including health care, immunosensors, environmental monitoring, and food analysis.

7.7.11.1 Biosensors for Health Care

The selective measurement of diverse blood analytes such as glucose, urea, lactate, uric acid, cholesterol, and so on is critical for the screening and treatment of a variety of disease. There are several publications in the literature on the immobilization of glucose oxidase in conductive polymers for glucose estimation. Ramanathan et al. immobilized GOD after modifying the pore size in polypyrrole to increase enzyme loading. The interchange of para-toluene sulfonate and ferricyanide ions in solution with a smaller ion like chloride has been used to make polypyrrole more porous, allowing for increased loading of GOD. Glucose oxidase (GO_X) has been covalently coupled to poly(o-amino benzoic acid) (PAB), a carboxy group functionalized polyaniline. For glucose concentrations, both mediated (with ferrocene carboxylic acid and tetrathiafulvalene) and unmediated systems have been used.

The majority of urea biosensors rely on the detection of NH_4^+ or HCO_3^- sensitive electrodes. Osaka et al. created a very sensitive and quick flow injection device for urea measurement using the composite sheet made of an electropolymerized inactive polypyrrole and a polyion complex. Gambhir et al. created a urea biosensor by co-immobilizing urease and glutamate dehydrogenase on electrochemically produced polypyrrole/polyvinyl sulfonate. Lactate biosensors are enzyme-based devices that use lactate oxidase and lactate dehydrogenase. Chaubey et al. presented a lactate biosensor based on polypyrrole/polyvinyl sulfonate composite sheets.

Cholesterol biosensors have gotten a lot of attention because they are such an important parameter in clinical diagnostics. The enzyme cholesterol oxidase serves as the basis for amperometric cholesterol biosensors. Cholesterol oxidase and ferrocene carboxylate were electrochemically immobilized in polypyrrole to explain the sensitivity of the resultant films. Kumar et al. used dodecylbenzene sulfonate-doped polypyrrole films for physical adsorption of cholesterol oxidase. These films were used to estimate cholesterol using ferricyanide mediation.

7.7.11.2 Organic Immunosensors

A number of conductive polymer-based immunosensors have been created. Sargent's group has developed a novel technique for reversible immunosensors based on pulsed electrochemical detection utilizing conductive polymers. The method has been used to identify organochlorine pesticides such as PCBs, atrazines, and chlorinated phenols. Porter studied electroplated conductive polymers as an antibody receptor in an immunosensor. They demonstrated that antibodies against conductive polymers (carbazole as a hapten) can react to modify the electrochemistry of the polymer. The antiserum reaction was discovered to alter the polymer electrochemistry via an amperometric response and so may be used as a sensor for a direct immunoacid. Ohmicron Corporation in Pennsylvania has also created an immunosensor that is a portable tool for estimating atrazine. The equipment allows for a total test duration of less than 15 min and can identify pesticides at the parts per billion level.

Immunosensors are a type of biosensor that recognizes antigens based on the capacity of antibodies or portions of antibodies to bind certain antigens (i.e., a molecule capable of evoking an immune response). The early organic immunosensors used a CP film on a metal substrate (for example, PANI or PPy on Au) to improve charge transfer between the biorecognition element and the electrode. The non-specific attachment of the analyte or interfering molecules to the CP surface and/or gate electrode is a significant characteristic that influences immunosensor sensitivity. Another team developed OTFTs using a similar blocking method (Fig. 7.8a). They immobilized procalcitonin-specific antibodies on a poly(3-hexylthiophene) (P3HT) active material and employed BSA as a post-treatment to minimize non-specific adsorption [21]. The detection limits of the devices were as low as 2.2×10^{-12} mol/L, which is consistent with clinical applications.

To retain immunological activity and lead to optimal device performance, antibodies must be immobilized on the device surface in a favorable orientation with

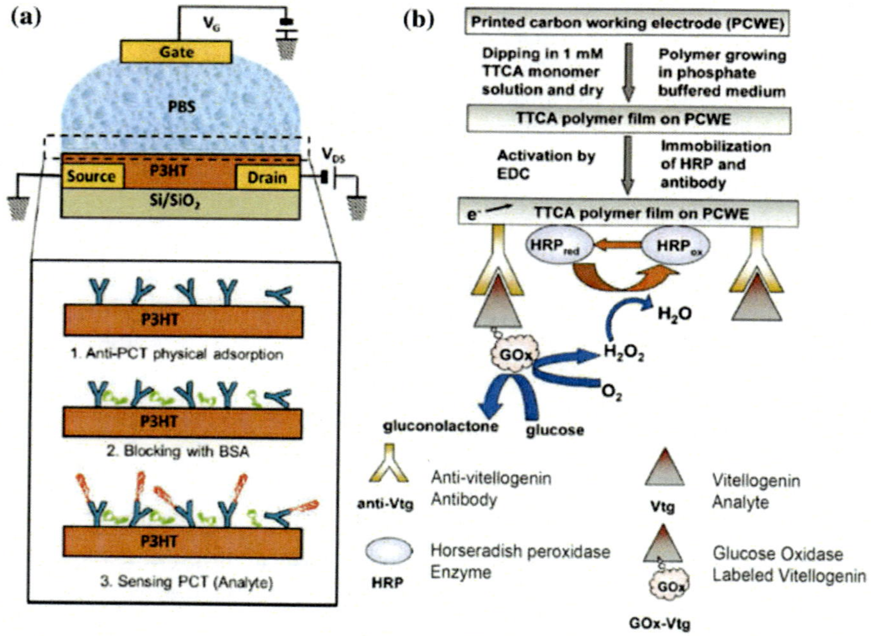

Fig. 7.8 **a** Immunosensor based on an OTFT having antiprocalcitonin (PCT) antibody physically adsorbed on poly(3-hexylthiophene) as sensing element, and bovine serum albumin as a blocking element. **b** Immunosensor using 1-ethyl3-(3-dimethylamino-propyl) carbodiimide mediated coupling to bind the antibody (anti-Vtg) to the carboxylic groups of terthiophene-based CP film (TTCA). Reprinted from ref. [30], copyright 2019, with permission from John Wiley and Sons

minimal structural alterations. Surface modification on already made CP films has been employed as a technique to immobilize the antibody via covalent bonding or electrostatic interactions. The researchers used 1-ethyl-3-(3-dimethylamino-propyl) carbodiimide (EDC) to enhance the formation of amide bonds between the carboxylic groups of a terthiophene-based CP and the exposed lysine residues of the selected antigen (Fig. 7.8b). The immunosensor identified vitellogenin precursor protein at concentrations as low as 0.09 ng mL^{-1} (around 5.6×10^{-13} mol/L). Covalent immobilization, on the other hand, might result in the loss of antibody function. Self-assembly techniques (e.g., host–guest interactions, as shown for PEDOT:PSS functionalized with ProLinker) provide a superior approach in this regard [3].

7.7.11.3 Enzyme-Based Biosensors

Analytical devices that use an enzyme as a biorecognition element are known as enzyme-based biosensors. To create a signal, the gadgets use the enzyme capacity to catalyze a specific biological reaction. The first organic, enzyme-based biosensors were created by entrapping the enzyme in the CP matrix during electropolymerization. To facilitate the passage of electrons from the enzyme to the CP, an electron acceptor was also introduced, either in solution or inside the CP film.

The use of PANI as an active substance resulted in OECTs that switched from OFF to ON in response to analyte detection. In this scenario, the reaction with the mediator resulted in the conversion of PANI from an insulating to a conducting state. The use of redox mediator additives for glucose detection in vivo is not recommended owing to probable leakage and toxicity. As a result, researchers created a nanojunction OECT for glucose monitoring that acts as an electron acceptor by utilizing naturally dissolved oxygen in solution.

Electropolymerization or crosslinking were used to entrap the enzyme in the polymer network. Other methods, like as inkjet printing and spin coating, have been utilized to create biosensing electrodes and devices with high sensitivity and stability. These approaches are straightforward and inexpensive, however, they all lack control over enzyme orientation and function preservation. To optimize the CP–protein interface and increase functioning, biofunctionalization techniques incorporating non-covalent or covalent interactions have been devised. Biofunctionalization methods for producing biosensors outperformed others in terms of selectivity, stability, and signal transduction. The researchers created a PEDOT/poly(allylamine) (PAH) composite biosensor in which the enzyme (GOx) is linked to the surface of the film via supramolecular interactions. Specifically, utilizing the crosslinker divinylsulfone, the main amine groups of PAH were employed to covalently bind mannose patterns. Then, using recognition-directed assembly, concanavalin A and GOx were immobilized on the surface. This method resulted in a 2.7-fold improvement in the bioelectrocatalytic signal when compared to pure Au platforms.

7.7.11.4 DNA Biosensors

DNA biosensors offer a wide range of applications, including clinical diagnosis of hereditary illnesses, fast detection of pathogenic infections, and screening of cDNA colonies in molecular biology. There have been a few instances of DNA interactions with conductive polymers. Livache et al. described a one-step electrodeposition of PPy sheets functionalized with a covalently attached oligonucleotide. Another option is to embed DNA probes in electropolymerised polypyrrole sheets and observe the current changes caused by hybridization. Gambhir et al. have reported the results of the studies relating to the characteristics of physically adsorbed DNA (Calf thymus) on conducting PPy/PVS films. The immobilization of DNA on a conductive polymer matrix allows for the detection of a signal (amperometric or potentiometric) created as a result of protein or drug interaction. Further research revealed that immobilizing the biorecognition element at the gate electrode improved sensitivity. The immobilization of DNA on a conductive polymer matrix makes it easier to detect a signal (amperometric or potentiometric) created as a result of protein or drug interaction with DNA.

The biorecognition element in DNA biosensors is a single-stranded DNA/RNA with a known base sequence. DNA pairing with a complementary strand (the analyte) has been proven to successfully transduce CPs in solution or film form into a quantifiable optical or electrochemical signal. Researchers presented the first example of a DNA biosensor based on an OECT in 2004. The sensing element of the described biosensor was inserted within an electropolymerized PEDOT active layer, allowing DNA detection at concentrations as low as 1×10^{-8} mol/L. The sensing mechanism was based on a change in the surface potential at the gate electrode caused by DNA immobilization and hybridization, which resulted in a change in the offset voltage. The sensitivity was 1×10^{-9} mol/L in steady-state conditions and 1×10^{-11} mol/L when electric pulses were applied to the gate electrode. The device stability in aqueous solutions allowed it to be integrated into microfluidic systems, while its operational stability retained its performance over many days.

7.7.11.5 Biosensors for Environmental Monitoring

Biosensors have the potential to supplement both laboratory-based and field-based environmental monitoring approaches. The conductive polymer-based gas sensor for environmental monitoring is based on the fact that when polymers are exposed to vapor, they undergo fast conductivity changes that are typically reversible at room temperature. The change in conductivity is most likely caused by a reversible decrease of the polymer caused by exposure to organic vapor, as well as a change in the moisture content of the film. A thin sheet of conductive polymer based on a microelectronic ammonia gas sensor device has been created, marking the first commercial application of conductive polymer in gas sensor devices. The device is a resistor element that exhibits huge fluctuations when extremely low concentrations of ammonia are present in the environment, allowing the appropriate action or alarm activation. The use of conductive polymers in impedance-type sensors allows for the construction of a low-cost, highly sensitive, and selective room temperature gas sensor. They described a number of organic vapor sensors based on PPy, poly-N-methyl pyrrole, poly(5-carboxy indole), and polyaniline. These materials, on the other hand, are favorable due to their lack of specificity; i.e., they respond to a wide variety of various gases and vapors. However, their selectivity is shown to be superior to that of inorganic material-based gas-sensitive resistors.

Thomas Fare et al. created an immunosensor based on conductive polymer for herbicide atrazine screening. Adeloju and Yuan report a novel method for flow amperometric biosensing of formate using single, double, and triple layered configurations of the ferrocyanide mediated PPy-based biosensor. The measurement of formate concentrations in the atmosphere, natural waterways, and sediments has been discovered to be relevant. This layered structure comprising formate dehydrogenase, NAD, and mediator has shown flow amperometric biosensing of formate in aqueous conditions with a detection limit of 2.5 mM. Lepsenyi et al. described a low-cost polyaniline-based ammonia gas sensor (1999). They used cyclic voltammetry to deposit a sensing film of polyaniline that had been electrochemically doped by cyclic voltammetry on thin and thick film electrodes where the isolation gap was produced by laser engraving. They tested the sensitivity, selectivity, temperature, dependence, and long-term behavior of the sensor [7].

7.8 Gas Sensor Based on Conductive Polymers

In most situations, the optical and electrical characteristics of conductive polymers alter when exposed to chemical vapors such as NH_3, HCI, $(NO)_x$, H_2S, $CHCl_3$, SO_2, CO_2, CCl_4, CH_2Cl_2, HSCN, methanol, acetone, ethanol, ethyl acetate, and so on. Jin et al. used absorption spectroscopy to investigate the ammonia gas sensing capabilities of the polyaniline film and discovered that the optical response time was less than 15 s and the rapid regeneration time was less than 2 min at ambient temperature. The sensors featured an ammonia detection limit of 1 ppm and a linear dynamic range of 180 to 18,000 ppm. Christie et al. conducted a study on remote detection of gaseous ammonia using the near infrared transmission properties of polyaniline because polyaniline films exhibit remarkable electrochromic properties as polyaniline films show multiple color changes depending on its oxidation states, yellow-green-dark blue-dark violet. The results demonstrated that at a wavelength of 1300 nm, the polyaniline sensing films reacted to ammonia concentrations as low as 6 ppm at 50% RH while remaining unresponsive to humidity fluctuations between 30 and 70% RH. The polyaniline sensing film has a lifespan of more than two years.

Penza et al. used the Langmuir–Blodgett technique to create a polypyrrole film-based gas sensor and tested its electrical resistance and capacitance in dry air with NH_3, CH_4, and CO. The results showed that the AC and DC resistance responses to NH_3 were extremely sensitive, whereas the responses to interfering gases like CH_4 and CO were negligibly weak. The sensitivity and selectivity of the NH_3

detector improved when the AC frequency decreased, and the optical operating condition in DC mode was reached. Poly(3-methoxyethoxythiophene) and its copolymers with aniline, thiophene, and pyrrole were chemically and electrochemically formed, according to Hossein et al. In their conducting condition, the copolymers are strongly conducting materials that are soluble in ordinary organic solvents. They studied the gas responses of the materials by measuring the mass change after exposure to hydrogen halides, hydrogen cyanide, halogens, monochloroacetic acid, 1-3-5 trichloromethylbenzene, methylbenzyl bromide, bromoacetone, and cyanogens bromide. These gas sensing materials outperform poly(aniline) blends with polyvinyl acetate, polystyrene, and poly(vinyl chloride) in terms of detection limit at room temperature, stability in the air, and gas selectivity.

Gallazzi et al. investigated polyalkoxythiophenes as sensing materials for electronic noses. The electrical resistance of the sensors after exposure to gases such as o-xylene, n-butanol, benzene, methanol, n-heptance, and 3-pentanone in a concentration range of 50–66,444 ppm in argon was investigated. Although a significant base line drift was observed in long time, it has not been proven to affect the sensitivity of the sensors. Sakurai et al. developed polythiophene (PTh) and poly(3-n-dodecylthiophene) (PDPTH) film gas sensors via electrochemical deposition method, and examined the response characteristics of these conductive polymer films to various gases over a wide temperature range. It was discovered that the resistance change of various polymer films varied greatly. The PTh film, in particular, responded to ammonia, but the PDPTH film clearly responded to hydrophobic gases including chloroform, methane, and ethanol. The reactions of these films to various gases were then evaluated using a pattern recognition algorithm to demonstrate that these modest gas sensors could operate together to differentiate the gases described in the research [15].

7.9 Perspectives

7.9.1 All-Solid-State Sensors

As a sensitive membrane, an electrochemically produced robust and ultrasensitive solid-state sulfate-selective electrode with Au wire-coated H_2SO_4-doped PANI was employed. Because of its low impedance of 36Ω, the electrode has high potentiometric stability and a rapid reaction time of 1 s. In 10^{-4}–1 mol/L Na_2SO_4 solutions, it exhibits a Nernstian response (26.22 mV decade^{-1}) to sulfate. The small electrode has been successfully built as a sensor for determining in situ sulfate content in the ocean.

7.9.2 Flexible Sensors

Flexible electronics have sparked a lot of interest for applications such as artificial intelligence, smart gadgets, electronic skins, soft robotics, and human–machine interaction systems. Over a concentration range of 5–1000 ppm, flexible NH_3 sensor based on PANI sheets demonstrate good long-term stability and repeatability. Furthermore, the mechanically robust sensor has great mechanical stability since it can be bent sharply without harm and still performs well after numerous bending cycles.

After 1000 bending cycles, a flexible PANI mesh sensor with an exceptional transparency of 88.4% at 600 nm wavelength exhibited high-performance resistive sensing of NH_3 gas without substantial deterioration. The MWCNT/PANI fabric sensor exhibits no significant change in base resistance even when bent in a wide range of 90°–270°, showing exceptional wearability. With a detection limit of 200 ppb, the sensor offers high sensitivity (92% for 100 ppm), outstanding selective rapid response (9 s), and recovery time (30 s). PANI serves a dual purpose by increasing flexibility and boosting sensor performance toward NH_3.

7.9.3 Stretchable Sensors

Stretchy electronics have applications in a variety of disciplines, including stretchable displays, wearable gadgets, stretchable energy storage devices, and implanted devices for human health monitoring. They have made strides in a number of applications for monitoring physical or environmental parameters. The ultrastretchable PANI composite fiber maintains fully constant conductivity after 5000 cycle stretching events of 200% strain due to the "twining spring" design and is used as an unique stretchy humidity sensor.

With a maximum strain of roughly 100%, the optimal conductivity and transparency of the PEDOT:PSS electrode may approach 0.14 $k\Omega\ sq^{-1}$ and 90%, respectively. The wrinkled PEDOT:PSS electrode-based triboelectric nanogenerator (WP-TENG) was able to accomplish high-performance active tactile sensing and may be employed as a human motion monitoring sensor to evaluate the bending angle of the elbow and joint.

7.9.4 Highly Sensitive Sensors

Three-scale nested surface wrinkling microstructures are used to create very sensitive pressure sensors based on hierarchically designed PPy sheets. Because of the multiscale nested microstructures that combine stimulus-responsive characteristics with the self-adaptive ability of wrinkling morphologies in PPy films, the as-fabricated piezoresistive pressure sensors have a high sensitivity (19.32 kPa^{-1}), a low detection limit (1 Pa), an ultrafast response (20 ms), and excellent durability and stability (more than 1000 cycles).

7.9.5 Highly Selective Sensors

A new electrochemical nanosensor based on p-nitrophenol (p-NP) imprinted polymer of PANI/GO nanocomposite for quantitative detection of p-NP in acetate buffer solution has been successfully developed (pH 5). In the presence of related phenol chemicals, the sensor demonstrated a low detection limit (0.02 mmol L^{-1}) and high selectivity for p-NP. The unique features of PANI and GO with large surface area and selective binding sites result in good stability and selectivity.

References

1. Bae J, Shin K, Kwon OS et al (2019) A succinct review of refined chemical sensor systems based on conducting polymer–cyclodextrin hybrids. J Ind Eng Chem 79:19–28
2. Correa D, Medeiros E, Oliveira J et al (2014) Nanostructured conjugated polymers in chemical sensors: synthesis, properties and applications. J nanosci nanotechnol 14(9):6509–6527
3. Darain F, Park DS, Park JS et al (2005) A separation-free amperometric immunosensor for vitellogenin based on screen-printed carbon arrays modified with a conductive polymer. Biosens Bioelectron 20(9):1780–1787
4. Deepshikha, Basu T (2011) A Review on Synthesis and Characterization of Nanostructured Conducting Polymers (NSCP) and Application in Biosensors. Anal Lett 44(6):1126–1171
5. El-Said WA, Abdelshakour M, Choi JH et al (2020) Application of conducting polymer nanostructures to electrochemical biosensors. Molecules 25(2):307
6. Gavgani JN, Hasani A, Nouri M et al (2016) Highly sensitive and flexible ammonia sensor based on S and N co-doped graphene quantum dots/polyaniline hybrid at room temperature. Sens Actuators B Chem 229:239–248
7. Gerard M, Chaubey A, Malhotra B (2002) Application of conducting polymers to biosensors. Biosens Bioelectron 17(5):345–359
8. Guo B, Ma Z, Pan LJ et al (2019) Properties of conductive polymer hydrogels and their application in sensors. J Polym Sci, Part B Polym Phys 57(23):1606–1621
9. Janata J, Josowicz M (2003) Conducting polymers in electronic chemical sensors. Nat mater 2(1):19–24
10. Jing X, Mi HY, Lin YJ et al (2018) Highly stretchable and biocompatible strain sensors based on mussel-inspired super-adhesive self-healing hydrogels for human motion monitoring. ACS Appl Mater Interfaces 10(24):20897–20909
11. Khalili N, Naguib H, Kwon R (2016) A constriction resistance model of conjugated polymer based piezoresistive sensors for electronic skin applications. Soft Matter 12(18):4180–4189
12. Klingstedt T, Nilsson KPR (2011) Conjugated polymers for enhanced bioimaging. Biochimica et Biophysica Acta (BBA)-General Subjects 1810(3):286–296
13. Li LL, Pan LJ, Ma Z et al (2018) All inkjet-printed amperometric multiplexed biosensors based on nanostructured conductive hydrogel electrodes. Nano lett 18(6):3322–3327

14. Li LL, Wang YQ, Pan LJ et al (2015) A nanostructured conductive hydrogels-based biosensor platform for human metabolite detection. Nano lett 15(2):1146–1151
15. Ma XF, Gao MJ, He XC et al (2010) Morphology tailoring of nano/micro-structured conductive polymers, composites and their applications in chemical sensors. Recent Pat Nanotechnol 4(3):150–163
16. Ma Z, Chen P, Cheng W et al (2018) Highly sensitive, printable nanostructured conductive polymer wireless sensor for food spoilage detection. Nano lett 18(7):4570–4575
17. Naveen MH, Gurudatt NG, Shim YB (2017) Applications of conducting polymer composites to electrochemical sensors: a review. Appl Mater Today 9:419–433
18. Pan LJ, Chortos A, Yu GH et al (2014) An ultra-sensitive resistive pressure sensor based on hollow-sphere microstructure induced elasticity in conducting polymer film. Nat Commun 5(1):1–8
19. Prajapati DG, Kandasubramanian B (2019) Progress in the development of intrinsically conducting polymer composites as biosensors. Macromol chem phys 220(10):1800561
20. Rahman M, Kumar P, Park D-S et al (2008) Electrochemical sensors based on organic conjugated polymers. Sensors 8(1):118–141
21. Seshadri P, Manoli K, Schneiderhan-Marra N et al (2018) Low-picomolar, label-free procalcitonin analytical detection with an electrolyte-gated organic field-effect transistor based electronic immunosensor. Biosens Bioelectron 104:113–119
22. Shim YB, Stilwell DE, Park SM (1991) Electrochemistry of conductive polymer X: Polyaniline-based potentiometric sensor for dissolved oxygen. Electroanalysis 3(1):31–36
23. Wang LL, Huang H, Xiao SH et al (2014) Enhanced sensitivity and stability of room-temperature NH_3 sensors using core–shell CeO_2 nanoparticles@ cross-linked PANI with p–n heterojunctions. ACS Appl Mater Interfaces 6(16):14131–14140
24. Wang T, Zhang Y, Liu QC et al (2018) A self-healable, highly stretchable, and solution processable conductive polymer composite for ultrasensitive strain and pressure sensing. Adv Funct Mater 28(7):201705551
25. Wang YM, Liu AP, Han YQ et al (2020) Sensors based on conductive polymers and their composites: a review. Poly Int 69(1):7–17
26. Wang ZW, Chen JC, Cong Y et al (2018) Ultrastretchable strain sensors and arrays with high sensitivity and linearity based on super tough conductive hydrogels. Chem Mater 30(21):8062–8069
27. Wu J, Wu ZX, Han SJ et al (2018) Extremely deformable, transparent, and high-performance gas sensor based on ionic conductive hydrogel. ACS Appl Mater Interfaces 11(2):2364–2373
28. Xia L, Wei ZX, Wan MX (2010) Conducting polymer nanostructures and their application in biosensors. J Colloid Interface Sci 341(1):1–11
29. Yu ZY, Li HJ, Zhang XM et al (2016) Facile synthesis of $NiCo_2O_4$@ Polyaniline core–shell nanocomposite for sensitive determination of glucose. Biosens Bioelectron 75:161–165
30. Zeglio E, Rutz AL, Winkler TE et al (2019) Conjugated polymers for assessing and controlling biological functions. Adv Mater 31(22):1806712
31. Zhang DZ, Wang DY, Li P et al (2018) Facile fabrication of high-performance QCM humidity sensor based on layer-by-layer self-assembled polyaniline/graphene oxide nanocomposite film. Sens Actuators B Chem 255:1869–1877